BIOCHEMISTRY, MOLECULAR BIOLOGY, AND PHYSIOLOGY OF PHOSPHOLIPASE A$_2$ AND ITS REGULATORY FACTORS

ADVANCES IN EXPERIMENTAL MEDICINE AND BIOLOGY

Recent Volumes in this Series

BIOCHEMISTRY, MOLECULAR BIOLOGY, AND PHYSIOLOGY OF PHOSPHOLIPASE A$_2$ AND ITS REGULATORY FACTORS

Edited by

Anil B. Mukherjee

National Institute of Child Health and Human Development/
National Institutes of Health
Bethesda, Maryland

PLENUM PRESS • NEW YORK AND LONDON

Library of Congress Cataloging in Publication Data

Biochemistry, molecular biology, and physiology of phospholipase A₂ and its regulatory factors / edited by Anil B. Mukherjee.

    p.   cm. — (Advances in experimental medicine and biology; v. 279)
Proceedings of a symposium held Sept. 19–20, 1989, in Bethesda, Md.
Includes bibliographical references.
Includes index.
ISBN-13: 978-1-4612-7910-5   e-ISBN-13: 978-1-4613-0651-1
DOI: 10.1007/978-1-4613-0651-1
    1. Phospholipase A₂ — Congresses. 2. Phospholipase A₂ — Mechanism of action — Congresses. 3. Phospholipase A₂ — Inhibitors — Congresses. I. Mukherjee, Anil B.
    [DNLM: 1. Phospholipases — antagonists & inhibitors — congresses. 2. Phospholipases — physiology — congresses. W1 AD559 v. 279 / QU 136 B6143 1989]
QP609.P553B56   1990
612'.01513 — dc20
DNLM/DLC   90-14266
for Library of Congress   CIP

Proceedings of a symposium on Biochemistry, Molecular Biology, and Physiology of Phospholipase A₂ and Its Regulatory Factors, held September 19–20, 1989, in Bethesda, Maryland

© 1990 Plenum Press, New York
Softcover reprint of the hardcover 1st edition 1990
A Division of Plenum Publishing Corporation
233 Spring Street, New York, N.Y. 10013

During the past decade there has been a dramatic expansion of our knowledge on phospholipases in general, and phospholipase A_2 (PLA_2) in particular. Progress in this field has been evident on many fronts, with novel information rapidly accumulating in the literature regarding the chemistry and molecular biology of this enzyme and its role in many important physiological processes. These include cellular signal transduction via the G-protein cycle, and in the generation of many cellular mediators, such as the platelet activating factor (PAF) and the eicosanoids that participate in the initiation and propagation of inflammation, to mention a few. This symposium was organized to obtain an overview of current investigations on this enzyme from the standpoint of its chemistry, molecular biology and physiology. Another important focus of this symposium concerns the regulation of PLA_2, including endogenous and synthetic inhibitors and activators of this enzyme. To review these important areas in PLA_2 research we invited scientists who made significant contributions in this field.

The papers in this volume are organized to emphasize the recent advances in several areas of investigation, including: (1) the structure and mechanism of action of PLA_2, (2) mechanism of activation of PLA_2, (3) molecular biology, physiology and endogenous inhibitors of this enzyme and finally, (4) clinical investigations emphasizing the pathophysiological role of this enzyme in human diseases. The first article in this volume is by Dr. Moseley Waite who very kindly agreed to review the research in the generalized field of phospholipases and provide a comprehensive overview of past, present and future directions in this broad field of research. The next article is by Drs. Ward and Pattabiraman who provided a detailed analysis of structural features of PLA_2 from different species. In the next three articles Dr. Heinrikson and his associates discuss the structure-function relationship of PLA_2 while the mechanism(s) of action of this enzyme is discussed by Dr. Dennis and his associates and Dr. Verheij and his colleagues.

The second group of articles focuses on the activation of PLA_2. Drs. Biltonen, Heimberg, Lathrop and Bell discuss the molecular aspects of PLA_2 activation and Dr. Cordella-Miele and colleagues describe a novel posttranslational modification of PLA_2 by transglutaminase that leads to its activation. The article by Dr. Hoffman and his colleagues describes the mechanism of activation of PLA_2 in human monocytes. These articles are followed by those of Drs. Miele and Baglioni and their associates who discuss the inhibitory properties of uteroglobin and antiflammin peptides. The work on the molecular biology of human placental PLA_2 is presented by Dr. Crowl and his colleagues. The physiological aspects of PLA_2 including the regulation of this enzyme via the G-protein cycle is discussed by Dr. Burch. Dr. Russo-Marie presents evidence that lipocortins inhibit cellular PLA_2 activity, whereas Drs. A. Hirata and F. Hirata discuss the biology of PLA_2 inhibitory proteins. Dr. Franson and his associates present data which suggest that PLA_2 activity may be inhibited by endogenous fatty acids and oligomers of prostaglandin Bl.

The last two articles are devoted to the clinical correlation of PLA_2 activation with pathophysiology of human diseases. Drs. Bomalaski and Clark discuss the activation of PLA_2 in rheumatoid arthritis and possible involvement of a novel protein called PLA_2 activating protein (PLAP) in this process. Drs. Pruzanski and Vadas present evidence for the role of soluble PLA_2s in human pathology. These articles provide an excellent overview of PLA_2 as a mediator of human inflammatory diseases such as rheumatoid arthritis.

I would like to express my deep appreciation to the scientists who have contributed to this symposium volume. Undoubtedly, great progress has been made in this field and much more remains to be done. I would like to emphasize that many outstanding scientists who have contributed in the field of PLA_2 research unfortunately could not be included in this symposium due to the constraints of time and money. I hope that our next symposium will be able to accommodate many more scientists in this field.

I would also like to thank Drs. Duane Alexander and Arthur Levine, Director and Scientific Director, respectively, of the National Institute of Child Health and Human Development, National Institutes of Health, for their support and appropriation of resources for organizing this minisymposium on PLA_2. Finally, I would like to thank the editors of Plenum Press for their enthusiastic support in publishing this volume.

Anil B. Mukherjee, M.D., Ph.D.

Section on Developmental Genetics
Human Genetics Branch
National Institute of Child Health
and Human Development
Bldg. 10, Room 9S242
Bethesda, Maryland 20892

CONTENTS

PHOSPHOLIPASES, ENZYMES THAT SHARE A SUBSTRATE CLASS

Moseley Waite

Department of Biochemistry
Bowman Gray School of Medicine of Wake Forest University
300 South Hawthorne Road
Winston-Salem, North Carolina 27103

INTRODUCTION

The title of this chapter was chosen to emphasize a point, namely, that PLs are classified as a group solely on the basis that they hydrolyze phospholipids. Beyond this commonality, this is a diverse group of enzymes, both in structure and in function.[1] One or more of these enzymes have been described in almost every, if not all organisms analyzed for their presence. The sites of hydrolysis and nomenclature for the PLs are given in Fig. 1.

Two general types of PLs exist, the acyl hydrolases (PLA_1, PLA_2, PLB),* and the phosphodiesterases (PLC, PLD). The functions of these enzymes vary considerably and more is being learned daily about their key role in cellular function. In general, three areas of function can be considered: 1) digestive, 2) membrane repair and remodeling, and 3) regulatory.

All types have been identified as digestive enzymes in a broad definition of digestive. For example, the best known examples of PLA_2s are derived from a multitude of venoms and from the mammalian pancreas whereas PLA_1 (lipases) are known to "digest" the phospholipids in circulating lipoproteins. Also, the beef pancreas contains a PLA_1 that is thought to be involved in digestion.[2] Both PLCs and PLDs have been identified in media of actively-growing bacteria. These were postulated to degrade extracellular materials for absorption, although a number are toxins.[1]

Membrane remodeling will not be considered here because of space constraints. Suffice it to mention that this essential role needs not

---

*Abbreviations used are: C18:0, stearic acid; C20:4, arachidonic acid; DG, diacylglycerol; f-MPL, formyl-methionyl-leucyl-phenylalanine; GPI, glycerophosphorylinositol; MDCK, Madin-Darby canine kidney; PA, phosphatidic acid; PL, phospholipase; PC, phosphatidylcholine; PE, phosphatidyl-ethanolamine; PEt, phosphatidylethanol; PG, phosphatidylglycerol; PI, phosphatidylinositol; PIP, phosphatidylinositol 4-monophosphate; PIP_2, phosphatidylinositol 4,5-bisphosphate; PKC, phospholipid-sensitive Ca^{2+}-dependent protein kinase C; PS, phosphatidylserine; LPI, lysophosphatidyl-inositol; LPL, lysophospholipase; TPA, tetradecanoyl-phorbol-acetate.

Biochemistry, Molecular Biology, and Physiology of Phospholipase A₂ and Its Regulatory Factors
Edited by A. B. Mukherjee, Plenum Press, New York, 1990

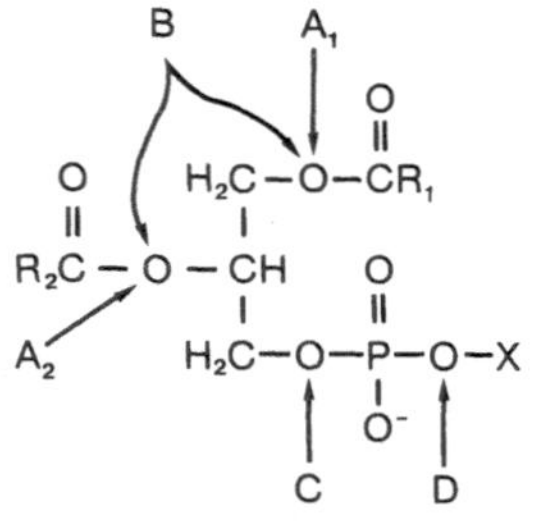

Fig. 1. Nomenclature for phospholipases

always involve complete deacylation-reacylation cycles since it is clear
that transacylations play a significant role in membrane remodeling.[3] It
is possible that PLs may function in transacylation systems and perhaps
even as transacylases.

Our knowledge of PLs in regulatory functions has emerged over the
past decade or so when it was recognized that PLA_2 action was the initial
step in the "arachidonate cascade".[4] Even before that, activation of the
PI-specific PLC during Ca^{2+} mobilization showed that this enzyme played a
central role in signal transduction in a number of systems.[5] PLCs with
other substrate specificity have now been demonstrated in signal
transduction[6] as well and quite recently PLD has been implicated as
another mammalial enzyme that can be stimulated during receptor occupancy.[7]
Fig. 2 shows some of the possible points at which PLs can function in
signal-coupling.

In this scheme a number of bioactive products are formed as the
result of membrane phospholipid catabolism. Without going into detail on
such a complex scheme, it is evident that PLs fulfill an essential role in
signal transduction. There are now suggestions that some events in this
scheme can exert their effects at the gene level including an involvement
in cell transformation. While the events depicted in Fig. 2 have been
documented clearly in a wide range of systems, the specific involvement of
individual reactions in a specific cell type and how those specific
metabolic events relate to the unique function of that cell often remain
to be established. Also, it has become evident that there is interplay
between some of the PLs, either directly or indirectly. For example, it
has been postulated that the Ca^{2+} mobilized by the action of PLC, via IP_3,
may activate a PLA_2 and, as a consequence, initiate the "arachidonate
cascade".[8] On the other hand, direct activation of the "arachidonate
cascade" by ionophore-induced PLA_2 action can subsequently activate PLC.[9]

CONSIDERATIONS IN THE ASSAY OF PHOSPHOLIPASES

A crucial consideration in the study of PLs is how they are measured.
Ideally, one would want to measure the specific PL of interest in its
native environment on its natural substrate. Indeed, this has been
possible for many of the digestive type of PLs and extrapolations have
been made to the regulatory and membrane remodeling PLs as attempts are
made to understand those enzymes. However, even with the digestive PLs,
differences in the method of assay have led to conflicting reports in the
literature. Since the natural substrates for PLs are water-insoluble and
since most PLs are activated by the lipid-water interface, most assays
employ lipid aggregates of one sort or another. Recent improvements in
understanding the organization of lipids in aqueous dispersion have aided

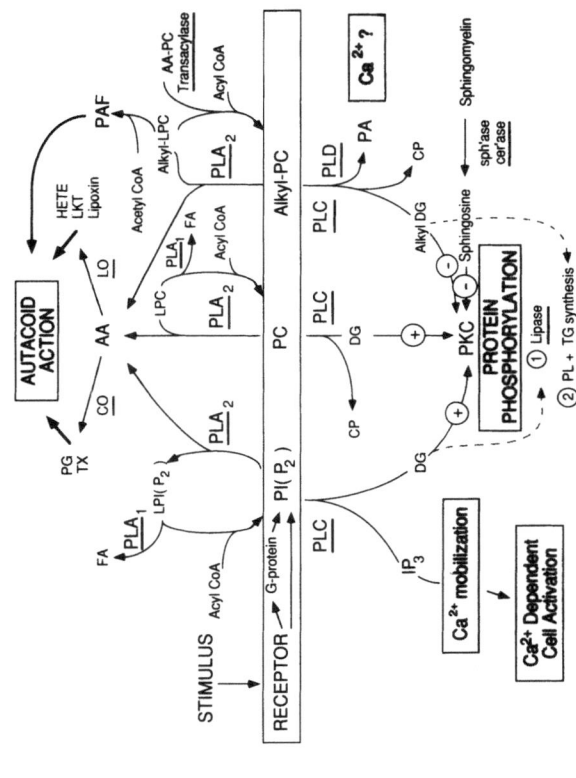

Fig. 2. The central role for phospholipases in signal transduction. The underlined terms are enzymes, and effectors or responses are in stippled boxes.

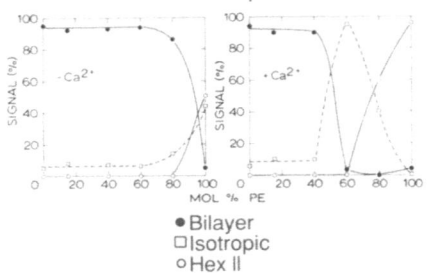

Bilayer, Isotropic, and Hex II Signals
31P-NMR Spectra

● Bilayer
☐ Isotropic
○ Hex II

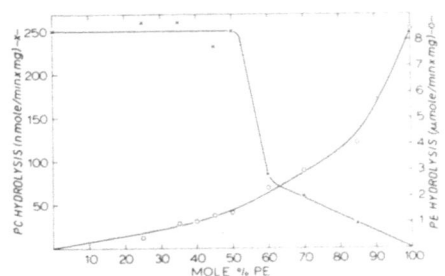

Fig. 3a. This figure demonstrates
the effect of Ca²⁺ on the
phases of PE:PC mixtures as
determined by NMR spectroscopy. (From Lenting et al.,
1988.[11])

Fig. 3b. Hydrolysis of the
PE:PC mixtures in Fig. 3a
by liver PLA₂ in the
presence of Ca²⁺. (From
Lenting et al., 1988.[11])

immensely in elucidating PL-substrate interaction. For example, the
development of methods to distinguish between lipid bilayers, hexagonal
arrays and micelles by NMR spectroscopy, differential scanning
calorimetry, and electron microscopy has permitted a systematic study on
this problem.[10] A good example of the effect of the physical state of
substrate on hydrolysis was shown with the liver mitochondrial PLA₂. As
shown in Fig. 3a, three phases can exist in PE:PC mixtures, depending on
the relative proportions of the two and the presence or absence of Ca²⁺.
Fig. 3b shows that the PLA₂ not only has a preference for PE over PC but
has a preference for PE in non-bilayer structures.[11]

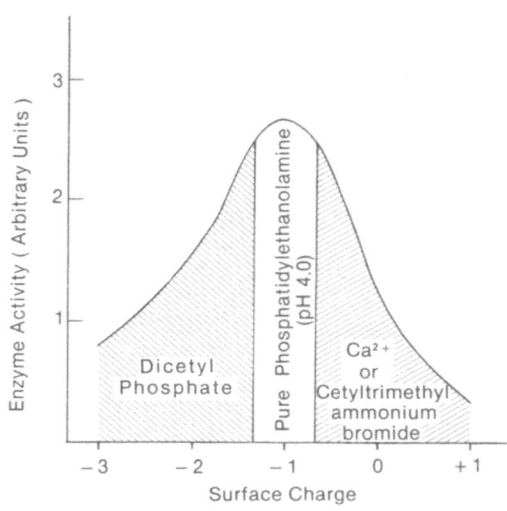

Striped areas indicate addition of Ca²⁺ or
amphipath to phosphatidylethanolamine

Fig. 4. Hydrolysis of PE by PLA₁ of rat liver
lysosomes. (From Robinson and Waite, 1983.[12])

Some other factors that can and often do regulate PL action are the charge at the interface (surface pH) and molecular packing. An example of the former was demonstrated with PLA_1 purified from rat liver lysosomes (Fig. 4).[12] In this case, surface charge could be regulated by the inclusion of charged amphipaths or by the addition of bivalent cations to anionic substrate.

Surface pressure effects of the substrate at the interface are most easily determined using monomolecular films of substrate.[13] Three factors that regulate PL action can be measured with this technique: 1) the rate of enzyme penetration into or absorption onto the interface, 2) activation by the interface, and 3) maximal surface pressure promoting hydrolysis. These events were studied using radiolabeled ε-amidated PLA_2 from pancreas (Fig. 5).[13] It is clear that pressures of a PC film greater than 10-11 dynes/cm limited binding, increased the time required to reach maximal velocity and decreased the maximal rate. However, the PLA molecules that did bind the interface were active since the specific enzymatic activity remained at a plateau.

While a number of other factors can regulate PL action, consideration of the physical properties of phospholipid is crucial since the phospholipids vary one from another. This becomes more evident if the goal is to describe either natural or pharmacologic regulators of PLs that may interact with and alter the substrate. With the recognition of regulator-substrate interactions, however, assay systems can be devised to study basic properties of PLs such as alterations in substrate specificity, and effects of modulators on enzyme-enzyme and enzyme-substrate binding. Some successfully used model systems include mixed substrate-detergent micelles, monolayer films, and bilayer vesicles, sometimes formed with inert amphipaths with substrates inserted.

The study of cellular PLs involved in regulation processes and in membrane remodeling present a more difficult task than that encountered

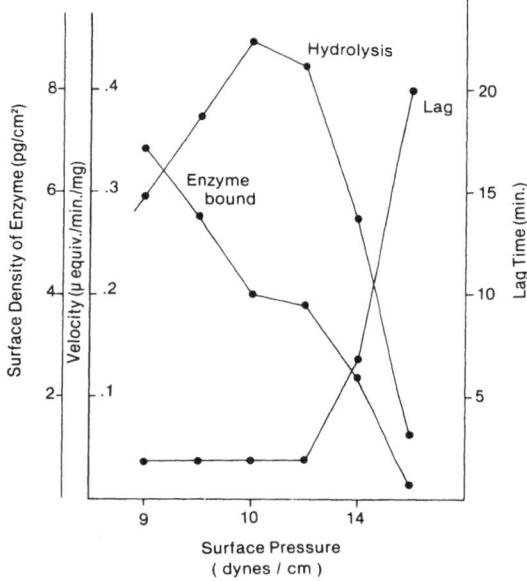

Fig. 5. Hydrolysis of PC monolayers by amidated porcine pancreatic PLA_2. (From Pattus and Slotboom, 1979.[13])

with digestive PLs even though digestive PLs have been studied successfully as general models for all PLs. The cellular PLs often are present in minute quantities and are sometimes membrane-bound. Also, their natural substrates are specific membrane sites, as opposed to digestive PLs that lack this specificity. To understand the function of these PLs, it is possible to use the models developed with digestive PLs to a certain point only. For the regulatory PLs we also need to know something of their specific molecular architecture and mode of action, localization within the cell, interaction with regulatory components of the cell, and the target membrane. The recognition of which PL within the cell is being measured, relative to the cellular function under investigation, is often difficult to discern. In some selected cases, progress is being made and within a few years many aspects of these goals will be resolved. Some advances with individual PLs or classes of PLs are taken up in the balance of the chapter.

1-ACYL HYDROLASES

Three types of PLs exist in this class of PLs: PLA_1, PLB, and LPL.[1] Often it has been found that these enzymes have overlapping substrate specificity, hence here they are considered as a single class. Both types of acyl hydrolases appear to proceed through an O-acyl cleavage mechanism, however, some 1-acyl hydrolases are thought to have a serine in the active site[14] whereas the PLA_2s studied thus far do not. Although the functions of these 1-acyl hydrolyases are poorly defined, at least some function as digestive enzymes. Since it has been noted that the 1-acyl group of phospholipids turns over as rapidly or more rapidly than that at position-2 in lipid remodeling,[15] the 1-acyl hydrolases may serve in this capacity as well. There is little information that would suggest PLA_1s are involved in stimulus-response coupling although our group has shown that arachidonate in the sn-3 position of lyso(bis)phosphatidic acid is released when alveolar macrophages are challenged with TPA.[16] This is a unique system, however, since this phospholipid is normally in very low quantities in most cells and because of its unusual stereochemistry.

Here, three 1-acyl hydrolases are described: PLB that hydrolyzes both acyl groups, hepatic lipase that removes only the 1-acyl group but has very broad substrate specificity, and heart cytosolic LPL-transacylase. While considerable information on bacterial 1-acyl hydrolases exists,[1] they are not described here.

PLB is a rather complex type of PL since it removes both acyl groups. One important factor in the study of PLB specificity was the purification of the enzyme to homogeneity, thereby eliminating the possibility that more than one enzyme is responsible for multiple hydrolyses.[17] The ratio of the activity on the two acyl chains is dependent upon the presence of detergent which inhibits LPL activity and increases hydrolysis of the diacyl substrate by the PLB purified from P. notatum[18] and from beef pancreas and liver.[19] Even though the P. notatum and beef enzymes share this property, they clearly are distinct. A mechanism was proposed for the P. notatum PLB for which there is substantial evidence: 1) substrate binding to Site I, 2) transfer of the 2-acyl group to catalytic site and transfer of LPL to Site II, 3) hydrolysis of 2-acyl group from catalytic site, and 4) transfer of 1-acyl group to catalytic site with subsequent hydrolysis.[20]

Some salient features of this proposal include acyl intermediates, initial attack at position-2 of the substrate, and the possibility of transacylation in the presence of an acceptor lysophospholipid since an enzyme-acyl group is postulated. Indeed, many enzymes of this general class are known to form diacyl lipids from two lysolipids. It is interesting to note that even though the initial attack is at position-2, its mode of action is quite distinct from PLA$_2$ from venom and pancreas.[1]

We have used a mixed micellar system to characterize the PLA$_1$ activity of hepatic lipase, an enzyme that also attacks monoacyl lipid, has transacylation activity and presumably an acyl-enzyme intermediate.[21] In this case, however, no attack at position-2 of the substrate could be detected. This enzyme has a broad substrate specificity and is not specific for phospholipids since neutral glycerides are also attacked. This, along with the lack of attack at position-2, clearly separates this enzyme from the P. notatum and liver PLB. We postulated that hepatic lipase has steric hindrance at the active site that precludes a large group remaining in the active site when the acyl-enzyme intermediate is thought to exist, unlike the PLBs and certain other 1-acyl hydrolases. Also, the mechanism of action of hepatic lipase appears to be distinct from the other 1-acyl esterases described here.

Two LPLs have been purified from rabbit heart cytosolic fraction with M_r of 23,000 and 63,000, similar in size to the two LPLs isolated from liver.[22,23] Comparison of the two heart LPLs is of interest since some differences in their mechanism of action appear to exist. The smaller of these enzymes was devoid of transacylase activity whereas the larger enzyme had both activities. The heart LPL catalyzed transacylation at very low substrate concentrations, 2 μM, near the cmc of the substrate. This is distinct from some other LPL that required high substrate concentrations to catalyze transacylations. Therefore, it appears that there might be three types of LPLs: those without transacylase activity, those with transacylase activity at high substrate concentrations, and those with transacylase activity with low substrate concentrations. The latter two may be distinguished by the heart enzyme's capacity to use monomeric substrate for transacylation.[23] This interesting possibility will require further study. Although the mechanism of the two heart LPLs is different, both were thought to prevent lysophospholipid accumulation and that inhibition of the LPL by acyl carnitines would account for lysophospholipid accumulation in ischemia.

PHOSPHOLIPASES A$_2$

The 2-acyl hydrolases certainly are the most completely studied group of PLs. These enzymes differ from the 1-acyl hydrolases in that they do not catalyze transacylations, are positionally specific, and hydrolyze only phospholipids, as far as is known. While some 1-acyl hydrolases met one or more of these criteria, as a group it appears that they do not. However, the PLA$_2$s are widely spread phylogenetically, as are the 1-acyl hydrolases, and are involved in digestive and remodeling functions like the 1-acyl hydrolases. It is well documented that PLA$_2$s are key enzymes in stimulus-response signaling as well. Several mammalian PLA$_2$s have been purified and their amino acid sequence shown to have a high degree of homology with venom or pancreatic enzymes. We do not yet have direct evidence that PLA$_2$s purified from mammalian cellular sources are those involved in signaling. However, it appears that structural differences between the PLA$_2$s purified thus far other than those involved in the catalytic mechanism will account for the remarkable differences in their functions. Comparison will be made in this section between the PLA$_2$s with M_r 12-15 kD with regard to their structure-function characteristics. Also, these will be contrasted with PLA$_2$s that have different properties. Details of the PLA$_2$s, their functions, structure and regulation, will be covered in subsequent chapters of this book.

Some of the features that appear to be invariant in the structure of the PLA$_2$s of this type are a high number of disulfide bonds (up to 7), a Ca^{2+} "binding loop" involving the residues at positions 28 (Tyr), 30 (Gly), 32 (Glu), and 49 (Asp), a helical region at the -NH$_2$ terminal region.[1] These enzymes require Ca^{2+} and are optimally active in the alkaline pH range. The catalytic mechanism for these PLAs is thought to be a proton-relay system that involves the bound Ca^{2+}, His-48, and Asp-99 (Fig. 6).[24]

A fundamental difference in the arrangement of the cysteines in the primary structure and therefore the cross-linking disulfide bonds formed between them has led to the creation of two classes of PLs. The first, Group I, comprised of the pancreatic, Elapidae, and Hydrophidae PLA$_2$s, have a unique disulfide bond between Cys-11 and Cys-77 whereas the Group II enzymes, comprised of the Viperidae and Crotalidae, have a unique disulfide bridge beween Cys-50 and the Cys at the C-terminal end of a peptide extended by six residues.[25]

Beyond these and other common but perhaps not invariant features of the PLs, subtle differences between these enzymes will undoubtedly account for a wide variation in their physiologic function. This is particularly true when recent information on the cellular PLA$_2$ is included. The recent recognition of cellular PLA$_2$s with structure nearly identical to the digestive PLA$_2$s demonstrates the conservation of this compact and stable enzyme class through evolution. These enzymes, localized within mammalian cells, are thought to be involved in a wide variety of physiologic and pathologic phenomenon and are under rather rigid control within the cell, without which cellular membranes would be degraded. Conceptually then, variations in protein structure may regulate membrane binding (penetration) to substrate, control by regulatory proteins, lipids or other cellular

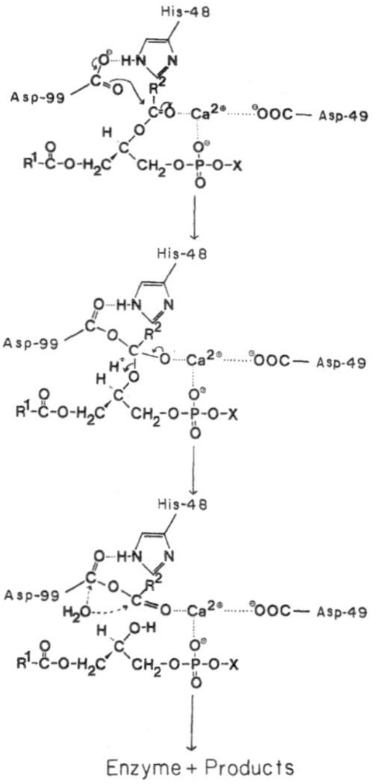

Fig. 6. Proposed proton relay mechanism for PLA$_2$ catalysis. (From Slotboom et al., 1982.[24])

components, and dictate substrate specificity that defines the PL's specific function. Examples of the last may include recognition of a specific acyl group at position-2 (arachidonate), or an ether vs. an acyl linkage at position-1. One of the big challenges in the field now is to establish structure-catalytic activity-function relationships for the PLA$_2$s. Recent advances in purifying and sequencing the mammalian enzymes bode well for meeting these challenges.

Recent evidence demonstrated that both groups of PLA$_2$s exist in a single tissue. As shown in Table 1, within the rat spleen both Group I (pancreas) and Group II (A. halys blomhoffii) enzymes have been identified, based on limited sequence data.[26,27] It is not yet clear if the soluble enzyme is synthesized as a proenzyme that can be activated by proteolysis. However, a similar PLA$_2$ from lung appeared to be encoded by a cDNA that suggested the PLA$_2$ was synthesized as a proenzyme.[28] If so, this provides a potential activation sequence for a cellular PLA$_2$. A Group I PLA$_2$ was described in the gastric mucosa as well.[29] The membrane-associated PLA$_2$ is clearly distinct from the soluble splenic enzyme yet was identical to the rat platelet and peritoneal PLA$_2$ (cf. Table 3), based on the information presently at hand, and differs only at residue-1 from the rat liver enzyme.[30] It is not clear at this point whether or not the two splenic enzymes are in the same or different cell types.

Comparison of five partial sequences of human PLA$_2$s shows that three fall into Group II (synovial fluid,[31] platelet,[32] and placenta[33]) whereas, like in the rat, the lung PLA$_2$[34] is in Group I and nearly identical to the pancreatic enzyme[35] (Table 2). While based on limited data, it is of interest that the three Group I PLA$_2$s have identical sequences. This raises the possibility that the three have a common origin in the platelet, although complete sequence data are required to establish this point. The question remains open as to the cellular origin of the lung PLA$_2$ and whether or not it is a secretory enzyme.

Comparisons made with the rat platelet and peritoneal fluid PLA$_2$s[36,37] indicate that the platelet may be the origin of the peritoneal fluid PLA$_2$, similar to the human enzymes. This, however, appears not to be the case for the rabbit since they are not identical (Table 3). In this case, the platelet[38] and peritoneal[39] PLA$_2$s differ at position 27 (H → S). The rabbit serum PLA$_2$ (personal communication),[40] sequenced through 20 residues, is common to both the platelet and peritoneal PLA$_2$s which leaves the door open as to the possibility that the serum enzyme may be derived from the platelet or the source of the peritoneal enzyme, but not both. Interestingly, the rabbit leukocyte PLA$_2$, while similar to the other three rabbit PLA$_2$s, is distinct and probably is not the source of either the serum or peritoneal PLA$_2$s (personal communication).[40]

Table 1.   Rat Spleen Phospholipases A$_2$

	1	5	10	15	20	25	30	35
A. halys blom-hoffii (basic)	H L L Q F R K M I K - K M T G K E P V I S Y A F Y							
Membrane[a]	X L L E F G Q M I L - F K T G K R A D V S Y G F Y G C X C G V G G R G S P							
Soluble[a]	A V W Q F R N M I K C T I P G S D P L R E Y N N Y G C Y C G L G							
Pancreas[b]	A V W Q F R N N I K C T I P G S D P L R E Y N N Y G C Y C G L G G S G T P							

[a] From Okamoto et al.[26,27] Identical to platelet and peritoneal PLA$_2$; differs from liver enzyme at residue 1 (van den Bosch and coworkers[30]).

[b] Gastric mucosal PLA$_2$ identical for first 15 residues.[29]

Table 2. Comparison of Human Phospholipases A$_2$

	1	5	10	15	20	25	30	35
Synovial fluid[a,c]	N L V N	F H R M I	K L T T G	K E A A L	S Y G F Y	G C X C G	V G G R G	
Platelet[b]	N L V N	F H R M I	K L T T G	K E A A L	S Y G F Y	G C H C G	V G G R G	
Placenta[c]	N L V N	F H R M I	K L T T					
Lung[d]	A V W Q	F R K M I	K C V I P	G S D P F	L E Y N N	Y G C T C	G L G G S	
Pancreas[e]	A V W Q	F R K M I	K C V I P	G S D P F	L E Y N N	Y G C V C	G L G G S	

[a] Inoue and coworkers[31]

[b] Kramer et al.[32]

[c] Lai and Wada[33]

[d] Seilhamer et al.[34]

[e] Verheij et al.[35]

A proposal has been made to relate these PLA$_2$s in an evolutionary tree.[41] Such relationships have proven useful in relating function of the PLA$_2$s as well as relationships between species. The challenge now is to incorporate new information on the cellular PLA$_2$s as well as the additional data on venom and pancreatic PLA$_2$s to update this dendrology. These comparisons will be particularly valuable when the cellular functions and regulatory mechanisms of PLA$_2$s are established.

The tools of site-directed mutagenesis have been applied to studies of the pancreatic PLA$_2$ with the purpose of elucidating the function of key residues in the enzyme. As will be detailed by Verheij later, some key observations include the essentiality of Asp-49 in Ca^{2+} binding and catalysis,[42] Glu-77 and Asp-66 in binding of the second Ca^{2+} ion,[43] and Tyr-69 in enzyme stereospecificity.[44] Further, removal of residues 62-66 (which are absent in Elapidae venom PLA$_2$) enhanced catalytic activity on zwitterionic substrates.[45]

While a great deal of information we have on cellular PLA$_2$s comes from those enzymes with M$_r$ = 13,000-15,000, it is clear that PLA$_2$s with different characteristics exist. For example, a PLA$_2$ was purified from sheep platelets that preferentially hydrolyzed substrates with an ether linkage in position-1.[46] This PLA$_2$ was purified as a dimer with an apparent M$_r$ = 58,000 and the suggestion has been made that the monomers are similar in size but are not identical. Another interesting

Table 3. Comparison of Rat and Rabbit Phospholipases A$_2$

	1	5	10	15	20	25	30	35
Rat peritoneal[a]	X L L E	F G Q M I	L F K T G	K R A D V	S Y G F Y	G C X C G	V	
Rat platelet[a]	S L L E	F G Q M I	L F K T G	K R A D V	S Y G F Y	G C H C G	V G G R G	
Rabbit platelet[a]	H L L D	F R K M I	R Y T T G	K G A T T	S Y G A Y	G C _S_ C G	V G G R	
Rabbit peritoneal[b]	H L L D	F R K M I	R Y T T G	K G A T T	S Y G A Y	G C _H_ C G	V G G R G	
Rabbit serum[b]	H L L D	F R K M I	R Y T T G	K G A T T				
Rabbit leukocyte[b]	A L L D	F R L M I	R Y T T G	K G A T				

[a] Inoue and coworkers[36,37,38]

[b] Elsbach and coworkers[39,40]

characteristic of this PLA_2 is its sensitivity to Ca^{2+}; maximal activity was observed with less than 1 μM Ca^{2+}. It will be of interest to establish if a family of enzymes of this type exists and how the properties of these enzymes compare in structure to the Groups I and II enzymes.

Considerable effort has been expended to understand the means by which PLA_2s are regulated. A major force behind this effort is the implication that PLA_2s are involved in a number of pathologies, either by their direct uncontrolled action on membranes or because of their central role in eicosanoid biosynthesis. A brief listing of some regulatory mechanisms includes: 1) proteins, either directly or following a phosphorylation event, 2) acylation, 3) dimerization, 4) amination, 5) fatty acids and/or eicosanoids, 6) translocation, 7) cellular uptake and degradation (for secretory PLA_2s), and 8) combinations of the above.

A number of these events will be covered elsewhere and only two examples will be described here since they are not subjects of this book. Quite recently, Inoue and coworkers have purified two proteins with homology to the complement component C3 that inhibit Group II but not Group I PLA_2s (personal communication).[47] The relationship of a PLA_2 inhibitor to the complement system would provide a type of feedback regulation in inflammation. The specificity of Group II inhibition is also attractive since these enzymes are those found in synovial and peritoneal fluids and secretory cells in circulation. While a number of proteins such as lipocortins have the capacity to inhibit PLA_2s as well as some other PLs, they appear to act by "substrate depletion," that is, the inhibitors bind to substrate rather than to the enzyme. This was initially established by showing that all inhibition was lost when substrate was present in a slight excess over inhibitor.[48] This does not appear to be the case with the inhibitors isolated from rat peritoneal exudates. The conditions for obtaining 50% inhibition of E. coli membrane hydrolysis are:

Inhibition of Rat Peritoneal PLA_2 by Purified Peritoneal Peptides[a]

| | Amount of Component | |
	Mass	Mol
PLA_2	2 ngm	160 fmol
Inhibitor	150 ngm	5 pmol
Substrate		50 nmol

[a]Conditions giving 50% inhibition; calculated from Ref. 47.

As can be seen, the inhibitor protein was present in a very small quantity, relative to the substrate when expressed either on a molar or weight basis. Further, a 4-fold increase in substrate did not overcome this inhibition, unlike the inhibition seen with lipocortin. This would argue against the inhibitory effect being exerted on the substrate. This is further supported by the observation that only the Group II PLA_2s are inhibited by these proteins. While other substrates need to be tested, these Group I related inhibitory proteins have considerable interest, especially since the Group I PLA_2s appear to be the more common cellular enzymes.

A second protein of considerable interest in PLA_2 activation and in the killing of bacteria by leukocytes is bacterial/permeability increasing protein (BPI).[49] This protein, found in the leukocyte granules, interacts with the PLA_2 in a specific manner that facilitates PLA_2 interaction with the E. coli membrane to initiate hydrolysis. Studies thus far strongly suggest that BPI interaction with PLA_2s requires a specific orientation of the amino residues of lysines and arginines in the $-NH_2$ terminal region of the enzyme. In this case, the regulatory protein, BPI, influences a digestive PLA_2 in one of several roles BPI plays as a bactericidal agent.

PHOSPHOLIPASES C

This class of PLs was initially described from bacterial sources[1] and, in fact, one of the first enzymes shown to degrade PC was found in El Tor Vibrios in 1901. One of the early interests in bacterial PLs stemmed from their action as toxins. For example, PLC from Clostridium perfringens was identified as the α-toxin active in gangrene. Interestingly, a single species of bacteria, Bacillus cereus, has three PLCs that differ in their substrate specificities: one with rather broad activity (PC, PE, PS), one that degrades the inositol lipids (PL, LPI), and a sphingomyelinase.[1] Perhaps the PLC whose molecular architecture is best known was purified from Bacillus cereus (PC-hydrolyzing). In addition to its requirement for Ca^{2+} in catalysis, Zn^{2+} is present that confers the required structure to the enzyme for activity.[50] Other PLCs have been studied in the Pseudomonas, Clostridium, Bacillus, and Streptomyces genera, primarily.[1]

The first PLCs recognized in mammalian systems were specific for PI or its phosphorylated derivatives PIP and PIP_2. As early as 1953, PLC was suggested to be present in brain tissue. Much of the interest in the PI-specific PLCs stemmed from their relationship to Ca^{2+} mobilization, an interest that remains current. It is currently hypothesized that when PLC acts on PIP_2 internal stores of Ca^{2+} are mobilized from the endoplasmic reticulum or other organelles that in turn can lead to a cascade of events that include an increased flux of Ca^{2+} from the extracellular pool. Likewise, the other product of the reaction, DG, is an activator of PKC which, in turn, sets off a sequence of protein phosphorylations with subsequent metabolic activation (Fig. 2; summarized in Ref. 1). Alternatively, the DG generated by PLC action may alter membrane properties and enhance membrane fusion events such as endocytosis.[51]

A number of PI-specific PLCs have been purified from diverse sources that include brain, heart, platelets, seminal vesicles, and liver. This work is thoroughly reviewed in Ref. 52. Comparison of structural properties of these PLCs using immunocross-reactivity and sequence analysis shows a surprising lack of homology. Also, many differ in their Ca^{2+} requirement, isoelectric point, and substrate specificity. An example of the last difference was demonstrated with three PLCs purified from brain where it was shown that their specific activities towards PI was the same at the optimal pH for each enzyme. However, at neutral pH the order was δ > γ > β when PI was the substrate but β > δ > γ when PIP_2 was used (see Table 4 for nomenclature). Identification of the PI-specific PLCs has been complicated by their susceptibility to proteolysis. Now that the molecular properties for several have been established, reasonable comparison and a unified nomenclature have been proposed (Table 4).

Thus far, at least four distinct proteins have been characterized even though one, the β, has been isolated in partially-modified forms, β-1, β-2 and β-3; all three forms are equally active. Rhee and coworkers also have compared the sequences of the four PLCs and found that three have conserved regions designated X and Y (Fig. 7). However, since PLC-α lacked these regions, it has been difficult to assign a specific role to the X and Y regions for substrate or Ca^{2+} binding, or catalysis. The PLC-γ uniquely contains three regions homologous to tyrosine kinases'

Table 4. Isozymes of PLC Purified from Different Tissues[a]

Source	Molecular mass		Nomenclature	
	SDS-PAGE (kD)	cDNA (kD)	Original report	Proposed
Rat liver	68		I	α
Sheep seminal vesicular gland	65		I	α
	~85		II	δ or ε
Bovine brain	150	138.2[b]	I	β-1
	140			β-2
	100			β-3
	145	148.4[b]	II	γ
	85	85.8[b]	III	δ
Guinea pig uterus	62	56.6[c]	II	α
Bovine brain	154	138.6	PLC-154	β-1
	88			δ
Rat brain	85		II	δ
	85		III	ε
Human platelets	61		mPLC-II	
Rat liver	87			

[a]From Rhee et al.[52]

[b]Calculated molecular mass based on amino acid sequence deduced from rat brain cDNA.

[c]Calculated molecular mass based on amino acid sequence deduced from rat basophilic leukemic cell cDNA.

regulatory domains A, B and C. This raises the possibility that PLC-γ may be regulated in a manner similar to these four kinases, probably through interactions with G-proteins. Also, evidence available suggests that both positive and negative regulation is dependent upon protein phosphorylation by PKC and cAMP-dependent kinases, respectively.

Recent work in our and other laboratories indicates that PLCs are active on phospholipids that do not contain inositol. Wolf and Gross[53] described a PLC in the cytosolic fraction of dog heart that degraded PC but not PI or sphingomyelin. Studies in whole cells or cell homogenates have been complicated somewhat, however, by the presence of the competing enzymes PLD and PA phosphatase that also produce DG from PG (Fig. 8). Further complicating the analysis of DG formation is the presence of DG kinase that will phosphorylate DG produced by PLC to yield PA. The criteria shown at the bottom of Fig. 8 indicate the type of experiments that are required to differentiate between PLC and PLD action. We showed that TPA stimulation of MDCK cells led to a rapid rise in DG with a subsequent formation of PA.[6] Also, phosphocholine was found to accumulate although free choline was also detected that may have arisen from PLD or phosphocholine hydrolase action. Further discussion of cellular PLDs is presented in the next section.

Since the DG produced by PLCs active on a variety of substrates are potential activators of PKC, it is important to differentiate between the potential substrates for PLCs. From such studies one can approximate the relative activities of the PI-specific PLCs and the PLC(s) that degrade

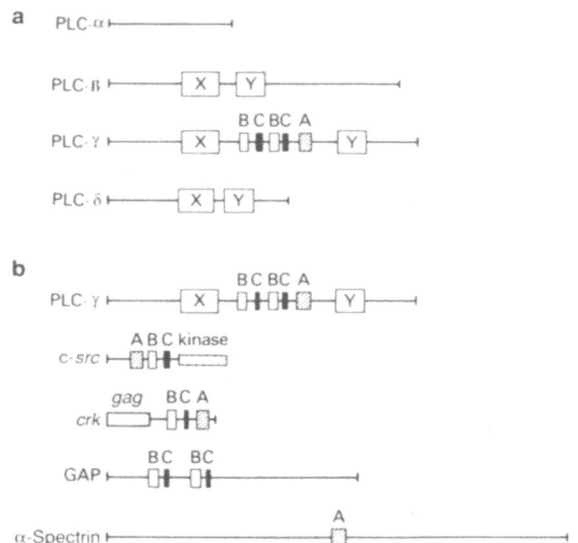

Fig. 7.   (a)   Linear representation of PLC isozymes.

Open boxes X and Y each denote the regions of ~150 and ~120 amino acids, respectively, of similar sequence found in PLC-β, PLC-γ, and PLC-δ. The conserved regions are not found in PLC-α.

(b)   Structural similarities between PLC-γ, pp60^{c-src}, p47$^{gag-crk}$, GAP, and α-spectrin.

The regions displaying sequence similarity are indicated by hatched box A, open box B, and black box C, which represent ~50, ~40, and ~15 amino acid residues, respectively.

(From Rhee et al., 1989.[52])

other PLs such as PC and PE. Since the acyl composition of a PL is somewhat characteristic for that PL, the acyl composition of DG produced upon cell stimulation and PLC activation can give some indication of the precursor. For example, PI has a very high content of C18:0 and C20:4, therefore, an increase in DG with a high amount of the C18:0 and C20:4 species would be consistent with the action of a PI-specific PLC whereas an increase in other species indicates other DG precursors are hydrolyzed. A detailed study of DG species from hepatocytes indicated that PC was a primary source of DG even though a large increase in C18:0 and C20:4 DG was produced when cells were challenged with vasopressin.[54] The other advantage of this approach was that the relative activation of PLCs by different agonists could be explored and, when coupled with the use of isotopic precursors, estimates could be made as to how rapidly a given PL pool turned over, relative to its availability to PLCs and PLDs.

Sphingomyelinases also fall into the PLC family and considerable interest has been generated in sphingomyelinases since a product of the reaction, sphingosine, has been implicated as a negative regulator of PKC in a variety of systems (Fig. 2; reviewed in Ref. 55). In this case, the

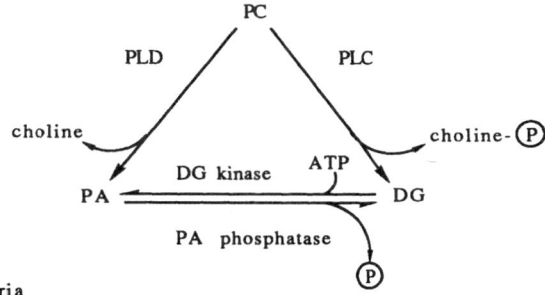

Criteria

PLC   1   Show stoichiometric DG + choline phosphate
       2   Kinetics; DG appears before PA

PLD   1   Phosphate retained on lipid; no ATP incorporated
       2   Kinetics; PA appears before DG

Fig. 8.   This scheme depicts pathways by which DG and
PA can arise.

sphingomyelinase is coupled to a ceramidase that degrades ceramide to
sphingosine. Determination of the amount of sphingosine formed in liver
plasma membranes suggests that this membrane, the proposed site of PKC
action, could provide the necessary lipid to block PKC action. Related to
that, DG, but not TPA, stimulates sphingosine production in pituitary
cells thereby providing a type of feedback system for PKC (Fig. 9).[56]

While this scheme is speculative, the effects of added sphingosine or
sphingomyelinase on cell characteristics such as TPA-induced differentiation
make this scheme attractive.[57]

Another class of PLCs under study with increased intensity are those
that liberate proteins from the cell surface by cleavage of a PI derivative,

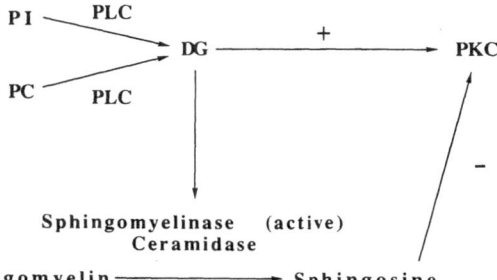

Fig. 9.   This scheme depicts pathways by which DG
produced from PLC action can modulate PKC
both positively (directly) or negatively
(indirectly) through sphingosine formation.
(From Merrill, 1989.[55])

15

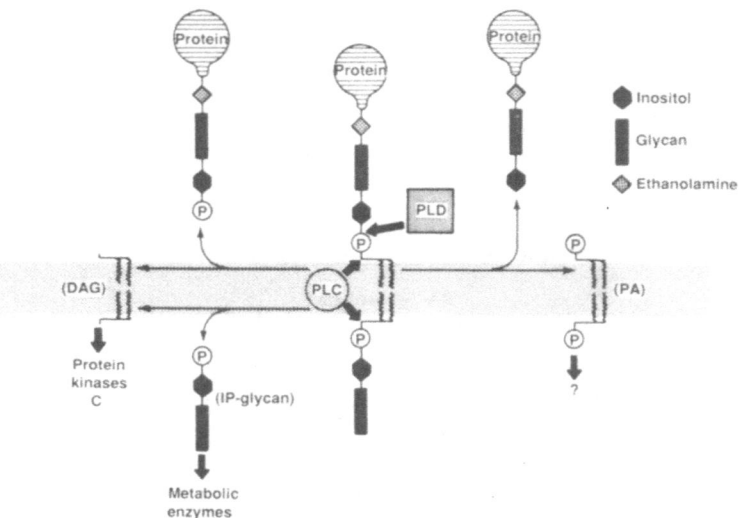

Fig. 10. Model for the release of glycosyl-PI anchored proteins
         by the action of PLC and PLD. (From Low and Saltiel,
         1988.[60])

glycosyl-PI, that covalently anchors the protein to the membrane (Fig. 10).
In a similar fashion PLD can liberate cell-surface anchored proteins.
Some proteins that are liberated in this manner are alkaline phosphatase,
5'-nucleotidase, and acetylcholinesterase. The first PI-specific PLC
shown to catalyze enzyme release was purified from Bacillus cereus[58];
later similar enzymes were isolated from a variety of bacteria (see
Refs. 59 and 60 for an extensive review on this subject). As pointed out
by Low in this review, it is not clear whether or not the bacterial PLC
functions to liberate these proteins in situ. Specific PLC for GPI
proteins have recently been identified. One of considerable interest,
from the pathophysiology prospective, was isolated from Trypanosoma
brucei. This PLC can remove a number of GPI-anchored proteins including
the variant surface glycoprotein from the tryposomal membrane. Unlike the
bacterial PLCs, this PLC does not degrade PI and is membrane-associated.[61]
A GPI-specific PLC has been purified from liver plasma membranes that can
degrade the insulin-sensitive GPI.[62] Undoubtedly, a number of other
similar PLCs and related PLDs will be discovered now that assay systems
suitable for their measurement have been developed. The physiological
impact of this class of PLCs and PLDs undoubtedly will be great.

PHOSPHOLIPASES D

     The history of PLDs is somewhat like that of PLCs in that for a
number of years they were thought to be absent from mammalian tissues. In
this case, however, the original sources of PLDs were plants such as
carrots and cabbages. One characteristic of the PLDs found in the earlier
works was that spinach leaves synthesized phosphatidylmethanol when
extracted with methanol. This led to the discovery that PLDs from plant
catalyze a "base exchange." In this example, PLD catalyzed the "exchange"
of the base group of the spinach phospholipids for methanol, in addition
to catalyzing hydrolysis to PA. The reaction is not a true base exchange
since the enzyme-bound intermediate is phosphatidate rather than the base
group. Correctly, therefore, PLD catalyzes a transphosphatidylation.[1]

In this mechanism a number of hydroxyls act as phosphatidate acceptors, including H_2O that produces PA. This mechanism appears to be correct for the bacterial (<u>Corynebacterium</u>, <u>Streptomyces</u>, <u>Haemophilus</u>, and <u>Vibrio</u>), animal, as well as plant PLDs. As will be discussed, this characteristic has been useful in differentiating PLDs from PLCs in a number of studies in whole cells or cell homogenates.[1]

Initial studies on PLDs in mammalian tissues were complicated by the presence of true base exchange enzymes that had no apparent hydrolytic activity. However, a hydrolytic system was identified in brain with catalytic properties different from those of the base exchange system that demonstrated that PLDs do exist in mammalian tissue.[63] A separate PLD was described in the mid-1970s that acted on lysophospholipid with an ether linkage in position-1; the presence of an ester group in position-2 drastically lowered hydrolysis.[64] This latter characteristic distinguishes this PLD from the others described thus far.

A number of studies have now shown that PLD can be stimulated by agonists or by the PKC activator TPA. As described above, the difficulty encountered when DG is measured is to determine the extent to which the stimulant is activating a PLC vs. a PLD. An example of one approach is given here since it takes advantage of the unique transphosphatidylation properties of the PLC to produce PEt when ethanol is present in the system.[7] Also, this approach differentiates between PA (PEt) formed by PLD from other sources of PA such as the <u>de novo</u> synthetic pathway. In this study, Pai and coworkers incubated HL-60 granulocytes with [³H]alkyl-[³²P]LPC that was acylated in the cells to form [³H]alkyl-[³²P]PC. Subsequent stimulation of the cells with the chemotactic peptide f-MLP in the presence of ethanol led to the formation of the expected products [³H]alkyl-[³²P]PA and [³H]alkyl-[³²P]PEt with the same isotopic ratio as the precursor [³H]alkyl-[³²P]PC (Fig. 11). This demonstrates the presence of PLD in these cells even though an estimate of the relative PLC activity was not given. An important aspect of this study was the demonstration that no [³²P]ATP was formed in the cells that could have given rise to [³H]alkyl-[³²P]PA by phosphorylation of [³H]alkyl-DG. While the stimulus-coupling mechanism with f-MLP to PLD is unclear, evidence exists for the involvement of guanine nucleotides in hepatocyte membranes.[65]

EVIDENCE FOR PLD ACTION IN HL-60 GRANULOCYTES

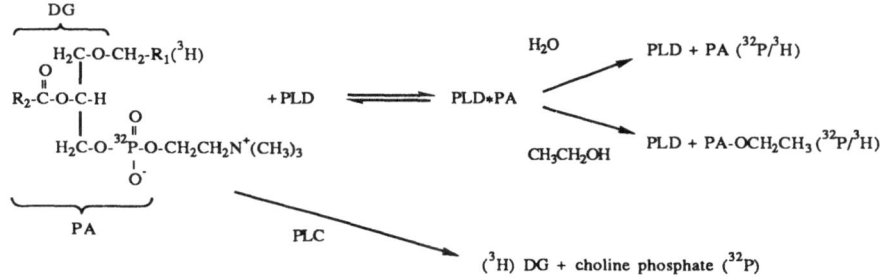

Fig. 11.   This scheme depicts how [³H]alkyl-[³²P]PC can be used to differentiate between PLC and PLD action. (From Pai et al., 1989.[7])

As shown in Fig. 10, PLDs also liberate glycosyl-PI anchored proteins on the cell surface. Such a PLD has now been purified to homogeneity from human plasma and shown to have M_r of 110,000.[66] It is activated by Ca^{2+} and since 1,10-phenanthroline inhibits, additional metals may be involved. Thus far the physiologic function is unclear since this PLD does not act on intact cells.

SUMMARY

Considerable work has gone into the study of PLs since the first suggestions of their existence nearly a century ago. This work has intensified enormously since the mid-1970s when their role in signal-coupling mechanisms and in pathophysiology was recognized. While much has been done to understand this diverse group of enzymes at the molecular and mechanistic levels, the discovery of new PLs has far outstripped our capacity to study them in sufficient detail to appreciate what makes each unique while perhaps having some common mechanisms of action and regulation. One would almost plead: No new PLs — Let us study those at hand! That is not the case in our field and the discovery of new PLs will continue. It is important, however, that an understanding be gained of these enzymes at the molecular level, how they interact with their substrates, and how regulatory factors can target the function of PLs in situ.

REFERENCES

1. M. Waite, The Phospholipases, in: "Handbook of Lipid Research," D. J. Hanahan, ed., vol. 5, 332 pages, Plenum Publishing Corporation, New York (1987).
2. H. van den Bosch, A. J. Aarsman, A. J. Slotboom, and L. L. M. van Deenen, On the specificity of rat liver lysophospholipase, Biochim. Biophys. Acta 164:215 (1968).
3. R. M. Kramer, C. R. Pritzker, and D. Deykin, Coenzyme A-mediated arachidonic acid transacylation in human platelets, J. Biol. Chem. 259:2403 (1984).
4. C. E. Walsh, M. Waite, M. J. Thomas, and L. R. DeChatelet, Release and metabolism of arachidonic acid in human neutrophils, J. Biol. Chem. 256:7228 (1981).
5. M. R. Hokin, and L. E. Hokin, The role of phosphatidic acid and phosphoinositide in transmembrane transport elicited by acetylcholine and other humoral agents, Int. Rev. Neurobiol. 2:99 (1960).
6. L. W. Daniel, M. Waite, and R. L. Wykle, A novel mechanism of diglyceride formation: 12-O-tetradecanoyl-phorbol-13-acetate stimulates the cyclic breakdown and resynthesis of phosphatidylcholine, J. Biol. Chem. 261:9128 (1986).
7. J.-K. Pai, M. I. Siegel, R. W. Egan, and M. Billah, Phospholipase D catalyzes phospholipid metabolism in chemotactic peptide-stimulated HL-60 granulocytes, J. Biol. Chem. 263:12472 (1988).
8. W. Siess, P. C. Weber, and E. G. Lapetina, Activation of phospholipase C is dissociated from arachidonate metabolism during platelet shape change induced by thrombin or platelet-activating factor: epinephrine does not induce phospholipase C activation or platelet shape change, J. Biol. Chem. 259:8286 (1984).
9. S. F. Rittenhouse, Activation of human platelet phospholipase C by ionophore A23187 is totally dependent upon cyclo-oxygenase products and ADP, Biochem. J. 222:103 (1984).
10. G. Lindblom, and L. Rilfors, Cubic phases and isotropic structures formed by membrane lipids - possible biological relevance, Biochim. Biophys. Acta 988:221 (1989).

11. H. B. M. Lenting, K. Nicolay, and H. van den Bosch, Regulatory aspects of mitochondrial phospholipase A_2: correlation of hydrolysis rates with substrate configuration as evidenced by ^{31}P-NMR, <u>Biochim. Biophys. Acta</u> 958:405 (1988).
12. M. Robinson, and M. Waite, Physical-chemical requirements for the catalysis of substrates by lysosomal phospholipase A_1, <u>J. Biol. Chem.</u> 258:14371 (1983).
13. F. Pattus, A. J. Slotboom, and G. H. deHaas, Regulation of phospholipase A_2 activity by the lipid-water interface: a monolayer approach, <u>Biochemistry</u> 18:2691 (1979).
14. M. C. Komaromy, and M. C. Schotz, Cloning of rat hepatic lipase cDNA: evidence for a lipase gene family, <u>Proc. Natl. Acad. Sci. USA</u> 84:1526 (1987).
15. T. Kuwae, P. C. Schmid, and H. H. O. Schmid, Assessment of phospholipid deacylation-reacylation cycles by a stable isotope technique, <u>Biochem. Biophys. Res. Commun.</u> 142:86 (1987).
16. F. R. Cochran, V. L. Roddick, J. R. Connor, J. T. Thornburg, and M. Waite, Regulation of arachidonic acid metabolism in resident and BCG-activated macrophages: role of lyso(bis)phosphatidic acid, <u>J. Immunol.</u> 138:1877 (1987).
17. N. Kawasaki, and K. Saito, Purification and some properties of lysophospholipase from <u>Penicillium</u> <u>notatum</u>, <u>Biochim. Biophys. Acta</u> 296:426 (1973).
18. N. Kawasaki, J. Sugatani, and K. Saito, Studies on a phospholipase B from <u>Penicillium</u> <u>notatum</u>, <u>J. Biochem.</u> 77:1233 (1975).
19. H. van den Bosch, Phospholipases, <u>in</u>: "Phospholipids," J. N. Hawthorne and G. B. Ansell, eds., vol. 4, pp. 313-357, Elsevier Biomedical, Amsterdam (1982).
20. K. Saito, and M. Kates, Substrate specificity of a highly-purified phospholipase B from <u>Penicillium</u> <u>notatum</u>, <u>Biochim. Biophys. Acta</u> 369:245 (1974).
21. M. Waite, and P. Sisson, Studies on the substrate specificity of the phospholipase A_1 of the plasma membrane of rat liver, <u>J. Biol. Chem.</u> 249:6401 (1974).
22. R. W. Gross, and B. E. Sobel, Rabbit myocardial cytosolic lyso-phospholipase: purification, characterization, and competitive inhibition by L-palmitoyl carnitine, <u>J. Biol. Chem.</u> 258:5221 (1983).
23. R. W. Gross, R. C. Drisdel, and B. E. Sobel, Rabbit myocardial lysophospholipase-transacylase: purification, characterization, and inhibition by endogenous cardiac amphiphiles, <u>J. Biol. Chem.</u> 258:15165 (1983).
24. A. J. Slotboom, H. M. Verheij, and G. H. deHaas, On the mechanism of phospholipase A_2, <u>in</u>: "Phospholipids," J. N. Hawthorne and G. B. Ansell, eds., vol. 4, ch. 10, pp. 359-434, Elsevier Biomedical, Amsterdam (1982).
25. R. L. Heinrikson, E. T. Krueger, and P. S. Keim, Amino acid sequence of phospholipase A_2-α from the venom of <u>Crotalus</u> <u>adamanteus</u>: a new classification of phospholipase A_2 based upon structural determinants, <u>J. Biol. Chem.</u> 252:4913 (1977).
26. H. Tojo, T. Ono, S. Kuramitsu, H. Kagamiyama, and M. Okamoto, A phospholipase A_2 in the supernatant fraction of rat spleen: its similarity to rat pancreatic phospholipase A_2, <u>J. Biol. Chem.</u> 263:5724 (1988).
27. T. Ono, H. Tojo, S. Kuramitsu, H. Kagamiyama, and M. Okamoto, Purification and characterization of a membrane-associated phospholipase A_2 from rat spleen: its comparison with a cytosolic phospholipase A_2 S-1, <u>J. Biol. Chem.</u> 263:5732 (1988).
28. O. Ohara, M. Tamaki, E. Nakamura, Y. Tsuruta, Y. Fujii, M. Shin, H. Teraoka, and M. Okamoto, Dog and rat pancreatic phospholipases A_2: complete amino acid sequences deduced from complementary DNAs, <u>J. Biochem.</u> 99:733 (1986).

29. H. Tojo, T. Ono, and M. Okamoto, A pancreatic-type phospholipase A_2 in rat gastric mucosa, <u>Biochem. Biophys. Res. Commun</u>. 151:1188 (1988).

30. A. J. Aarsman, J. G. N. deJong, E. Arnoldussen, F. W. Neys, P. D. van Wassenaar, and H. van den Bosch, Immunoaffinity purification, partial sequence, and subcellular localization of rat liver phospholipase A_2, <u>J. Biol. Chem</u>. 264:10008 (1989).

31. S. Hara, I. Kudo, K. Matsuta, T. Miyamoto, and K. Inoue, Amino acid composition and NH_2-terminal amino acid sequence of human phospholipase A_2 purified from rheumatoid synovial fluid, <u>J. Biochem</u>. 104:326 (1988).

32. R. M. Kramer, C. Hession, B. Johansen, G. Hayes, P. McGray, E. P. Chow, R. Tizard, and R. B. Pepinsky, Structure and properties of a human non-pancreatic phospholipase A_2, <u>J. Biol. Chem</u>. 264:5768 (1989).

33. C.-Y. Lai, and K. Wada, Phospholipase A_2 from human synovial fluid: purification and structural homology to the placental enzyme, <u>Biochem. Biophys. Res. Commun</u>. 157:488 (1988).

34. J. J. Seilhamer, T. L. Randall, M. Yamanaka, and L. K. Johnson, Pancreatic phospholipase A_2: isolation of the human gene and cDNAs from porcine pancreas and human lung, <u>DNA</u> 5:519 (1986).

35. H. M. Verheij, J. Westerman, B. Sternby, and G. H. deHaas, The complete primary structure of phospholipase A_2 from human pancreas, <u>Biochim. Biophys. Acta</u> 747:93 (1983).

36. H. W. Chang, I. Kudo, M. Tomita, and K. Inoue, Purification and characterization of extracellular phospholipase A_2 from peritoneal cavity of caseinate-treated rat, <u>J. Biochem</u>. 102:147 (1987).

37. M. Hayakawa, K. Horigome, I. Kudo, M. Tomita, S. Nojima, and K. Inoue, Amino acid composition and NH_2-terminal amino acid sequence of rat platelet secretory phospholipase A_2, <u>J. Biochem</u>. 101:1311 (1987).

38. H. Mizushima, I. Kudo, K. Horigome, M. Murakami, M. Hayakawa, D.-K. Kim, E. Kondo, M. Tomita, and K. Inoue, Purification of rabbit platelet secretory phospholipase A_2 and its characteristics, <u>J. Biochem</u>. 105:520 (1989).

39. S. Forst, J. Weiss, and P. Elsbach, Structural and functional properties of a phospholipase A_2 purified from an inflammatory exudate, <u>Biochemistry</u> 25:8381 (1986).

40. G. C. Wright, C. E. Ooi, J. Weiss, and P. Elsbach, Purification of a cellular (granulocyte) and an extracellular (serum) phospholipase A_2 that participate in the destruction of <u>Escherichia</u> <u>coli</u> in a rabbit inflammatory exudate, <u>J. Biol. Chem</u>., submitted (1989).

41. M. J. Dufton, and R. C. Hider, Classification of phospholipases A_2 according to sequence: evolutionary and pharmacological implications, <u>Eur. J. Biochem</u> 137:5454 (1983).

42. C. J. van den Bergh, A. J. Slotboom, H. M. Verheij, and G. H. deHaas, The role of aspartic acid-49 in the active site of phospholipase A_2: a site-specific mutagenesis study of porcine pancreatic phospholipase A_2 and the rationale of the enzymatic activity of [lysine^{49}]- phospholipase A_2 <u>Agkistrodon</u> <u>piscivorus</u> <u>piscivorus</u>' venom, <u>Eur. J. Biochem</u>. 176:353 (1988).

43. C. J. van den Bergh, A. C. A. P. A. Bekkers, H. M. Verheij, and G. H. deHaas, Glutamic acid 71 and aspartic acid 66 control the binding of the second calcium ion in porcine pancreatic phospholipase A_2, <u>Eur. J. Biochem</u>. 182:307 (1989).

44. O. P. Kuipers, R. Dijkman, C. E. G. M. Pals, H. M. Verheij, and G. H. deHaas, Evidence for the involvement of tyrosine-69 in the control of stereospecificity of porcine pancreatic phospholipase A_2, <u>Protein Engr</u>. 2:467 (1989).

45. O. P. Kuipers, M. M. G. M. Thunnissen, P. deGeus, B. W. Dijkstra, J. Drenth, H. M. Verheij, and G. H. deHaas, Enhanced activity and altered specificity of phospholipase A_2 by deletion of a surface loop, <u>Science</u> 244:82 (1989).

46. L. A. Loeb, and R. W. Gross, Identification and purification of sheep platelet phospholipase A_2 isoforms: activation by physiologic concentrations of calcium ion, J. Biol. Chem. 261:10467 (1986).

47. Y. Suwa, I. Kudo, M. Okada, A. Imaizumi, Y. Suzuki, H. W. Chang, and K. Inoue, Novel proteinous inhibitors of phospholipase A_2 purified from rat inflamed sites, submitted (1989).

48. F. F. Davidson, E. A. Dennis, M. Powell, and J. Glenney, Inhibition of phospholipase A_2 by "lipocortins" and calpactins: an effect of binding to substrate phospholipids, J. Biol. Chem. 262:1698 (1987).

49. P. Elsbach, and J. Weiss, Phagocytosis of bacteria and phospholipid degradation, Biochim. Biophys. Acta 947:29 (1988).

50. K. Aalmo, L. Hansen, E. Hough, K. Jynge, J. Krane, C. Little, and C. B. Storm, An anion binding site in the active centre of phospholipase C from Bacillus cereus, Biochem. Int. 8:27 (1984).

51. D. P. Siegel, J. Banschbach, D. Alford, H. Ellens, L. J. Lis, P. J. Quinn, P. L. Yeagle, and J. Bentz, Physiological levels of diacylglycerols in phospholipid membranes induce membrane fusion and stabilize inverted phases, Biochemistry 28:3703 (1989).

52. S. G. Rhee, P.-G. Suh, S.-H. Ryu, and S. Y. Lee, Studies of inositol phospholipid-specific phospholipase C, Science 244:546 (1989).

53. R. A. Wolf, and R. W. Gross, Identification of neutral active phospholipase C which hydrolyzes choline glycerophospholipids and plasmalogen selective phospholipase A_2 in canine myocardium, J. Biol. Chem. 260:7296.

54. G. Augert, S. B. Bocckino, P. F. Blackmore, and J. H. Exton, Hormonal stimulation of diacylglycerol formation in hepatocytes: evidence for phosphatidylcholine breakdown, J. Biol. Chem., in press (1989).

55. A. H. Merrill, Lipid modulators of cell function, Nutr. Rev. 47:161 (1989).

56. R. N. Kolesnick, 1,2-Diacylglycerols but not phorbol esters stimulate sphingomyelin hydrolysis in GH_3 pituitary cells, J. Biol. Chem. 262:16759 (1987).

57. R. N. Kolesnick, Sphingomyelinase action inhibits phorbol ester-induced differentiation of human promyelocytic leukemic (HL-60) cells, J. Biol. Chem. 264:7617 (1989).

58. H. Ikezawa, M. Yamanegi, R. Taguchi, T. Miyashita, and T. Ohyabu, Studies on phosphatidylinositol phosphodiesterase (phospholipase C type) of Bacillus cereus. I. Purification, properties and phosphatase-releasing activity, Biochim. Biophys. Acta 450:154 (1976).

59. M. G. Low, Degradation of glycosyl-phosphatidylinositol anchors by specific phospholipases, in: "Glycosylphosphatidylinositol Membrane Protein Anchors and Cell Signalling Events," A. J. Turner, ed., ch. 2, Ellis Horwood Publ., U.K., in press (1989).

60. M. G. Low, and A. R. Saltiel, Structural and functional roles of glycosylphosphatidylinositol in membranes, Science 239:268 (1988).

61. R. Bulow, and P. Overath, Purification and characterization of the membrane-form variant surface glycoprotein hydrolase of Trypanosoma brucei, J. Biol. Chem. 261:11918 (1986).

62. J. A. Fox, N. M. Soliz, and A. R. Saltiel, Purification of a phosphatidylinositol-glycan-specific phospholipase C from liver plasma membranes: a possible target of insulin action, Proc. Natl. Acad. Sci. USA 84:2663 (1987).

63. J. N. Kanfer, The base exchange enzymes and phospholipase D of mammalian tissue, Can. J. Biochem. 58:1370 (1980).

64. R. L. Wykle, and J. M. Schremmer, A lysophospholipase D pathway in the metabolism of ether-linked lipids in brain microsomes, J. Biol. Chem. 249:1742 (1974).

65. S. B. Bocckino, P. F. Blackmore, P. B. Wilson, and J. H. Exton, Phosphatidate accumulation in hormone-treated hepatocytes via a phospholipase D mechanism, J. Biol. Chem. 262:15309 (1987).

66. M. A. Davitz, J. Hom, and S. Schenkman, Purification of a glycosyl-phosphatidylinositol-specific phospholipase D from human plasma, J. Biol. Chem. 264:13760 (1989).

COMPARATIVE ANATOMY OF PHOSPHOLIPASE A2 STRUCTURES

K. B. Ward and N. Pattabiraman

Code 6030, Laboratory for the Structure of Matter
Naval Research Laboratory, Washington D.C. 20375

INTRODUCTION

For a number of years our laboratory has been interested in phospholipases A2, especially those that are neurotoxic. Within the last year or two we have begun to direct our interest to nontoxic phospholipases that are of importance in human health. The goal of our presentation is to set the stage for you, and to introduce you to many of the terms and concepts that will be given in other presentations of this symposium. Many of these will deal with enzyme mechanisms, enzyme regulation, enzyme activation and so forth, and will be given by researchers actively involved in these areas. A full appreciation of their presentations will be facilitated by an understanding of the detailed structural anatomy of phospholipases A2.

We review here some important results of primary structure analysis of phospholipase A2; a more extensive discussion will be found in Waite's presentation. We summarize both secondary and tertiary structures not only because of their importance for enzymatic activity, but also their impact on mechanisms of the so called pharmacological activities, that is, effects like myotoxicity, anti-coagulant properties, and hemolytic effects that may or may not be related to PLPA2 activity. Finally we will discuss the quaternary structure of a number of phospholipases A2, and the role this level of organization may play in activity. Some major sources of phospholipases A2 are shown in Table 1. They are essentially found in two areas. There are a number of mammalian sources, some of which are given here for cellular phospholipases A2. In addition there is a very common extracellular, and therefore soluble, phospholipase A2 that is found in mammalian pancreas. There are also a variety of non-mammalian sources, particularly venoms. Less detail is known about the bee venom enzyme, whereas a good deal is known, particularly from the

Biochemistry, Molecular Biology, and Physiology of Phospholipase A₂ and Its Regulatory Factors
Edited by A. B. Mukherjee, Plenum Press, New York, 1990

23

Table 1. Sources of Phospholipase A2 (from Waite(1987))

```
A.  Mammalian:
    1.    Intracelluar:
          Neutophils, Brain, Platelets, Macrophages, Erythro-
          cytes, Liver mitochondria, Lung exudate, Ascites
          hepatoma cells, Intestine (lumenal and mucosal),
          Heart mitochodria and lysosomes and spleen

    2.    Extracellular:
          Human, Bovine, and Porcine pancreas

B.  Non mammalian:
    1.    Bee venoms
    2.    Snake venoms
          a. Elapidae, Hydrophidae (Cobras and Sea Snakes)
          b. Viperadae, Crotalidae (Vipers and Pit vipers)
```

structural standpoint, about those phospholipases found in snake venom. In particular, there are two classes of phospholipases found in snake venoms according to whether the snakes are in the cobra or sea snake families as opposed to the viper or pit viper families.

Structurally, phospholipases A2 are called α-β proteins. They contain the typical building blocks one find in proteins: α-helices, β-pleated sheets and disulphide bonds which play a particularly important role in these molecules. There are also regions of the peptide chain that are for the lack of a better term called "random coil". They are almost certainly not random; if we knew enough about structure and function we would know why they are disposed in space the way they are.

THE ARCHETYPICAL PHOSPHOLIPASE A2 STRUCTURE

In presentations in this symposium, you will be shown a number of three-dimensional pictures of phospholipases A2. If you are not expert at interpreting those structural representa-tions, it may be useful to begin by studying the archetypical structure shown in figure 1. This is the "standard view" of the molecule that we use in our laboratory. We therefore consider it "God's view" of the enzyme, but workers in other laboratories are used to viewing the molecule from some other angle, and this can be confusing for non-structural researchers who seek information from different structural papers. Because many researchers interested in structure analysis results have difficulty viewing stereo-images, we have chosen to present structures using ball-and-stick-like images with somewhat exaggerated prospective. In these drawings, the sizes of atoms and bonds decrease with distance from the viewer; the resulting depth illusion is often more easily preceived than when other methods are used.

Figure 1 shows an α-carbon backbone tracing of bovine pancreas phospholipase A2, the first molecule of this class whose structure was determined in Drenth's laboratory (Dijkstra *et al.*,1978, Dijkstra, *et al.*, 1981a; Dijkstra,*et al*, 1981b) using single crystal x-ray analyses. All helices are shown with solid black bonds, whereas random chain is shown using dashed bonds. Near the center of this view, and nearest to the viewer, lies the N terminus, which initiates the A helix. A short bend lies between this and the short B helix which follows. The chain then meanders away from the viewer following a "random" course ending near the rear of the molecule. At this point, the long C helix begins its course back towards the front by proceeding down and to the left. The chain then turns abruptly toward the front and spirals toward

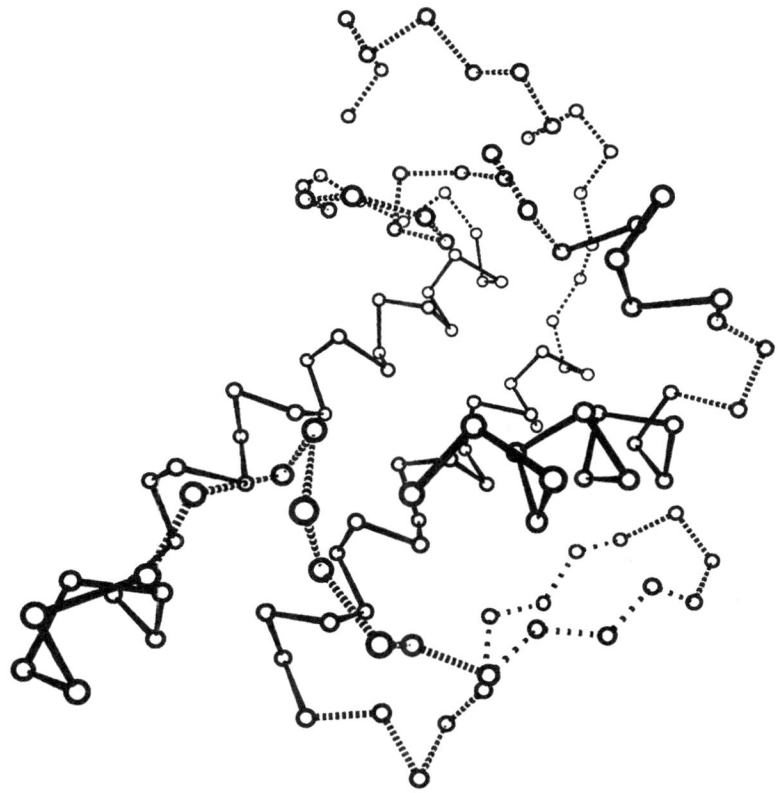

Figure 1. An archetypical view of the trace of α-carbon atoms of bovine pancreatic phopholipase A2.

the viewer as the poorly formed D helix. A length of random coil then stretches from left to right, first up, then down, across the front of the molecule until there begins the single example of antiparallel β-sheet structure (shown with bonds containing fewer dashes), which lies below the A helix. Following this "β-wing" the chain turns down and leftward before forming the long, final E helix which proceedes away from the viewer, ending in the rear of the molecule. A random coil "tail" then snakes up along the back of the protein, turning leftward before ending at the C terminus. This simplified structural summary may well be useful when considering more detailed structures in other presentations.

Figure 2 is a similar α-carbon diagram which emphasizes the location of disulfide bonds in the enzyme. Note how these physically pin the various secondary structural features of the molecule together, and make it very rigid. With few exceptions

Figure 2. α-carbon atom tracing for the Group I Phospholipase from bovine pancreas. Disulfide bridges shown with dashed lines.

these 7 disulfides occur in all phospholipases A2 and are thus obviously very important in maintaining the structure which conveys the functions in which we are all interested.

There exist 53 sequences of phospholipase A2 and homologues that occur in mammalian pancreas and venom sources (van den Bergh, et al., 1989). There are 20 residues that are always conserved and therefore must play crucial structure-function roles. Half of these comprise cystines. So there are 10 half cystines always present and there are 10 other residues that are also vital. In addition, there are 10 more residues that are "usually conserved", i.e. they are present in almost all sequences examined. The importance of the last 10 is that when they are replaced, the proteins invariably exhibit altered function. In fact it has been from the examination of these proteins that so much has been learned about the functions of these residue positions. The side chains of these residues are located in the central part of the molecule referred to as the "structural core" of phospholipase A2.   They are shown in figure 7, with invariant residues being shown in yellow, whereas "usually conserved" residues are in magenta, attached to a white α-carbon backbone similar to figure 2. Various regions of that core will be referred to often in other presentations in this symposium.

GROUP I AND GROUP II STRUCTURES

Heinrikson et al. (1987) were the first to classify the primary structures of phospholipases A2 into two major groups, and here we illustrate the corresponding differences in tertiary structure. The Group I phospholipases A2 are found in cobra and sea snake venoms and also in the mammalian pancreas exemplified by the bovine phospholipase A2 used as our structural archetype (figures 1 and 2).   The Group II phospholipases are found in crotalid (pit viper) and viper venoms and are best characterized by a phospholipase A2 from the western diamond back rattlesnake *Crotalus atrox* whose structure was determined by Keith, *et al.*(1981) and described by Brunie, *et al.* (1985). A simplified structure of this molecule is shown in figure 3. Group I phospholipases A2 always contain a disulfide bridge between CYS-11 and CYS-76; this bridge is missing in Group II enzymes. Despite this "missing" disulfide, members of Group II also have 7 disulfides, because they contain an "additional" bridge connecting the middle of the C helix to the C terminus of an extra long (by 7 residues) tail which also characterizes the Group II proteins (compare figure 2 to figure 3). The fourth and final distinguishing characteristic of Group II phopholipases is the absence of the D helix (the so called "elapid loop", from the cobra family, *Elapidae*), present in Group I.

Not withstanding these structural differences which distinguish Group I from Group II enzymes, the overall tertiary structure of the two classes is remarkably similar (Renetsedes, et al., 1985). The importance of maintaining this underlying structure is implied by the structural comparisions shown in figure 4. The (CYS-11; CYS-77) disulfide bridge serves to tie the β-wing to the A helix. This tethering function is accomplished in Group II enzymes (which lack the disulfide) by a salt-bridge between LYS-11 and GLU-77. This suggests an important functional role for the close proximity of these two structural features.

CALCIUM ION BINDING SITE

A part of the central core of sidechains conserved in phospholipase A2 consists of residues involved in binding calcium

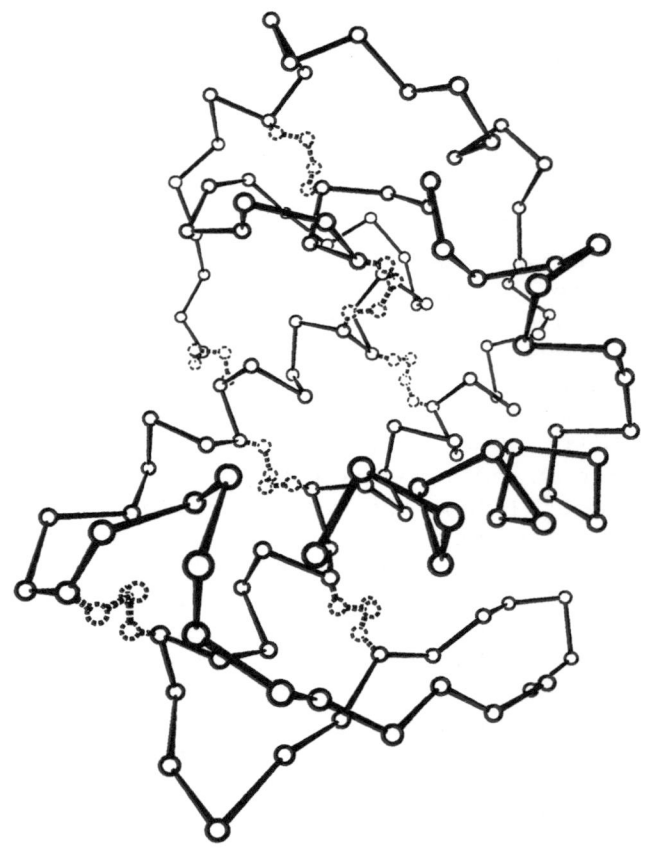

Figure 3. α-carbon atom tracing for a Group II (C. Atrox) phospholipase A2. Disulfide bridges shown with dashed lines.

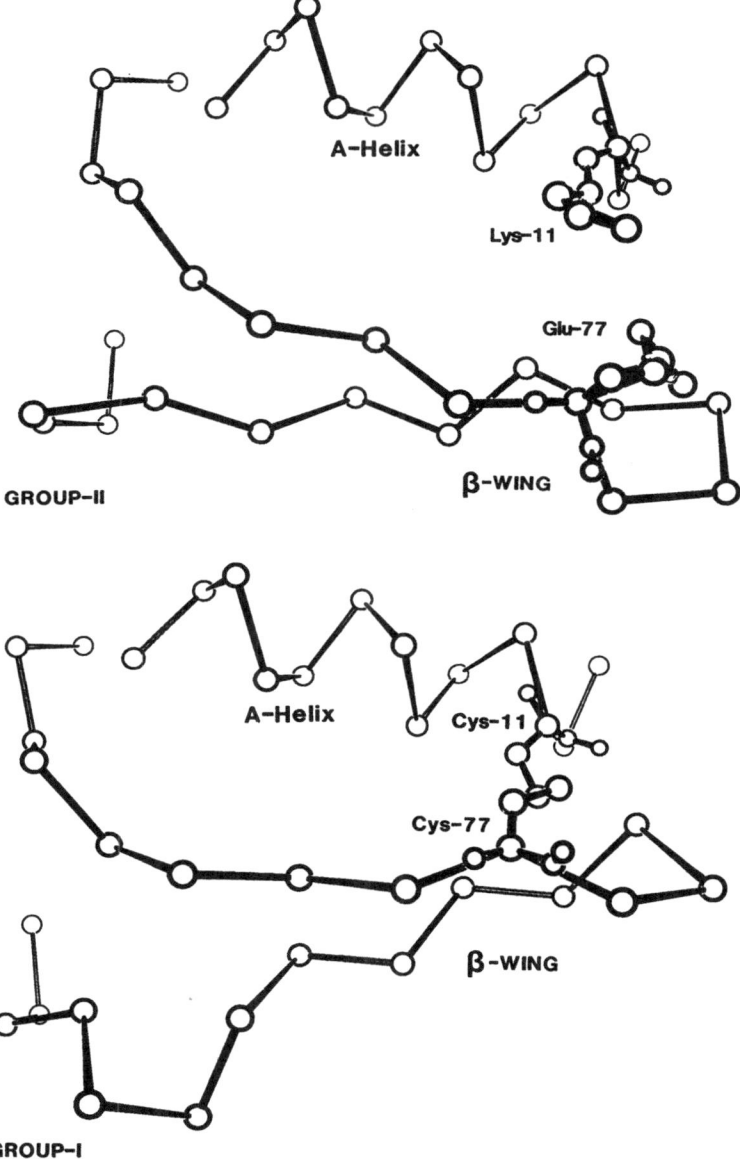

Figure 4. Tethers between the A helix and the β-wing in Group I (CYS-11 and CYS-77) and Group II (LYS-11 and GLU-77) phospholipase A2.

ion required for activity. Figure 5 shows the location of this region in a Group I enzyme, and a more detailed illustration of this "calcium binding loop", from a different perspective, is given in figure 6. Three ligands for the obligatory cation are provided by main chain carbonyl oxygens of residues 28,30 and 32 of the binding loop. The carboxylate side chain of conserved ASP-49, located on the C helix, provides two additional ligands, and additional liganding sites are occupied by tightly bound water molecules. Other presentations in this symposium will present further details about the important functional role played by this region of the enzyme.

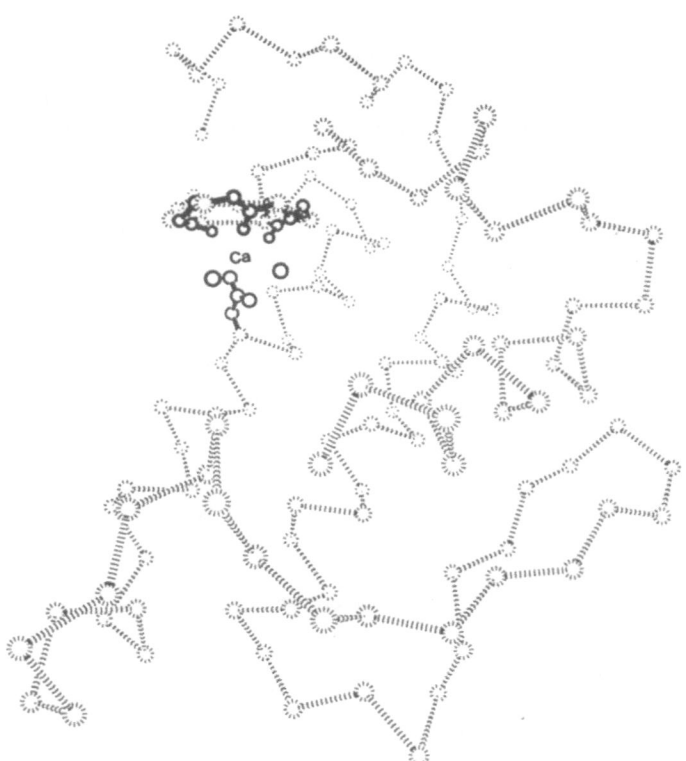

Figure 5. Location of the calcium ion binding site in bovine pancreatic phospholipase A2. Liganding groups shown in solid lines.

SOLVENT ACCESSIBILITY

    The ball and stick models of the α-carbon backbone in previous figures can mislead one into a false sense of the openness of such structures, even when they are expanded to include all atoms in the structure. Using atomic coordinates derived from an x-ray structure data base (Bernstein, *et al.*, 1977), we can construct a surface covering these molecules, which gives a better measure of the exterior "appearance" of the protein. This surface is computed by rolling a spherical "solvent" molecule around the exterior of the protein molecule, maintaining contact with the van der Waals surfaces of atoms near the exterior. The locus of points on the surface of the solvent molecule as it "kisses" the protein during this journey is called the "solvent accessible surface" (Langridge *et al.*, 1981; Connolly, 1981).

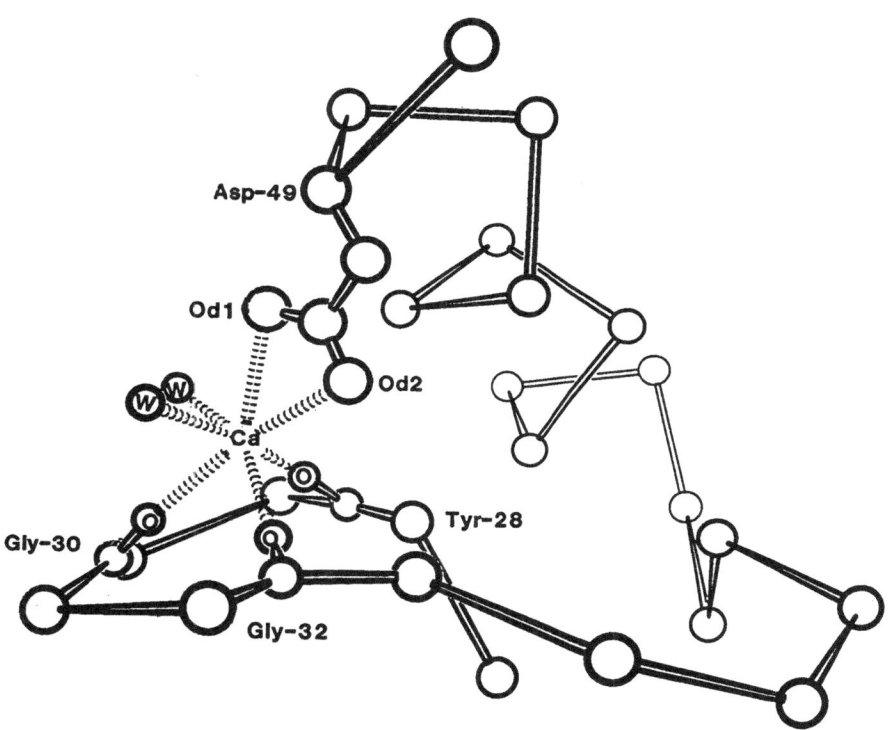

Figure 6. Detailed view of calcium ion binding site, showing the coordination sphere. Water molecules ligands shown as W.

SURFACE ELECTROSTATICS

Weiner *et al.*(1982) and Conrad (1988) have extended this concept by color coding the points on this surface according to the value of the electrostatic potential calculated 1.4 Å away from the point along a line perpendicular to the surface . Figures 8 , 9 and 10 illustrate the dramatic qualitative structural information that such surfaces provide. In these figures, neutral

Figure 7. A view of the invariant (yellow) and "usually conserved" (magenta) residues attached to a white α-carbon backbone of phospholipase A2 structure.

Figure 8. A view of the solvent accessible surface (represented by dots) of the bovine pancreatic (left) and C.atrox (right) phospholipase A2. Anionic electrostatic potentials are represented by magenta dots, cationic electrostatic potentials by cyan dots, and electrostatic potentials near zero by white dots.

Figure 9. The electrostatic surface viewed from a prospective at right angles from that in figure 8. Refer to figure 8 for color coding.

Figure 10. A view of a part of the electrostatic molecular surface of the nontoxic (C. atrox, right) and the highly neurotoxic phospholipase A2 subunit of mojave toxin (left). Refer to figure 8 for color coding.

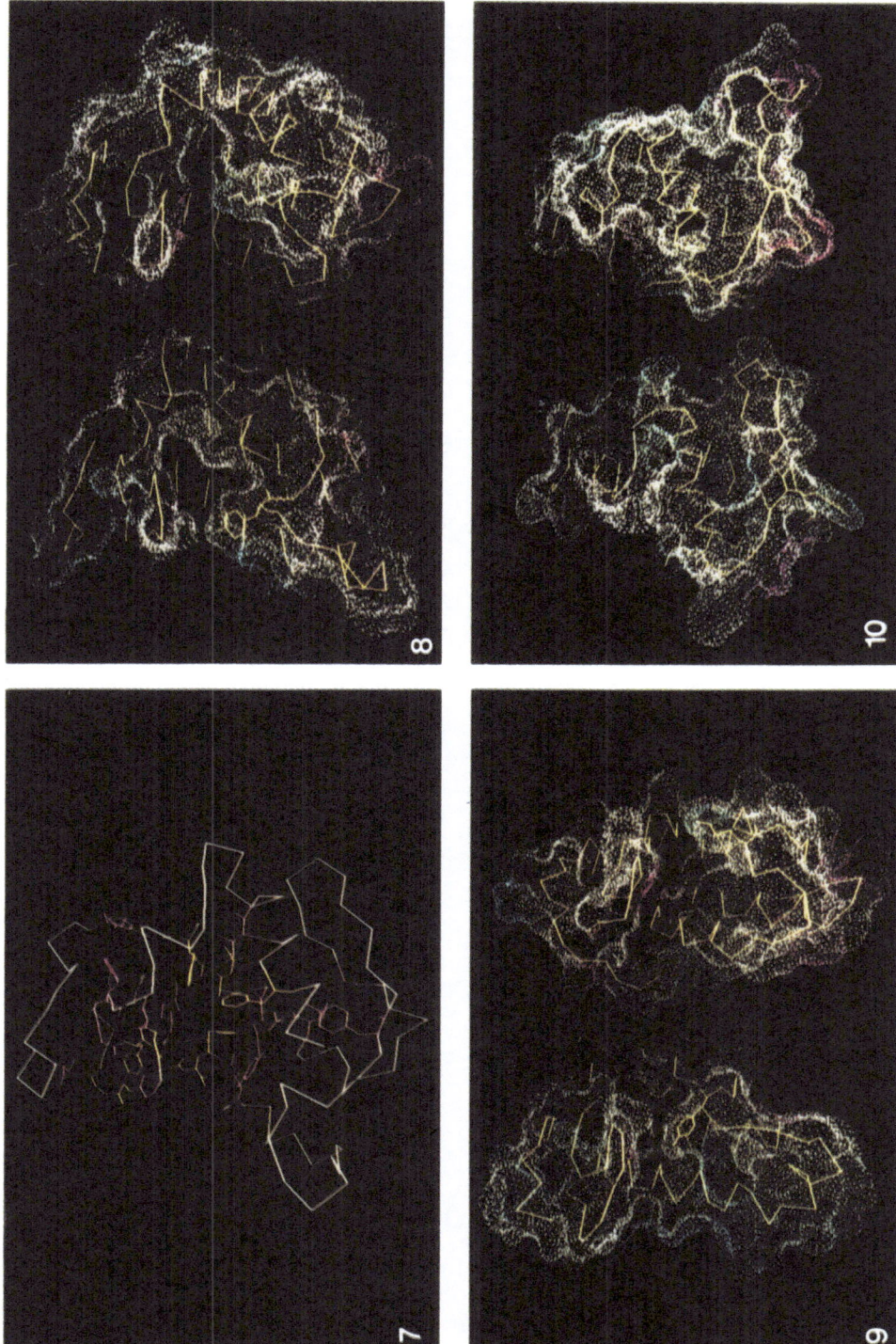

or only slightly charged regions are shown in white, points near anionic regions are colored magenta (purplish-red) and the cationic points are shown in cyan (greenish-blue).

Figure 8 shows a Group I phopholipase A2 (left, bovine pancrease) and a member of Group II (right,*C. atrox*), each in the "standard position" of figures 2 and 3, and both adorned with their solvent accessible coats of several colors. Differences in surface features become apparent in this display. Note the surface pockets leading to the site of activity . The tyrosine-69 of the porcine molecule contrasted with the lysine-69 at the equivalent position dramatically alters the nature of access to this pocket. The apparently more open nature of this pocket in the *C. atrox* structure is misleading because the left side is blocked by a dimeric mate (not shown) in this phospholipase.

If we face the molecules in figure 8 and command both to execute a precise left face, we are presented with the view shown in figure 9. This allows another detailed comparison of the surface clefts. Note the large cationic (cyan) surface patch on the porcine structure (left) compared to *C. atrox* (right), a feature which is likely related to the propensity of the *C. atrox* protein to form dimers. Finally, note the marked dipolar character (cationic top combined with anionic bottom) of the *C. atrox* molecule. Whether this is a general feature of functional importance must await evidence provided by future structure analyses on other members of both classes.

NEUROTOXICITY

In order to understand the structural correlates of presynaptic neurotoxicity of some phospholipases A2, we have recently determined the structure of the first presynaptic neurotoxin using x-ray diffraction methods (Collins, *et al.*, 1989). This protein is mojave toxin obtained from *C. scutulatus scutulatus*, the mojave rattlesnake, and closely related to the similar crotoxin from a South American rattlesnake. These proteins are dimeric and contain a phospholipase subunit and a "chaperone" subunit of undetermined function.

Figure 9 compares the charged surface features of the nontoxic *C. atrox* protein (right) with the highly neurotoxic phospholipase subunit of mojave toxin (left). Tsai *et al.*(1987) have argued that neurotoxic phospholipases are distinguished by cationic residues at certain positions in the primary structure. These correspond to the ARG-65 and LYS-69 located at the bottom of the molecule in this view. The resulting cationic (cyan) regions contrast dramatically with the anionic (magenta) surface patches on the *C. atrox* molecule (right).

QUATERNARY STRUCTURE

Both toxic and nontoxic phospholipases are known to exist in a wide varity of quaternary forms as shown in table 2. The quaternary state of the enzyme while interacting with aggregated substrate has not yet been clarified. Whereas some presynaptic neurotoxins such as ammodytoxin (Aleksiev and Tchorbanov, 1976) are monomeric, *N. naja naja* venom (Deems and Dennis, 1975) contains dimers and the venom of *Vipera berus orientale* (Boffa *et al.*, 1976) contains homotrimers. The importance of these quaternary forms for activity is not clear.

For heteroligomer toxins, however, much evidence points to an active role for the non-phospholipase subunits. The covalently bound B chain of β-bungarotoxin shows no phospholipase activity, and Kini and Iwanaga (1986) suggest it is responsible for neurotoxicity, apparently acting at the same site as dendrotoxin (Black, *et al.*, 1988), which is not a phospholipase.

Table 2   Quaternary Structures of Phospholipase A2
Containing Proteins

α = Phospholipase A2 subunit; β = chaperone subunit; γ = glycoprotein subunit; δ = non phospholipase, covalently attached	
Quaternary	**Examples**
1. α	Procine Pancreatic, *A. piscovorous,* ammodytoxin, notexin
2 αβ	Crotoxin, mojave toxin
3. αδ	β-bungarotoxin
4. $α_2$	*C. atrox, N. naja naja*
5. $α_3$	*Vipera berus orientale*
6. αβγ	Taipoxin

In mojave toxin and crotoxin, the phospholipase subunit is accompanied by a nontoxic "chaperone" subunit which apparently insures that this potent toxin only attacks specific targets of high binding affinity (Bon et al., 1979).

One of the most potent snake venom toxins, taipoxin, exists as a heterotrimer. Only the α-subunit is toxic, whereas the other two, one of which is glycosylated, serve an ancillary role; all three have closely related primary structures.

CONCLUSION

The molecular architecture of phospholipases A2 has been revealed in several instances by x-ray diffraction analyses. These results continue to play an important role in our understanding of enzymatic mechanisms. Many questions concerning mechanisms of toxicity must await similar analyses of heteroligomeric toxins containing both phospholipase and non-phospholipase subunits.

ACKNOWLEDGMENTS

This work was supported by the United States Army Medical Research and Development Command. Black and white chain tracings were generated using SCHAKAL88B (Keller, 1988), and color pictures were generated using MIDAS (Ferrin 1988a,b).

REFERENCES

Alekiev, B., and Tchorbanov, B., 1976, Toxicon 14:477.
Bernstein F.C., Koetzle T.F., Williams G.J.B., Meyer E.F., Brice M.D., Rodgers J.R., Kennard, O., Shimanouchi T., Tasumi, M, 1977, J. Mol. Biol., 112:535.
Block, A. R., Donegan, C. R., Denng, B. J., and Dolly, J.O.,1988, Biochemistry 27:6814.
Boffa, G. A., Boffa, M.-C., and Winchenne, J.-J. 1976, Biochim. Biophys. Acta 429:828.
Bon, C., Changeux, J. P., Jerg, T. W. and Fraenkel-Conrat, H. 1979, Eur. J. Biochem. 99:471.
Brunie, S., Bolin, J., Gewirth, D. and Sigler, P. B, 1985, J. Biol.Chem., 260:9742.
Collins, D., Hardman, K. D., Norden, T. N., Pett, V. B., Ward,K. B., Aird, S. D. and Kaiser, I. I, 1989, Toxicon 27:38.
Connolly, M.1981, MS, a program to calculate a solvent accessible molecular surface , Users manual, University of California, San Francisco (1981).

Conrad H., 1988, ESP, a program to calculate electrostatic potential surface, Users manual, University of California, San Francisco.

Deems, R. A. and Dennis, E. A., 1975, J. Biol. Chem. 250:9008

Dijkstra, B. W., Drenth, J., Kalk, K. H., and Vandermaelen, P. J., 1978, J. Mol. Biol. 124:53.

Dijkstra, B. W., Kalk., K. H., Hol, W. G. J. and Drenth, J. 1981a, J. Mol. Biol., 147:97.

Dijkstra, B. W., Drenth, J. and Kalk, K. H, 1981b, Nature 289:605.

Eaker, D., 1978, in "Versatility of Proteins", pp 413, Li, C. H. ed, Academic Press, New York.

Ferrin, T. E., Huang, C., Jarvis, L. E. and Langridge, R., 1988a, J. Mol. Graphics., 6:2.

Ferrin, T. E., Huang, C., Jarvis, L. E. and Langridge, R., 1988b, J. Mol. Graphics., 6:13.

Heinrikson, R. L., Krueger, E. T., and Keim, P.S., 1977, J. Biol. Chem. 252:4913.

Keith, C., Feldman, D. S., Deganello, S., Glick, J., Ward, K. B., Jones, E. O. and Sigler, P. B., 1981, J. Biol. Chem. (1981) 256:8602.

Keller, E., 1988 SCHAKAL88B a program for the graphics representation of molecular and crystallographic models, Kristallographisches Institut der Universitaet, Freiburg, FRG.

Kini, R.M. and Iwanaga, S., 1986, Toxicon, 24:527.

Langridge, R., Ferrin, T. E., Kuntz, I.D., Connolly, M. L., 1981, Science 211:661.

Renetseder, R., Brunie, S., Dijkstra, B. W., Drenth, J. and Sigler, P. B, 1985, J. Biol. Chem., 260:11627.

Tsai, I.-H., Liu, H.-C., and Chang, T., 1987, Biochem. Biophys. Acta, 916:94.

Weiner P.K., Langridge, R., Blaney, J. M., Schaefer, R. and Kollman, P.A., 1982, Proc. Natl. Acad. Sci. USA, 99:3754.

Waite, M., 1987, in "The Phospholipases", Chapters 7 and 9, Plenum Press, New York.

van den Bergh, C. J., Slotboum, A. J., Verheij, H.M., and de Hass, G. H., 1989, J. Cell Biochem. 39:379.

A NOVEL BIFUNCTIONAL MECHANISM OF SURFACE RECOGNITION

BY PHOSPHOLIPASE A$_2$

Robert L. Heinrikson and Ferenc J. Kézdy

Biopolymer Chemistry Unit
The Upjohn Company
Kalamazoo MI 49001

INTRODUCTION

The dedication of yet another volume to the subject of the phospholipases A$_2$ (PLA$_2$) attests to the continued high level of interest in these enzymes. As will be evident from the articles contained herein, a major focus of interest is in regard to the involvement of PLA$_2$ in the initiation of a metabolic cascade leading to the production of potent mediators of inflammatory diseases. Accordingly, an active search is underway for modulators or inhibitors of enzyme activity that might be useful in the treatment of inflammation, asthma, etc.. Furthermore, from a purely mechanistic point of view, the PLA$_2$ have always been prime subjects for the study of interfacial catalysis. These enzymes are small, water soluble, stable, and work alone with only a requirement for calcium in their hydrolysis of water-insoluble phospholipids. Also, in contrast to the substrates of other lipolytic enzymes, phospholipids lend themselves ideally to such studies in that they form a variety of stable and structurally well-defined surfaces in micelles, reverse micelles, monolayers, planar bilayers, and curved vesicles with which to study catalysis at the boundary between two phases.

Although from a structural standpoint, the PLA$_2$ rank among the best characterized enzymes (for reviews see 1-3), the physical and chemical mechanisms of their catalysis remain conjectural. Nevertheless, two phenomena have emerged as hallmarks of interfacial enzyme catalysis by PLA$_2$. These appear to occur always and only at phospholipid surfaces, be they pure or mixed micelles, monolayers, or bilayers. The first of these features unique to interfacial enzyme catalysis is commonly referred to as "interfacial recognition"; it consists of a higher reactivity of the enzyme toward a substrate in any of the aggregated states than toward the same substrate in homogeneous solution. With PLA$_2$ this difference in reactivity amounts, at most, to two orders of magnitude, although in the past, comparison of the rate of hydrolysis of one substrate in aqueous solution with that of a different substrate in an aggregated state led to misleadingly high estimates of the interfacial rate enhancement. The second phenomenon, often observed in the PLA$_2$-catalyzed hydrolysis of aggregated substrates and proposed to be a distinctive feature of interfacial catalysis, is "interfacial activation". This process consists of a readily observable latency phase which follows the addition of PLA$_2$ to a solution in contact with a phospholipid surface (4,5).

Biochemistry, Molecular Biology, and Physiology of Phospholipase A₂ and Its Regulatory Factors
Edited by A. B. Mukherjee, Plenum Press, New York, 1990

37

Unlike interfacial recognition, however, interfacial activation varies considerably with the type of PLA$_2$ used; several PLA$_2$ have a lag phase of as much as 5 to 15 minutes in the hydrolysis of monomolecular layers of phospholipids at the air/water interface, while enzymes from other species show no measurable lag periods (6).

In spite of its lack of predictability, interfacial activation has been considered to be the necessary, direct cause of interfacial recognition. According to one popular view, interfacial activation is the slow penetration into and subsequent tight anchoring of the enzyme within the monomolecular layer of phospholipid (4,7). This model which links, causally, the activation and recognition processes, predicts that tight binding to the surface renders the enzyme processive; interfacial recognition is then simply the increase in catalytic efficiency due to processivity. In fact, however, a number of equally plausible physical models have also been put forth to explain the rate enhancement seen in the hydrolysis of aggregated -- as opposed to monodisperse -- substrates (8). High local concentration of the substrate at the interface, product desorption facilitated by the interface, partial dehydration and/or conformational restriction of the scissile ester function at the interface, and conformational changes induced either at the molecular level by the substrate or more broadly by the anisotropic, amphiphilic environment of the interface -- the "quality of the interface" -- (4) have all been invoked at one time or another to explain interfacial recognition. In retrospect, the major reason for the generation of such a large body of seemingly contradictory theories lies in our assumption that catalysis of all PLA$_2$ should be the same, down to the most minute detail, irrespective of the source of the enzyme and the chemical nature and physical state of the substrate. We now recognize that, although the essential features of the catalytic mechanism of PLA$_2$ are universal, many details are particular to one given experimental system and occur only with a certain class of enzymes, a certain type of substrate, a given assay system, and a given set of physical conditions. It is therefore unwarrented to extrapolate results from studies of a single PLA$_2$ and a single assay system to all enzymes and all systems.

Investigations into the surface recognition and chemical mechanism of PLA$_2$ have also been confounded by the extreme catalytic efficiency of the enzyme. Measurable rates of reaction are produced by such insignificant amounts of enzyme as to render any protein chemistry of the reaction mixture impossible. With other enzymes, such as the serine proteases, "non-specific substrates" proved to be the ideal tool for understanding the mechanism of the enzymatic reaction, not only because with these substrates the whole catalytic process is slowed down to an experimentally manageable time scale, but also because the large concentrations of enzyme permissible only in these slow reactions allowed one to observe, identify, and analyze the chemical nature of reaction intermediates.

The progress we wish to summarize in the following is largely the result of our efforts to study the protein chemistry and enzymology of as large a number of different PLA$_2$ as possible and to develop non-specific substrates for detailed kinetic study of the reaction catalyzed by a variety of PLA$_2$.

THE PLA$_2$ ACTIVE AT THE INTERFACE IS A DIMER

We have long been intrigued as to how PLA$_2$ become fully potentiated for catalysis, whether that process involve interfacial recognition or activation. For many years, our studies of these enzymes were based, exclusively, upon the PLA$_2$ from rattlesnake venoms. As it turns out, these enzymes are distinct from most of the venom and pancreatic PLA$_2$ in that they are dimeric in structure. Moreover, these enzymes are only active in the dimeric state and at the extreme dilutions at which their catalytic activity can be measured, they partially dissociate into inactive monomers (9). Thus, instead of interfacial activation, they seem to display an "interfacial inactivation", which we proved to be due solely to the readjustment of the monomer-dimer equilibrium. This experience led us to think about the PLA_2 mechanism in terms of quaternary as well as tertiary structure, and we wondered whether all PLA_2, be they monomers or dimers in solution, might function as dimers at the interface (10). Indeed, one of the several mechanisms proposed (8) consists of the attachment of one monomeric PLA_2 to the phospholipid surface followed by a conformational change leading to the binding of a second protomer to yield an activated, dimeric species. The proposal was suggestive but had no compelling experimental support. In fact, since the kinetics of interfacial activation are strictly first-order with respect to the enzyme (4), dimerization of the enzyme at the surface could not be the rate limiting step of the reaction. Moreover, in view of the extreme structural rigidity of the PLA_2 molecule, the possibility of conformational changes being responsible for activation seems to be remote. We had access to high resolution X-ray crystallographic structures of a typical monomeric PLA_2 from bovine pancreas complexed to Ca^{++} (11), and of the apo-enzyme form of our dimeric PLA_2 from the venom of *Crotalus atrox* (12), but neither provided insights as to a relationship between dimerization and function. Nevertheless, it was intriguing that the world of PLA_2 seemed to be divided between clearly dimeric and clearly monomeric enzymes at the extremes and, in between, a range of enzymes which showed intermediate tendencies toward dimerization in solution. Under no condition did the rattlesnake venom PLA_2 dimers show the latency phase characteristic of the monomeric pancreatic enzyme acting on phospholipid monolayers (6). However, in spite of the absence of any detectable interfacial activation, the *C. atrox* and *C. adamanteus* enzymes still showed the usual surface recognition characteristic of all PLA_2, i.e. the rate enhancement toward aggregated as opposed to monomolecular phospholipids. The most parsimonious hypothesis accounting for all these observations was that interfacial recognition is an intrinsic property of, and only of the dimeric PLA_2. In that case, however, interfacial activation of the monomeric enzymes should consist of dimerization, an idea seemingly contradicted by kinetic evidence. It was with the use of non-specific substrates that we were able to resolve this conundrum.

Recently, we have developed a series of water-soluble, chromogenic non-specific substrates for PLA, the 3-(acyloxy)- 4-nitrobenzoic acids (13). We first showed that a variety of PLA_2 hydrolyzed these substrates by the same mechanism as employed in the hydrolysis of natural phospholipids. This was judged by the kinetics, the Ca^{++}-dependency, and the pH- dependency of catalysis, and by the inhibition of the reaction by alkylation of the histidine at the active site of the enzyme. We also observed that hydrolysis of the acyloxy-nitrobenzoic acids in aqueous solution by monomeric -- but not by dimeric! -- enzymes is accompanied by a rate increase during the reaction, very much reminiscent of the kinetics of interfacial activation (14,15). We demonstrated that the rate increase is irreversible and that it is due to a chemical derivatization of the enzyme. Several monomeric PLA_2, but not the dimeric *C. atrox* enzyme, undergo substrate- level acylation during the course of hydrolysis of these water-soluble non-specific substrates (14). Extending these investigations to natural, long-chain phospholipids we were also able to detect a slow acyl transfer from substrate to enzyme. In this process, the acyl group normally hydrolyzed from the substrate is, instead,transferred -- in a small fraction of the catalytic events -- to form a covalent link at epsilon amino groups in the enzyme. The autoderivatized enzymes were isolated from the reaction mixture and shown to be fully capacitated for interfacial recognition. At the same time, these enzymes no longer displayed any detectable interfacial activation.

Insofar as autoacylation leads to a fully activated PLA_2, it also leads to compulsory dimerization (14,15). For example, with the basic monomeric PLA_2 from the venom of *Agkistrodon piscivorus piscivorus* [AppD49, (16)], two sites of acylation are required for activation of the enzyme, at both lysines 7 and 10. Neither monoacyl-Lys-7-, nor monoacyl-Lys-10-AppD49 is activated, nor do these derivatives show any tendency to dimerize in solution. The bis-acyl-Lys-7:Lys-10-AppD49, however, forms a stable dimer that can be resolved from AppD49 and monomeric derivatives by gel filtration. The pure dimer shows a rate enhancement of greater than 100-fold over the native enzyme in the hydrolysis of phospholipid monolayers and no interfacial activation. The chemistry associated with the activation of another PLA_2, that from porcine pancreas, was much simpler in that only one lysine at position 56 was found to undergo autoacylation (15). The end result, however, was the same; monoacyl-Lys-56 pancreatic PLA_2 was dimeric in solution, and showed a rate enhancement of at least two orders of magnitude over the native enzyme toward both water soluble and aggregated substrates. Neither of these acylated, activated PLA_2 species showed the lag-phase kinetics of interfacial activation in the hydrolysis of phospholipid monolayers, characteristic of the native forms. In other words, both their quaternary structural organization in solution and their catalytic properties were identical to those of the naturally occurring *C.atrox* dimer.

These findings demonstrate that it is a chemical, and not a physical event that is fundamental to the process of interfacial activation. The chemical derivatization process leads to compulsory PLA_2 dimerization. At least in the few cases we have examined thus far, the monomeric enzymes are all susceptible to interfacial activation and at the same time they all can be converted to activated, dimeric derivatives by simple self-acylation concurrent with the hydrolysis of substrates. On the other hand, our sequencing data (1) exclude the possibility that the dimeric rattlesnake enzymes would already be acylated in the native state. Thus considered, the *C.atrox* PLA_2 would appear to have evolved to a dimeric species that is independent of the acylation mechanism and is fully potentiated for interfacial recognition in the native form. Since it is not practical to investigate every monomeric PLA_2, we can only infer that acylation is a necessary prerequisite for dimerization of all *monomeric* enzymes. In fact, however, there are some PLA_2, deemed to be monomeric enzymes, that show some moderate tendency to dimerize in solution. Under the influence of an anisotropic, amphiphilic environment these enzymes could, perhaps, dimerize at the interface spontaneously, in a manner equivalent to that of the *C.atrox* enzyme, instead of through covalent derivatization. We would propose that if such were the case it still would not be necessary to invoke a conformational change to explain the phenomenon.

The general conclusion which emerges from the above discussion is that interfacial recognition in all PLA_2- catalyzed reactions is expressed by, and only by, PLA_2 dimers, but the means of achieving dimerization is diverse. We feel that, by and large, all relevant kinetic data are consistent with this proposal. This, of course, does not imply that monomeric PLA_2 would have no activity whatsoever toward aggregated substrates. In fact, such an activity is probably necessary in monomeric enzymes in order to effect self-acylation. The enzymatic activity of truly monomeric PLA_2 toward monomolecular solutions of phospholipids is well documented and it is preserved even when the enzyme is covalently linked to a solid support (17). In the latter system, of course, interfacial recognition is abolished.

AUTOACYLATION AND INTERFACIAL ACTIVATION

Sites of modification

Much remains to be done to work out the details of autoacylation and its implications for PLA_2 function. At the present time, however, there are a number of aspects of the phenomenon that must be considered. Thus far, we have only documented sites of modification at epsilon amino groups; it is perfectly reasonable that other side chains such as those of serine or tyrosine might undergo autoacylation, but further speculation in this regard is unwarranted. In point of fact, we are already faced with a kind of dilemma in explaining the variability seen thus far in the number and positions of lysine residues modified. It is usually the case that residues crucial for function are highly conserved in the phylogeny of a protein. When one compares the sequences of the dozens of PLA$_2$ available in the literature (see e.g., 2,8) one sees no invariant lysines, so perhaps it is not surprising that the snake venom AppD49 PLA$_2$ requires acylation at lysines 7 and 10, and that the pancreatic enzyme is activated by a single modification at lysine 56 Nevertheless, some might argue that autoacylation is a capricious event and not necessarily functionally relevant. We must, therefore, look further into structure to find a basis for this variable pattern of sites modified in the autocatalytic activation of these enzymes.

Years ago when we first proposed the classification of PLA$_2$ into two groups based upon particular distinctive structural elements (1), we were aware that this separation was elementary. Indeed, it was a bit like separating living forms into plants and animals. Even then it was clear that there existed distinctions among classes of PLA$_2$ in each group that could have served for further subdivision of enzyme species. Armed now with a focus on the distribution of lysines among particular sets of PLA$_2$, we see patterns that support the idea that what we have documented thus far will apply to certain other closely related enzymes. We would predict, for example that all pancreatic PLA$_2$ will be autoacylated at lysine 56; this inference is easy since all have lysine 56. When one turns to other group I elapid venom PLA$_2$, however, there is no exact counterpart to lysine 56, but there are lysines in the vicinity of this residue that could serve as sites of modification. These enzymes often have a lysine at position 6 that could undergo autoacylation in a manner reflective of what is seen with the group II AppD49 enzyme. Most, but not all, group II PLA$_2$ have a lysine in the N-terminal alpha helix, and we would propose that those that do will undergo modification in this amphiphilic helix. Broadly considered, the group I and II PLA$_2$ are distinguished by the presence in the former of a disulfide bridge that is absent in the latter. Moreover, the group I enzymes have an insertion of amino acids in the vicinity of lysine 56, something we referred to a decade ago as the "elapid loop" (1) that is missing in group II PLA$_2$.

We feel that these structural distinctions hold the key for understanding the pattern of autoacylation by monomeric PLA$_2$. Group II enzymes are modified in an N-terminal region unencumbered by a restrictive disulfide bridge, possibly to stabilize an interaction fostered by that structure in the group I PLA$_2$. We would predict that group I enzymes will be activated by modifications in the center of the molecules near that region of hypervariability in sequence we have defined as the elapid loop. In either case, however, the acylation leads to dimerization, and this is the important aspect of the phenomenon that must be explained. We cannot, at present, establish a causal relationship between acylation and dimerization, but it is of interest that the modifications documented in these group I and group II PLA$_2$ occur in a region of structure variously referred to as a hydrophobic ring since " the residues around the entrance to the active site are variable, but with few exceptions, they are hydrophobic." (2). Although this hydrophobic area has been assumed to play a role in facilitating the binding of PLA$_2$ to lipid surfaces, we feel it could also define the contact area between protomers in the process of dimerization. Thus considered, it would be easy to explain the diversity in the number and types of lysines modified insofar as positive charges in this area would negate the possibility of dimerization. Lysines 7 and 10 in AppD49 lie on the same face of the helix, and interpose into this hydrophobic realm; it is necessary to acylate both in order to achieve dimerization and activation. Lysine 56 in the pancreatic enzyme also imposes what could be an unfavorable charge in the hydrophobic face; a single modification at this position is sufficient to enable dimerization and activation. This enzyme also has a lysine at 10, but it is adjacent to a disulfide bridge and therefore, presumably, out of the picture as concerns dimerization potential.

Chain length specificity

These reflections bring us to a consideration of the question as to whether activation and dimerization result from abolition of a charge, or from insertion of a fatty acyl side chain. If the role of acylation is to induce dimerization by creating a hydrophobic interaction of the fatty acyl group of one protomer with either the fatty acyl group or hydrophobic side chains of the other, then the chain length of the acyl group should be paramount in determining the efficiency of surface recognition. This doesn't seem to be the case; octanoylated and oleoylated enzymes are equally capable of surface recognition and even dibutyryl lecithin in aqueous solution is an activator of the pancreatic enzyme. This apparent chain length independence also precludes the possibility that the interaction of the covalently linked acyl group would serve to constrain the substrate molecule to a position optimal for interaction with the catalytic groups. By default, then, the dimerization of the native protomers should be prevented by local electrostatic interactions, even though the latter are globally important in stabilizing the dimer form of dimeric PLA_2 (18). If indeed the ε-ammonium groups interfere with dimerization, then the monomeric enzymes should show, progressively, some surface recognition in the pH region where Lys groups become deprotonated. The pancreatic enzyme does in fact show such a tendency; at pH = 10.5 its surface activation is five times faster than at pH = 8 (4).

Kinetics of self-acylation

When measured in the presence of phospholipid monolayers, the kinetics of surface activation are strictly first order with respect to the enzyme (4). This then indicates that the rate limiting step is not dimerization per se, but rather a slow transformation of the protomer into an active species. The experimental flexibility of the system consisting of pancreatic PLA_2 and 4-nitro-3-octanoyloxybenzoate allowed us to add some details to the pathway of dimer production. We found (15) that in this system also the rate of activation is first order with respect to the enzyme protomer. One should then conclude that the occasional acylation, parallel to hydrolysis, occurs intramolecularly. In other words, in the catalytic reaction the ε-amino group of the target lysine should be able to substitute for water, the normal acyl acceptor. After this slow, intramolecular acylation step, rapid formation of a hybrid acylated protomer:native protomer and a subsequent rapid intradimer acylation generate the activated dimer. The kinetics of the later phases of the reaction show that the activated dimer is, in fact, a potent activator of the remaining native protomer, presumably by facile exchange of one acylated protomer with a non-activated one.

Irrespective of the mode of acyl transfer, the most important conclusion one reaches from the above considerations is that the catalytic mechanism is not restricted to water being the sole possible acyl acceptor. The enzyme is able to catalyze acyl transfer to other acceptors as well. Thus, the puzzling failure of repeated attempts to observe PLA_2-catalyzed acyl transfer to a variety of acceptors is not due to some unique mechanistic properties of the water molecule, but rather to restricted access to a highly constrained active site pocket.

When one considers the structural organization of enzymes involved in hydrolysis of large substrates, e.g., ribonuclease, lysozyme, trypsin, etc., to name but a few, it is always the case that these enzymes, be they ever so small, display a multi-domain character with the active sites in clefts between domains. The multi-domain construction of the cleft confers a definite advantage to the enzyme in terms of flexibility and hence adaptability to the shape of the macromolecular substrate without sacrificing the rigidity of the catalytic apparatus necessary for optimal intracomplex reactivity. And yet, one is hard put to perceive any such domain structure in any PLA$_2$ monomer; these molecules are rigidly defined by an array of alpha helices and, almost always, seven disulfide bridges in such a manner as to prevent any part of the molecule to move independently of the rest of the structure. If multidomain structure is indeed advantageous for enzyme catalysis, then it becomes appealing to think of these enzymes as maximally active in the dimeric form, where each protomer becomes a domain and forms half of the cleft for substrate binding. A striking precedent for such a dimeric enzyme is the case of the aspartyl protease from the human immunodeficiency virus HIV-1 (19). Unlike other enzymes of this class, which are all monomeric, double-domain proteins, this enzyme is dimeric, where each protomer is a single domain protein, and the active site cleft is constructed by equal contributions from the two protomers.

We see in the taxonomy of venomous reptiles a satisfying correlation between morphology and state of evolution. Rattlesnakes are considered to be the most highly evolved reptiles, and they, alone, among venomous snakes examined thus far, have dimeric PLA$_2$ exclusively in their venoms. We have found recently (20), that *Lahesis muta*, the bushmaster snake and also a highly evolved crotalid, also has only a single dimeric PLA$_2$. Dimers are seen, along with an abundance of monomeric species in members of the evolutionarily advanced *Agkistrodon* and *Bothrops* families, but are not evident in more primitive species characterized by the elapids and hydrophids. Thus, it might be argued that a dimeric PLA$_2$ has evolved to produce an enzyme that is fully potentiated for function and, therefore, that dimerization is a fundamental element in the enzyme function. From the analogy with the aspartyl proteases it is then tempting to speculate that in some organism, or in the evolutionary future, one will find a 25 to 30 kd PLA$_2$ which arose by gene duplication and which possesses intrinsic surface recognition.

Surface recognition through autoacylation and dimerization is not, a priori, the only possible pathway for optimizing interactions between a water-soluble enzyme and an aggregated substrate. Many lipases engaging in heterogeneous catalysis require, or at least are aided by, cofactor proteins with intrinsically high surface affinities. Pancreatic lipase uses the colipase; lipoprotein lipase, apoC2; lecithin:cholesterol acyl transferase (LCAT), apoA1, and so on. One could imagine that the surface-bound cofactor serves as a temporary anchor to which the enzyme binds and thus becomes processive. It has been shown, however, that the cofactor apoA1 is not specific for LCAT and that, in fact, any amphiphilic peptide could fulfill the role of activator (21). On the other hand, both LCAT (22) and pancreatic lipase (23) undergo substrate-level, autocatalytic acylation at more than one site, a feature difficult to reconcile with a mechanism of acyl transfer occurring through the formation of a high energy acyl-enzyme intermediate. Perhaps these enzymes also are activated by irreversible acylation.

Regulation of enzyme activity is of paramount importance in living systems, and PLA$_2$ can be considered to be highly dangerous enzymes.

Dimerization as a requisite for fully activated PLA$_2$ could serve a crucial function here, helping to mitigate the activity of membrane-bound and digestive enzymes as well. It is true that the pancreatic PLA$_2$ is elaborated as a zymogen with low activity, but any adventitious activation of this enzyme outside the intestinal tract would have disastrous consequences, and a second line of regulation by acylation would provide another safeguard against premature activation.

A BIFUNCTIONAL ENZYME: SURFACE RECOGNITION AND CATALYSIS

The arguments provided above all constitute valid, although not compelling, reasons to support the hypothesis that dimerization is a sine qua non for enhanced PLA$_2$ catalysis at surfaces. What is reasonable is not always, or even often, correct, however, and it is crucial that the dimer pathway also be validated from the mechanistic point of view. Specifically, it should be shown that dimerization confers a mechanistic advantage which is not found in the monomeric enzyme. This brings us, then, to the subject of the molecular mechanism of interfacial recognition. In the following, we will propose that a dimeric enzyme is a more efficient surface catalyst than the monomer because it can bind two substrate molecules simultaneously.

Dimerization could, a priori, result in surface recognition through the generation of an enzyme species with high surface affinity and hence, processivity. According to this model, the enhanced surface affinity of the enzyme is not due to an increased active-site:substrate interaction but rather to an increased interaction of the enzyme as a whole with the surface as a whole. For example, the activated enzyme could have an exposed hydrophobic domain which would enable it to bind to lipid surfaces. But, neither the PLA$_2$ from the venom of *C. atrox*, nor the one from the venom of *C. adamanteus* -- both dimeric -- is processive toward lecithin single bilayer vesicles (6) or lipoproteins (24,25) nor have they an unusually high affinity toward lipid surfaces. Thus, the mechanism of surface recognition cannot be processivity, nor can it involve an increase of several orders of magnitude in affinity for phospholipid surfaces.

On the other hand, dimeric PLA$_2$ possess two substrate binding sites and thus have the potential to bind simultaneously two substrate molecules. Since phospholipid surfaces are the only constructs wherein two phospholipids can be juxtaposed optimally to bind the two proximal binding sites in an enzyme dimer, we considered that surface recognition might simply be the consequence of such a bifunctional mode of binding to the substrate.

Bifunctional binding to phospholipid surfaces is by no means a novel phenomenon. In the binding of Ca^{++} to phospholipid surfaces the only experimentally observable adduct is a diphospholipid:Ca^{++} complex (26,27). In other words, in the environment of a phospholipid monolayer the bifunctional complex is much more stable than any monofunctional complex. Thus, a significant gain in binding efficiency is achieved when the active site of the enzyme can accommodate two substrate molecules simultaneously. The exact value of the contribution of this effect to catalysis is, however, difficult to estimate from existing data. First, even if K_m is a true equilibrium binding constant, direct comparison of the K_m of the homogeneous reaction for a given substrate with that of the surface reaction does not reflect the efficacy of the bifunctional interaction alone, since the local environment of the phospholipid surface exerts several forces on the enzyme molecule itself.

These forces are not involved in the catalytic process *per se*, yet they will influence the rate of the enzymatic reaction favorably or negatively, depending on the amino acid sequence, the isoelectric point, the molecular dipole, and the conformational details of the individual enzymes. Second, the affinity of Ca^{++} toward lecithin in water is rather low, about two orders of magnitude less than that of the $PLA_2:Ca^{++}$ complex. It has been shown, however, that in mixed organic:water media the affinity of Ca^{++} toward lecithin increases significantly with decreasing water content (28,29). We feel that the environment of the active site of PLA_2 is hydrophobic enough to constitute a medium more conducive than water to lecithin:Ca^{++} complexation. But then, the high sensitivity of the complex to the local environment would again make its stability dependent on a variety of extraneous factors. In conclusion, bifunctional binding of the substrate at the interface to a dimeric enzyme should definitely be more efficient than monofunctional binding to either a monomeric or a dimeric enzyme. At the present time we can only guess that this binding contributes most of the observed two orders of magnitude increase in catalytic efficiency, but further experimentation will be necessary to establish this point.

When two phospholipid molecules bind simultaneously to the dimeric enzyme, an additional modest increase in catalytic efficiency may be expected from the entropic effects of the substrate:substrate interaction. Indeed, in the case of natural phospholipids with 16 to 18 carbon atom fatty acids, the two acyl chains of the reactive substrate molecule should interact more intimately with the acyl chains of the second substrate molecule than with any possible hydrophobic side chain on the enzyme. This then should result in an additional loss of mobility and therefore increase of the entropy of activation.

Some rearrangement of the active site must occur during dimerization, as indicated by the two to three fold increase in activity of the AppD49 enzyme toward 4-nitro-3-octanoyloxy benzoate (14) and a 200 fold increase with the porcine pancreatic enzyme (15). The detailed mechanism of this increase in catalytic power is not known at the present time. However, we should emphasize again that dimerization is not necessary for enzyme activity at the interface. The monomeric form is already catalytically active and we should like to conclude that dimerization does not create de novo a catalytic site, as is the case, for example, with alkaline phosphatase, glutamine synthetase, and alcohol dehydrogenase.

In principle, the kinetic hallmark of bifunctional surface catalysis should be that the surface reaction is second order with respect to the substrate at low substrate concentration. Experimentally, however, the time course of the reaction of a dimeric enzyme with either plasma lipoproteins (24,25) or single bilayer phospholipid vesicles (6) at low concentrations is first order with respect to the substrate. Since in these reactions the lysolecithin product remains, by and large, on the surface, the density of the phosphate head groups does not change during the reaction. Thus, it is quite likely that lysolecithin can easily assume the role of the second, or helper substrate in the formation of the bifunctional complex. A variety of observations do, indeed, support this hypothesis (30).

Finally, our proposed bifunctional mechanism of surface recognition provides a new framework within which to address fundamental questions concerning PLA_2 catalysis. These include the quaternary structure of the activated enzyme:Ca^{++} complex, the exact location of the two substrate molecules in that structure, the chemical mechanism of catalysis, and the specificity of the enzyme, to cite but a few.

ENVOI

The mechanism of surface recognition by bifunctional enzymes, proposed in this paper, is consistent with all salient features of PLA$_2$-catalyzed surface reactions and is validated experimentally for a few representative enzymes. For the time being, it is based only on a patchwork of experimental observations, select reports from the literature, and a lavish number of extrapolations. We feel that, as with any hypothesis, its main merit resides in the experimental testability of its predictions and we do hope that it will stimulate further investigation into this fascinating problem.

ACKNOWLEDGMENTS

The results described in this paper represent the work of postdoctoral associates and graduate students in our laboratories. The contributions of Drs. S. Yokoyama, W. Cho, A. Tomasselli, J. Hui, M.A. Markowitz, B.W. Shen, and J.T. Seykora, is gratefully acknowledged. Last, but not least, the authors wish to thank Dr. John H. Law for his long standing input and support.

REFERENCES

1. Heinrikson, R.L., Krueger, E.T., and Keim, P.S. (1977) *J.Biol.Chem.*, **252**, 4913-4921.
2. Verheij, H.M., Slotboom, A.J., and de Haas, G.H. (1981) *Rev.Physiol.Biochem.Pharmocol.*, **91**, 91-203.
3. Renetseder, R., Brunie, S., Dijkstra, B.W., Drenth, J., and Sigler, P.B. (1985) *J.Biol.Chem.*, **260**, 11627-11634.
4. Verger, R., Mieras, M.C.E., and de Haas, G.H. (1973) *J.Biol.Chem.*, **248**, 4023-4034.
5. Romero, G., Thompson, K., and Biltonen, R.L. (1987) *J.Biol.Chem.*, **262**, 13476-13482.
6. Kupferberg, J.P., Yokoyama, S., and Kézdy, F.J. (1981) *J.Biol.Chem.*, **256**, 6274-6281.
7. Verger, R., and de Haas, G.H. (1976) *Ann. Rev. Biophys. Bioeng*, **5**, 77-117.
8. Dennis, E.A. (1983) *The Enzymes*, **16**, 307-353.
9. Shen, B.W., Tsao, F.H.C., Law, J.H., and Kézdy, F.J. (1975) *J.Am.Chem.Soc.*, **97**, 1205-1208.
10. Heinrikson, R.L. (1982) *Proteins in Biology and Medicine*, (Bradshaw, R.A., Hill, R.L., Tang, J., Liang, C., Tsao, T.C., and Tsou, C.L., eds.) pp.131-152, Academic Press, N.Y.
11. Dijkstra, B.W., Drenth, J., Kalk, K.H., and Vandermaelen, P.J. (1978) *J.Mol.Biol.*, **124**, 53-60.
12. Brunie, S., Bolin, J., Gewirth, D., and Sigler, P.B. (1985) *J.Biol.Chem.*, **260**, 9742-9749.
13. Cho, W., Markowitz, M.A., and Kézdy, F.J. (1988) *J.Am.Chem.Soc.*, **110**, 5166-5171.
14. Cho, W., Tomasselli, A.G., Heinrikson, R.L., and Kézdy, F.J. (1988) *J.Biol.Chem.*, **263**, 11237-11241.
15. Tomasselli, A.G., Hui, J., Fisher, J., Zurcher-Neely, H., Reardon, I.M., Oriaku, E., Kézdy, F.J., and Heinrikson, R.L. (1989) *J.Biol.Chem.*, **264**, 10041-10047.
16. Maraganore, J.M., Merutka, G., Cho, W., Welches, W., Kézdy, F.J., and Heinrikson, R.L. (1984) *J.Biol.Chem.*, **259**, 13839-13843.
17. Lombardo, D., and Dennis, E.A. (1985) *J.Biol.Chem.*, **260**, 16114-16121.
18. Wells, M.A. (1971) *Biochemistry*, **10**, 4074-4078.
19. Wlodawer, A., Miller, M., Jaskolski, M., Sathyanarayana, B.K., Baldwin, E., Weber, I.T., Selk, L.M., Clawson, L., Schneider, J., and Kent, S.B.H. (1989) *Science*, **245**, 616-621.
20. Heinrikson, R.L. unpublished observations.
21. Yokoyama, S., Fukushima, D., Kupferberg, J.P., Kézdy, F.J., and Kaiser, E.T. (1980) *J.Biol.Chem.*, **255**, 7333-7339.

22. Jauhiainen, M., and Dolphin, P.J. (1986) *J.Biol.Chem.*, **261**, 7032-7043.
23. Guidoni, A., Benkouka, F., DeCaro, J., and Rovery, M. (1981) *Biochim. Biophys. Acta* **660** 148-150.
24. Pattnaik, N.M., Kézdy, F.J., and Scanu, A.M. (1976) *J.Biol.Chem.*, **251**, 1984-1990.
25. Aggerbeck, L.P., Kézdy, F.J., and Scanu, A.M. (1976) *J.Biol.Chem.*, **251**, 3823-3830.
26. Houser, H., and Dawson, R.M.C. (1968) *Biochem. J.,* **109**, 909-916.
27. Altenbach, C., and Selig, J. (1984) *Biochemistry*, **23**, 3913-3920.
28. Wells, M.A. (1974) *Biochemistry*, **13**, 2258-2264.
29. Yabusaki, K.K., and Wells, M.A. (1975) *Biochemistry* **14**, 162-166.
30. Yokoyama, S. and Kezdy, F.J., submitted to *J. Lipid Res.*

ACTIVATION, AGGREGATION, INHIBITION AND THE MECHANISM

OF PHOSPHOLIPASE A$_2$

Theordore L. Hazlett, Raymond A. Deems, and Edward A. Dennis

Department of Chemistry
University of California, San Diego
La Jolla, California 92093

I. INTRODUCTION

Phospholipids comprise a large portion of cell membranes and serve to form the hydrophobic barrier between the internal and external environments. Their regulation is crucial for membrane integrity and cell viability. Phospholipid catabolism plays an important role in this regulatory process. It has recently been found to have another, equally important, cellular function. Many of the breakdown products of phospholipid catabolism are important components in a number of regulatory pathways. These include the inositol phosphate second messengers, which are degradation products of phosphatidylinositol, and the eicosanoids, which are products of arachidonic acid metabolism (1). Phospholipid catabolism is carried out by four basic types of enzymes that hydrolyze the four ester bonds found in phospholipids. These enzymes are the phospholipases A$_1$, A$_2$, C, and D (2). Phospholipases A$_1$ and A$_2$ are responsible for cleaving the fatty acid ester bonds while phospholipases C and D cleave at the headgroup phosphoester bonds.

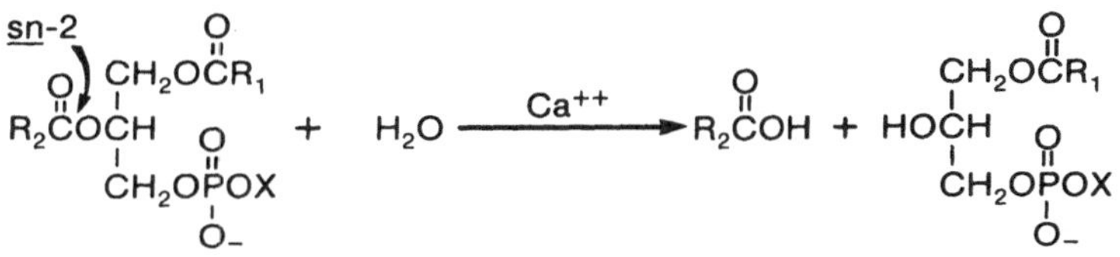

Figure 1: The reaction catalyzed by phospholipase A$_2$. The enzyme, in the presence of water and Ca^{2+}, cleaves the *sn*-2 fatty acid chain of phospholipids.

Biochemistry, Molecular Biology, and Physiology of Phospholipase A₂ and Its Regulatory Factors
Edited by A. B. Mukherjee, Plenum Press, New York, 1990

49

This report will focus on the extracellular phospholipase A_2s, the most extensively studied of these four enzymes. The reaction catalyzed by phospholipase A_2 is shown in Figure 1. Historically, these enzymes have been the focus of phospholipase A_2 research because of their stability and availability. The extracellular phospholipase A_2s can be found in large amounts in the mammalian pancreas as well as in various snake and insect venoms. Isolation of relatively pure preparations is straightforward. The enzyme is very stable and can tolerate harsh treatments, commonly used in modification studies, with little loss in activity. The stability appears to be common to all of the extracellular phospholipases A_2s and is thought to be due to the seven disulfide bonds which cross-link the protein. The quantities obtained from routine purifications are on the order of milligrams; enough for kinetic studies and for the more "protein-expensive" structural studies.

The study of the extracellular phospholipase A_2 has proven important, not only in its own right, but also because these enzymes can be used as membrane probes, as models for the less trackable intracellular phospholipase A_2s, and as the route to understanding the unique properties and problems that phospholipid interfaces present. The extracellular phospholipase A_2s have proven useful in solubilizing membrane proteins (3,4) and in studies of membrane asymmetry (5,6), their usefulness deriving from the efficiency with which they hydrolyze interfacial phospholipids. Other studies have indicated that phospholipase A_2 is sensitive to the nature of the interface and can distinguish between different phospholipid pools on the surface of natural membranes (7). Membrane structure and protein-lipid interactions continue to be of great interest. Phospholipase A_2 and its interaction with lipids is a convenient model system for studying these areas.

In addition, phospholipase A_2 has become important in biomedical research where it has been implicated in the regulation of prostaglandin release (1). The prostaglandin precursor, arachidonic acid, is esterified to the *sn*-2 fatty acid position of cellular phospholipids. The action of a phospholipase A_2, which specifically cleaves this bond, would be the most direct path for arachidonic acid release. It has also been shown that certain phospholipase A_2 inhibitors can inhibit arachidonic acid release in various cell lines (8). The extracellular phospholipase A_2s have proven to be useful models for guiding both the search for and the studies of the intracellular enzymes.

This report will cover three aspects of phospholipase A_2 research and will discuss their implication for the enzyme's catalytic mechanism. These three aspects are i) the enzyme's quaternary structure, ii) its inhibition, and iii) its activation. The first aspect deals with the aggregation state of the active enzyme. Of the four well characterized extracellular phospholipase A_2s, two are monomers, one is a dimer, and the fourth can be either a monomer or a dimer, depending on the conditions. The significance of these differences in quaternary structure are not clear; although, there is evidence that they may play a role in the activation of these enzymes. The second aspect deals with the activation of these enzymes which is manifested in two distinctly different ways. The porcine pancreatic, cobra venom (*Naja*), and rattlesnake venom (*Crotalus*) enzymes exhibit interfacial activation. They all hydrolyze monomeric phospholipid at a relatively slow rate which increases dramatically when phospholipid interfaces are present (9,10). These enzymes can also be activated by phospholipid. The cobra enzyme is activated by phosphorylcholine containing lipids (11,12) and the pancreatic enzyme by negatively charged species (13,14). Whether these two activations are different faces of the same phenomenon or are completely separate processes, has yet to be determined. The last aspect deals with the

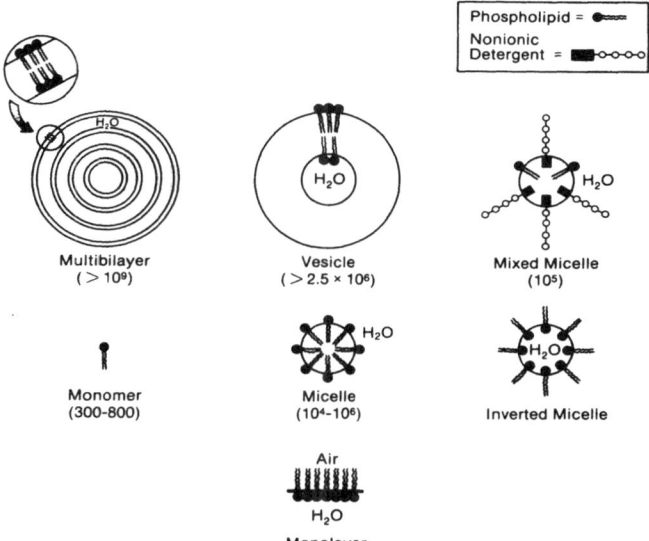

Figure 2: Schematic representation of possible aggregation states of phospholipids and apparent molecular weights. Adapted from (2).

inhibition studies that have yielded important information about what amino acid residues are important for the phospholipid/enzyme interaction.

Before delving into these subjects in greater detail, it is important to discuss several points about the environment in which these enzymes operate. This environment sets these enzymes apart and presents the researcher with a unique set of problems.

II. LIPID AGGREGATION STATES

Most naturally occurring phospholipids have a very low solubility in aqueous solutions. It has been reported that the monomer concentration of long chain phospholipids is of the order of 10^{-9} M (15-17). Due to the amphiphilic nature of these molecules, they aggregate into large ordered structures. Some possible structures for phosphatidylcholine (PC) are represented schematically in Figure 2. The headgroup and the hydrophobic chain determine which of these structures will be formed in an aqueous environment. Only synthetic short chain lipids, such as dibutyryl PC, exist in significant concentrations as monomers. Lipolytic enzymes acting on these large aggregates are subject to two surface phenomena that do not exist for enzymes acting on soluble substrates. The first is that the enzymes interact not only with each individual substrate molecule but also with the aggregate as a whole. The second factor is that these enzymes exhibit "surface dilution kinetics". Both of these phenomena must be kept in mind when analyzing any experiments involving lypolytic enzymes.

A. Phospholipase A₂ Interaction with the Lipid-Water Interface

Any time a lipolytic enzyme binds to a phospholipid which is sequestered in a lipid-water interface a significant portion of its exterior must come in contact with both the interface and other lipid molecules. These interactions may cause conformational changes in the enzyme.

In addition to its effect on the enzyme, the aggregation process also affects the substrate molecules directly. A phospholipid monomer floating free in solution is a very different species than one that is sequestered in a phospholipid interface. The interaction of an enzyme with these two substrate species is also very different. The first difference is that the orientation of the interaction is determined by the organization of the phospholipid molecules in the surface. A large portion of the lipid molecule, the hydrophobic tail, is buried in the interface and is not directly accessible to the enzyme. Therefore, the enzyme must approach the headgroup first. Another difference is that the phospholipid is no longer surrounded by water but is immersed in a sea of lipid. This difference in milieu can affect the conformation of the lipid molecule. Any change in the aggregated structure can change the mobility of the substrate in the surface and the packing pressure. Both of these factors can affect how easily the enzyme can access the individual lipid molecules. While all of these effects are rather obvious they are often overlooked when analyzing experiments dealing with lipolytic enzymes and aggregated lipids.

B. Surface Dilution Kinetics

The second aspect that interfaces bring to lipid enzymology is the phenomenon of surface dilution kinetics. Lipid enzymologists discovered early on that the activity of these enzymes depends on both the bulk and the surface concentrations of the substrate(18). Deems et al. (19) developed a kinetic model that accounted for this dependence. The explanation of this phenomenon arises from the fact that any phospholipid surface is two-dimensional and that any enzyme that interacts with the surface and then undergoes a bimolecular collision with the substrate while on the surface will "see" the phospholipid's surface concentration. As the total concentration of the molecules in the surface is increased, the relative surface concentration of the substrate decreases, even if the bulk substrate concentration is held constant. The model predicts that the activity of the enzyme should decrease as the substrates surface concentration decreases. Implicit in this model is the assumption that the enzyme stays sequestered at the interface long enough to undergo these bimolecular collisions. The *N. naja naja* phospholipase A_2 (19) exhibits surface dilution kinetics as do the membrane-bound enzymes phosphatidylserine decarboxylase (20) and the amnionic lysophospholipase (21).

III. PHOSPHOLIPASE A_2 AGGREGATION

The oligomeric states of the extracellular phospholipase A_2 are varied. The pancreatic phospholipases are found to be primarily monomeric, the crotalid phospholipases, in contrast, remain dimeric under most conditions, while the cobra venom enzymes show ligand-dependent and concentration-dependent aggregation. All of these enzymes have monomer molecular weights of approximately 14,000. They all have a large amount of sequence homology, approaching 80% (22-24). The X-ray crystal structure of the pancreatic and rattlesnake enzymes are very similar (2,25-28). Considering the similarity in sequence and crystal structure, one might have expected to find a more uniform oligomeric state. This raises the following questions: What is the catalytically relevant oligomeric state of these enzymes? Are they the same for all of the enzymes? Does the substrate or the interface affect the oligomeric state?

These questions are complicated by the substrates own aggregation behavior, which is independent of the phospholipase's aggregation. The presence of lipid aggregates interferes with many hydrodynamic techniques used for the determination of protein size. The binding of the enzyme to large lipid aggregates can drastically change the apparent size, density, and shape of the protein. In addition, none of these techniques can distinguish between a dimer that has a single catalytic site, with inactive monomers, and two catalytically active monomers that happen to be aggregated. In the next three sections, we discuss the oligomeric nature of the pancreatic, the rattlesnake venom, and the cobra venom phospholipases. In the fourth section, we will discuss the one technique, radiation inactivation, that offers the best hope of determining what the catalytically relevant oligomeric state of these enzymes is in the presence of aggregated phospholipid.

A. Pancreatic Phospholipases A_2

The bovine and porcine pancreatic phospholipase A_2s appear to be monomeric enzymes. Ultracentrifugation and gel filtration experiments have shown that these enzymes are monomeric, even at enzyme concentrations of several mg ml^{-1}. These phospholipases have little tendency to aggregate (29). In addition, the X-ray crystal structure has been solved for these enzymes (28), and the asymmetric unit appears to be a monomer.

Attempts to measure the aggregation state under assay conditions, i.e. in the presence of substrate and the cofactor Ca^{2+}, have also been reported. Active enzyme ultracentrifugation experiments provided direct evidence that the monomeric enzyme was active (30). Samples were run in the presence of monomeric substrate and Ca^{2+}. The sedimentation coefficient obtained was consistent with a monomeric enzyme. While this demonstrated that the monomer was catalytically competent, it did not resolve the issue since phospholipids below their cmc values, as used above (30), are 100 to 1000 fold poorer substrates than interfacial lipids (9). Monomer hydrolysis may not reflect the true active state of the enzyme. Viewed another way, dimerization may be responsible for interfaction activation.

In more recent work (31), investigators have observed that the binding of lipid analogs to the porcine pancreatic phospholipase A_2 causes enzyme aggregation. The n-alkylphosphocholine analogs, at concentrations above their cmc's, cause aggregation of the pancreatic phospholipase while concentrations below the cmc did not. Incorporation of the phospholipase into the micelles caused a rearrangement of the micellar structure. The protein-lipid complex formed contained approximately 2 protein molecules and 100 lipid molecules and had a smaller hydrodynamic radius than a pure lipid micelle (31). Whether or not the phospholipases were associated within the complex was not established.

De Haas and coworkers investigated the binding properties of the negatively charged n-alkyl sulfates (detergents) and 1,2-diacylglycerol-3-sulfates (synthetic substrates) (32,33). They found that at premicellar detergent concentrations the enzyme formed a large protein/detergent aggregate. The same was seen for the sulfate-containing substrates. The presence of the negatively charged headgroup and hydrophobic tail are thought to neutralize the charge repulsion between enzyme molecules and to strengthen the hydrophobic interaction thus stabilizing the aggregate. Enzyme within this aggregate displayed a higher than normal specific activity, indicating that the complex is beneficial and protein-protein interactions may be important for a fully active phospholipase A_2 (34,35).

B. Crotalid Phospholipases A_2

The phospholipases A_2 from *Crotalus adamanteus* and *C. atrox* are examples of dimeric phospholipase A_2s. In the absence of ligands, these enzymes have a molecular weight of 30,000 determined from ultracentrifugation, gel filtration, and nondenaturing electrophoresis studies (36). The enzymes, when denatured, dissociate into two 15,000 dalton subunits.

The evidence that the active form of the *C. adamanteus* enzyme is a dimer are three fold. First, there is a correlation between dimer dissociation and activity loss in increasing urea concentrations (36). These data have been criticized since urea may affect more than the quarternary structure, and the inactivation may be due to a different denaturing effect occurring simultaneously with dimer dissociation. Secondly, an active-enzyme-ultracentrifugation experiment in the presence of monomeric substrate, similar to that later carried out on the pancreatic enzyme, showed that the active enzyme was a dimer (37). Third, modification studies suggested a dimer model and half site reactivity (38). The enzyme was found to be inactivated by the incorporation of 0.5 moles of ethoxyformic anhydride (ETA) per mole of enzyme. Renaturation experiments after 50% and 100% inactivation yielded specific activities consistent with an active dimer. Some questions still remain about the specificity of ETA, but no new work has been reported.

C. Cobra Venom Phospholipases A_2

The aggregation states of the cobra venom phospholipase falls somewhere in between those of the pancreatic and rattlesnake enzymes. Aggregation of this enzyme depends on enzyme concentration and on the presence of phospholipids (39).

Very early in our studies of the cobra venom enzyme, we discovered, by gel permeation chromatography and ultracentrifugation (40), that the enzyme undergoes a concentration dependent aggregation. At concentrations between 100 ug ml^{-1} and 500 ug ml^{-1}, the enzyme is a dimer. At higher concentrations, the phospholipase aggregates still further (40). The formation of an active dimer for this enzyme seemed plausible since the *C. adamanteus* phospholipase is a dimer. However, under routine assay conditions, the enzyme concentration is between 0.1 and 10 ug ml^{-1}, where the enzyme is a monomer.

The possibility remained that Ca^{2+} or phospholipid could shift the equilibrium toward the dimer. We employed superimidate crosslinking (39) and a photoactivatable crosslinking reagent (41) to confirm the concentration dependent aggregation and to show that monomeric substrate does induce enzyme aggregation. We have also used fluorescence polarization to investigate this question. The extrinsic probe, fluorescein isothiocyanate, was coupled to the enzyme and steady state flourescence polarization measurements were made to determine changes in the aggregation state of the fluorescently tagged enzyme. Addition of monomeric PL analogs increased the polarization (42,43). The results bolster the idea that under assay conditions the phospholipase A_2 exists in an aggregated form and that this aggregation is linked to enzyme activation. Glutarylaldehyde crosslinking and sucrose gradient centrifugation studies (44) also support this conclusion. We now know that the *N. naja naja* phospholipase A_2 undergoes a concentration dependent aggregation and that the monomeric PC substrate will cause the enzyme to aggregate under assay conditions. Furthermore studies on immobilized enzyme argue for a connection between aggregation and activation (10,45).

Studies on the other cobra venom phospholipases, *N. melanoleuca* and *N. naja oxiana*, have also been reported. The *N. melanoleuca* venom enzyme aggregates in the presence of n-alkyl phosphocholine lipid analogues (46). However, the aggregation does not appear to be a simple monomer to dimer transition. The interaction was investigated with light scattering, gel filtration, and equilibrium dialysis and appears to represent the formation of a lipid and protein complex containing approximately four enzyme molecules and 36 lipids. It is not known if these complexes contain dimers or not.

In a study on the *N. naja oxiana* phospholipase A_2, it was found that in the presence of a short chain substrate, dibutyryl PC, the enzyme would form a dimer as observed with electrophoretic methods (47,48). The cmc of dibutyryl PC is at least 75 mM (49) and therefore would not be likely to form a phospholipid/phospholipase A_2 complex similar to that suggested for the *N. melanoleuca* phospholipase A_2 and lipid analogue system.

D. Radiation Inactivation

The numerous studies outlined above adequately document the fact that both the pancreatic and the cobra venom enzymes form aggregated structures in the presence of aggregated phospholipids. What remains unclear is the aggregation state of the enzyme in these structures. Each report places more than one enzyme molecule in each aggregated structure. But we are still left with the question of whether they are dimers or not.

In the presence of large lipid aggregates, molecular weight determinations are extremely difficult to perform with hydrodynamic procedures, e.g. ultracentrifugation, gel permeations, light scattering, or sucrose density centrifugation. This is even more tenuous when the enzyme is bound to these large structures. Determining whether the molecular weight of the enzyme is 13,000 or 26,000 daltons when it is bound to a 100,000 dalton or larger phospholipid aggregate is a difficult task indeed. All of these techniques yield information about the physical aggregation of the enzyme but do not give any information about whether this aggregation is catalytically relevant. The possibility exists that even as a dimer both subunits are still catalytically active and that aggregation is not required for catalysis. This brings us to the crux of the problem. What is the aggregation state of the catalytically relevant unit in the presence of monomeric and micellar substrate?

The one procedure that can potentially answer this question is radiation inactivation (50,51). In this procedure, the activity of the enzyme is measured as a function of radiation dosage. Target theory (50) relates the loss of enzyme activity directly to the mass of the catalytic unit. Thus, if the monomer is indeed the functional unit, its molecular size determined by radiation inactivation will be that of the monomer even if the enzyme is aggregated. Another advantage of this technique is that the surrounding material does not interfere with the determination. Therefore, the molecular mass of the protein can be determined even in the presence of large lipid aggregates. A third advantage is that the correlation of loss of enzyme activity is related directly to its mass and does not depend on any secondary hydrodynamic property. We are currently employing this technique to investigate the size of the catalytically relevant unit of several phospholipase A_2s under various assay conditions: enzyme alone, and in the presence of Ca^{2+}, monomeric substrate analogues, and aggregated substrate analogues.

IV. PHOSPHOLIPASE A₂ INHIBITION

A. Histidine

Classical chemical modification studies have identified several amino acid residues whose modification affects enzyme activity. One of the first of these studies was reported by Volwerk et al. (52). It showed that the modification of a histidine (His-48) in the pancreatic enzyme with p-bromophenacyl bromide (BPB) gave stoichiometric inhibition. Modification of a histidine (also His-48) was found to inhibit the *N. naja naja* phospholipase A₂ as well (53). The importance of this residue in phospholipid binding has been shown through BPB modification and direct micellar binding studies. BPB-modified pancreatic enzyme could neither bind monomeric dodecanyl phosphatidylcholine nor monomeric D-lecithin. Furthermore the enzyme could no longer bind micellar lipid. It was argued by de Haas and coworkers that the BPB group sterically inhibited phospholipid binding. In agreement, evidence of a group with a pKa of 6.25, the pKa which might be expected for a histidine group, was shown to be important in micelle binding (52,54). This is in contrast to what has been found for the *N. naja naja* phospholipase A₂ which still binds to mixed micelles when modified with BPB (53). This is probably due to the fact that the cobra venom enzyme has two phospholipid binding sites. See the discussion of the dual phospholipid model below.

B. Tyrosine

Modification of porcine, bovine, and equine pancreatic phospholipases with tetranitromethane revealed an exposed and invariant Tyr-69 residue (55,56). When nitrated at this residue, 87% of the activity was lost. Reduction of this group to an amine resulted in the restoration of much of the activity. Further modification of the amine with a dansyl group increased the affinity of the enzyme for the interface. The authors concluded that the tyrosine was not a catalytically important residue (55,56). The activity loss and attraction of the dansylated enzyme to the interface suggested that this residue is important for interfacial binding. Yang and coworkers (57) modified the Tyr-3 and Tyr-62 residues in the phospholipase A₂ from *N. naja atra* and *N. nigricollis* venoms. The enzyme derivative lost the ability to bind 8-anilonaphthalene sulfonate (ANS). Since earlier studies suggested that ANS might bind the active site (58,59) the authors concluded that tyrosine residues are present at the substrate binding site.

C. Lysine

Up to four lysine residues on the *N. naja naja* phospholipase A₂ can be modified by various anhydrides, but the enzyme retained as much as 60% of its original activity toward micellar lipid (Darke and Dennis, unpublished). We have recently found that the marine natural product, manoalide, and several of its analogs are also potent inhibitors of cobra venom phospholipase A₂ (60-62). Manoalide also modifies four lysine residues and inhibits the enzyme by 60 to 80%. These experiments also suggested that manoalide may be inhibiting the enzyme by binding near the activator site and blocking activation by PC.

From modification studies on the phospholipases from *N. melanoleuca* (63), *N. nigricollis* (59), and *N. naja atra* (58) venom it was concluded that the lysine residues are not essential for catalysis but may play a role in interfacial binding. The *N. naja oxiana* phospholipase loses little activity if the E-NH₂ groups are modified with acetic anhydride, o-methylisourea, or the N-carboxyanhydride of o-nitrophenylsulphenyl-glycine (64). If the investigators labeled instead with succinic anhydride, which introduces negative charges on the same residues, the enzyme was inactivated toward micellar substrate. It was concluded that the lysine residues are not important for catalysis but are important for an ionic interaction with the surface charge of a phospholipid interface.

Essential lysines have also been reported for the *C. adamanteus* (38) and *Bitis gabonica* (65) venom phospholipases. In neither case was the site of modification clearly established. The *B. gabonica* enzyme was treated with pyridoxal phosphate followed by reduction with borohydride which modified one lysine residue per mole of phospholipase. However, four different modified residues were isolated and enzymatic inhibition was not complete. Lysine modification of the *C. adamanteus* venom phospholipase with 0.5 mole of ethoxyformic anhydride per mole of enzyme inactivated the phospholipase. The modification inhibited dimer formation, believed to be essential for this enzyme's activity, and led to the proposal of a half-sites mechanism. However, no half-sites mechanism has been definitively proven and further evidence on the *C. adamanteus* venom phospholipase has not been forthcoming.

D. α-NH$_2$ group

Modification of the α-amino terminal inactivates the pancreatic and the snake venom phospholipases (66-68). The enzymes are inactivated toward micellar substrate, but not monomer substrate. Direct binding studies demonstrated that the modified pancreatic phospholipase A$_2$ was no longer able to bind micellar lipids. Semisynthesis studies of the n-terminal region have also indicated that the overall secondary structure of the n-terminal region is important for "interfacial" binding of this enzyme (67,69,70).

It is interesting that of all of these inhibitors only the BPB completely inhibited the enzymes by modifying a single specific residue. All the others produced partial inhibition, modified several residues, and affected monomeric and micellar hydrolysis differently. Most prevented interfacial activation. This implies that the residues modified are not involved in catalysis directly but are in some way linked to activation or interfacial binding.

V. PHOSPHOLIPASE A$_2$ ACTIVATION

Several extracellular phospholipases exhibit two types of activation. The first and more general is the dramatic activation that occurs when a monomeric substrate aggregates. The other activation is due to an interaction between specific phospholipids and the enzymes. The cobra venom enzyme is activated by phosphorylcholine containing phospholipids while the pancreatic enzyme is activated by negatively charged species. These two activations are being intensely studied. A brief review of these studies is presented below.

A. Interfacial Activation

One of the earliest works in this field was on phospholipase A$_2$ from *Crotalus durissus terrificus* (71). Although this was a preliminary report with an impure enzyme, the study showed that monomeric short chain lipids could act as substrates and that the requirement for an interface was not absolute. It was also shown that

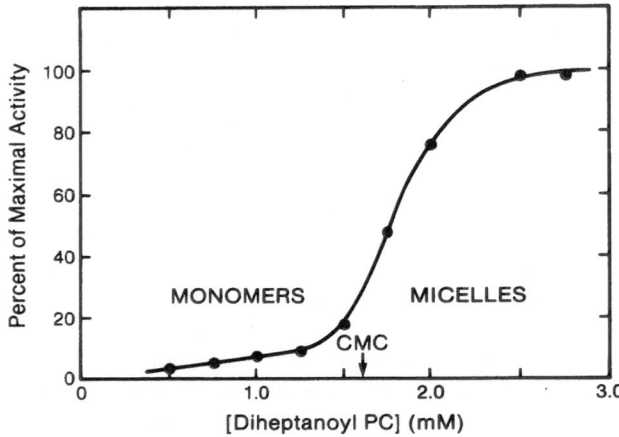

Figure 3: Interfacial activation of cobra venom phospholipase A$_2$.

the addition of detergent or lysophospholipid micelles stimulated the rate of lipid hydrolysis, an effect now known as "interfacial activation".

Subsequently several enzymes have been found that exhibit interfacial activation. The porcine phospholipase (9) enzyme hydrolyzes monomeric diheptanoyl PC, but as the concentration of substrate reaches and exceeds the cmc there is a 100-fold increase in the enzyme's specific activity. While the lipid concentration is increased by only 10% over the cmc range, the effect on activity is dramatic. A similar effect was observed in our laboratory for the *N. naja naja* venom phospholipase A$_2$ (Figure 3).

Investigators have also shown evidence for the importance of an interface for the porcine lipase (72,73). Lipase activity toward tripropionin is 100-fold greater if siliconized beads are present. The hydrophobic surface of the beads creates an artificial interface to which the soluble tripropionin, normally monomeric, can bind. The lipase then acts on the substrate at the interface, resulting in the extraordinarily high activity.

Theories on interfacial activation have been grouped into four general categories (2). The first groups together hypotheses explaining the activation effect in terms of an *enzyme conformational change* induced by the interface. The ionic and hydrophobic character of the interface optimizes the active site and enhances catalysis.

The second category deals with *substrate effects*. These substrate theories analyze the interfacial activation in terms of lipid orientation, lipid conformation, hydration of the polar headgroups, surface charge, surface pH, and surface lipid packing. These factors presumably orient the lipid and enzyme to induce the observed rate enhancement. We have demonstrated that the conformation of the phospholipid's *sn*-2 group is different when the phospholipid is monomeric or aggregated (74-77). More recently, we (78) and others (79) have looked at a cyclopentono-phospholipid analog in which the *sn*-2 chain is locked into a conformation similar to that found in aggregated natural phospholipids. It was found that these compounds did not show interfacial activation. These results suggest that substrate conformation may, in fact, contribute to the activation.

The third category deals with *product effects*. These theories deal primarily with product release. Release of the hydrophobic portion of the amphipathic products lysophospholipid and fatty acid into an aqueous environment would not be a favorable step. The presence of an interface could provide a suitable environment into which products could then be easily released. In this way, the overall rate of hydrolysis would increase, assuming that product release is the rate-limiting step. We have studied this possibility with the cobra venom enzyme using ^{13}C NMR techniques (80,81). We found that the rate limiting step was not product release when either monomeric or micellar phospholipids were present. Thus, product release is probably not a major factor in this activation.

The fourth category, *reaction effects*, refers to the availability of substrate. Sequestered at the interface, phospholipase A_2 would have an extremely high local substrate concentration, much higher than a similar concentration of monomerically dispersed lipid. Under these conditions, the bound enzyme would be closer to saturation and to maximal activity, thus explaining the activation. Increased substrate for the reaction, of course, could be a consequence of binding the enzyme to a substrate in the interface and thus may relate to an enzyme conformational change.

These four groups of theories cover most of the current explanations for the interfacial activation. The actual cause of the activation may involve more than one of the above theories.

B. PC-Activation

The *N. naja naja* phospholipase A_2 hydrolysis of phosphatidylethanolamine (PE) is activated by phosphocholine-containing lipids, such as lecithin (PC) and sphingomyelin (11,12,82). PE is a 10- to 20-fold poorer substrate in the absence of activator. To explain the data, two lipid binding sites have been postulated, a *catalytic* site and an *activator* site. The concept of two sites fits in well with the surface dilution model. The first binding site, which depends on the bulk lipid concentration, could be the activator site. Binding to this site sequesters the enzyme to the surface, causes a conformational change in the enzyme, and allows the second productive phospholipid binding. The latter is dependent on the surface concentration of the substrate. One would expect that non-activator substrates, such as PE, would not demonstrate interfacial activation. We have some evidence that PE alone does not show interfacial activation with the cobra venom enzyme (83).

One possible mechanism for this activation is that the activator lipids alter the quality of the interface and activate PE hydrolysis in this way. To test this, monomeric dibutyryl PC was used to activate the enzyme toward micellar PE. Since dibutyryl PC does not enter the interface (82), it cannot alter the interface as a longer chain PC might. It was found that the monomeric substrate did indeed activate micellar PE, and strengthened the proposal for two lipid sites.

We have also found that the presence of an interface is necessary for the PC activation of PE (83). Monomeric activators will not activate monomeric PE hydrolysis, but micellar activators will. This suggests that interfacial activation may be related to PC activation.

The pancreatic enzyme undergoes a similar activation; although, it is activated by negatively charged species. We have shown that the activity of the enzyme toward PC/Triton X-100 micelles depended upon how much fatty acid was present in the micelle (13). Volwerk et al. (14) subsequently examined a series of detergent phospholipid micelles and found that the enzymatic activity was dramatically enhanced when either negatively charged detergents or phospholipids were present.

VI. CONCLUSION

Interfacial activation, phospholipid activation, and the inhibitor results have been molded into a model for the cobra venom phospholipase A_2 (2,39,84). The basic tenet of this model, the "dual phospholipid model", is that the cobra venom enzyme

Two Site Single Subunit Model

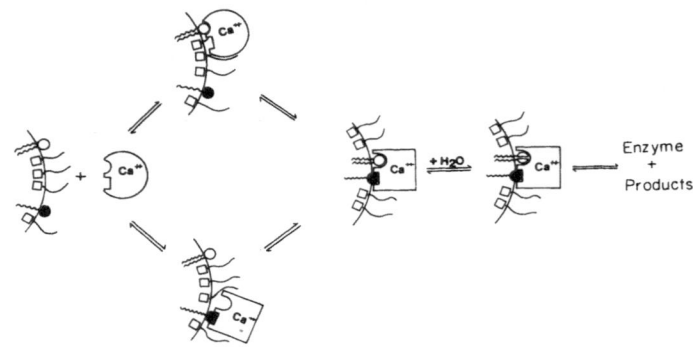

Two Site Dimer Model

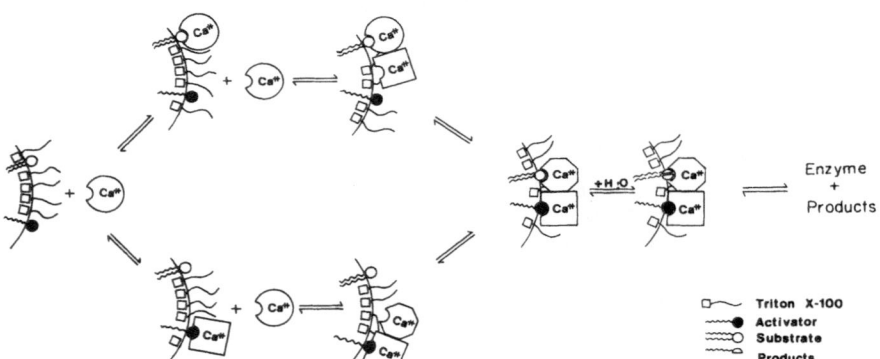

Figure 4: Two of several possible models for the hydrolysis of micellar lipid by phospholipase A_2. Both models indicate that the active phospholipase unit contains two lipid binding sites and undergoes a conformational change upon binding phosphocholine-containing lipids (activators). The single subunit model places both lipid binding sites on a phospholipase monomer. PC-activation occurs when an activator lipid, represented with a single acyl chain and filled polar headgroup, binds to the activator site. The dimer model proposes an asymmetric dimer to be the activatable phospholipase unit. Binding of an activator lipid then activates the second site on the opposite subunit. Adapted from (84).

has two phospholipid binding sites: an *activator site* and a *catalytic site*. Activation occurs when a phosphorylcholine containing phospholipid is bound to the activator site. Note that these same conditions also produce interfacial activation. The possibility that these two activations are the same, at least for the cobra venom enzyme, is currently being investigated in our laboratory. Two working models that incorporate the dual phospholipid model are given in Figure 4. In the two-site single-subunit model, both sites are present in an enzyme monomer. Binding of an activator lipid to the activator site then causes a conformational change resulting in an activated enzyme. In contrast, the dimer model predicts that the monomeric enzyme aggregates at the surface of a substrate interface and each subunit has only one site. Binding of activator lipid results in a conformational change, activating the second site on the associated subunit. Since the size of the catalytically relevant enzyme unit is not known, we cannot distinguish between these two models. Hopefully, the radiation inactivation experiments that are currently being conducted will answer this question.

The three phospholipase A_2s from the pancreas, cobra venom, and rattlesnake venom have marked sequence homology, have very similar X-ray crystal structures, all exhibit interfacial activation, and all react similarly with the inhibitors studied, and yet, their aggregation states seem to differ dramatically as does their response to various activating compounds. When more is known about these enzymes, will these apparent differences disappear? Or, on the other hand, are these differences caused by the few deviations in sequence and three-dimensional structure that do exist? The answers to these intriguing questions are being vigorously pursued in our laboratory.

ACKNOWLEDGEMENT

Work in our laboratory on phospholipase A_2 has been supported by NSF grant DMB88-17392 and NIH grant GM 20,501.

REFERENCES

1. Dennis, E. A. (1987) *BIO/TECHNOLOGY* (Nature Publishing Co.) 5, 1294.
2. Dennis, E. A. (1983) in *The Enzymes, Third Edition, Vol. 16,* (Boyer, P., Edit.) Academic Press, New York, 307-353
3. Rivas, E.A., Le Maire, M., and Gulik-Krzywicki, T. (1981) *Biochim. Biophys. Acta. 644*, 127.
4. Durkin, J.P., Pickwell, G.U., Trotter, J.T., and Shier, W.T. (1981) *Tocicon 19*, 535.
5. Adamich, M., and Dennis, E.A. (1978) *J. Biol. Chem. 253*, 5121.
6. Zwaal, R.F.A., Roelofsen, B., and Colley, C.M. (1973) *Biochim Biophys. Acta 300*, 159.
7. Shukla, S.D., and Hanahan, D.J. (1981) *Arch. Biochem. Biophys. 209*, 668.
8. Lister, M. D., Glaser, K. B., Ulevitch, R. J., and Dennis, E. A. (1989) *J. Biol. Chem. 264*, 8520.
9. Pieterson, W.A., Vidal,, J.C., Volwerk,, J.J., and de Haas, G.H. (1974) *Biochemistry 13*, 1455.
10. Lombardo, D., and Dennis, E. A. (1986) in *Enzymes of Lipid Metabolism,* (Freysz, L. and Gatt, S., Edit.), Plenum Press, New York, 133-138
11. Adamich, M., Roberts, M.F., and Dennis, E.A. (1979) *Biochemistry 18*, 3308.

12. Roberts, M.F., Adamich, M., Robson, R.J., and Dennis, E.A. (1979) *Biochemistry 18*, 3301.
13. Plückthun, A., and Dennis, E. A. (1985) *J. Biol. Chem. 260*, 11099.
14. Volwerk, J.J., Jost, P.C., de Haas, G.H., and Griffith, O.H. (1986) *Biochemistry 25*, 1726.
15. Tausk, R.J.M., Karmiggelt, J., Oudshoorn, C., and Overbeek, J.T.G. (1974) *Biophys. Chem. 1*, 175.
16. Smith, R., and Tanford, C. (1972) *J. Mol. Biol. 67*, 75.
17. Martin, F.J., and Macdonald, R.C. (1976) *Biochemistry 15*, 321.
18. Dennis, E.A. (1973) *J. Lipid Res. 14*, 152.
19. Deems, R.A., Eaton, B.R., and Dennis, E.A. (1975) *J. Biol. Chem. 250*, 9013.
20. Warner, T.G., and Dennis, E.A. (1975) *Archiv. Biochem. Biophys. 167*, 761.
21. Jarvis, A.A., Cain, C., and Dennis, E.A. (1984) *J. Biol. Chem. 259*, 15188.
22. Dufton, M.J., Eaker, D., and Hider, R.C. (1983) *Eur. J. Biochem. 137*, 537.
23. Dufton, M.J., and Hider, R.C. (1983) *Eur. J. Biochem. 137*, 545.
24. Davidson, F. F., and Dennis, E. A. (1989) Submitted.
25. Brunie, S., Bolin, J., Gewirth, D., and Sigler, P.B. (1985) *J. Biol. Chem. 260*, 9742.
26. Pasek, M., Kieth, C., Feldman, D., and Sigler, P.B. (1978) *J. Mol. Biol. 97*, 395.
27. Kieth, C., Feldman, D.S., Daganello, S., Glick, J., Ward, K.B., Jones, E.O., and Sigler, P.B. (1981) *J. Biol. Chem. 256*, 8602.
28. Dijkstra, B.W., Drenth, J., Kalk, K.H., and Vandermaelen, P.J. (1978) *J. Mol. Biol. 124*, 53.
29. Volwerk, J.J., Dedieu, A.G.R., Verheij, H.M., Dijkman, R., and De Haas, G.H. (1979) *Recl. Trav. Chim. Pays-Bas 98*, 214.
30. Slotboom, A.J., Verheij, H.M., and de Haas, G.H. (1982) in *Phospholipids* (J.H. Hawthorne and G.B. Ansell, eds), Elsevier Biomedical Press.
31. Hille, J.D.R., Donne-Op den Kelder, G.M., Sauve, P., de Haas, G.H., and Egmond, M.R. (1981) *Biochemistry 20*, 4068.
32. Hille, J.D.R., Egmond, M.R., Dijkman, R., van Oort, M.G., Jirgensons, B., and de Haas, G.H. (1983) *Biochemistry 22*, 5347.
33. Hille, J.D.R., Egmond, M.R., Dijkman, R., van Oort, M.G., Sauve, P., and de Haas, G.H. (1983) *Biochemistry 22*, 5353.
34. Van Oort, M.G., Dijkman, R., Hille, J.D.R., and de Haas, G.H. (1985) *Biochemistry 24*, 7987.
35. Van Oort, M.G., Dijkman, R., Hille, J.D.R., and de Haas, G.H. (1985) *Biochemistry 24*, 7993.
36. Wells, M.A. (1971) *Biochemistry 10*, 4073.
37. Smith, C.M., and Wells, M.A. (1981) *Biochim. Biophys. Acta 663*, 687.
38. Wells, M.A. (1973) *Biochemistry 12*, 1086.
39. Roberts, M.F., Deems, R.A., and Dennis, E.A. (1977) *Proc. Nat'l. Acad. Sci. USA 74*, 1950.
40. Deems, R.A., and Dennis, E.A. (1975) *J. Biol. Chem. 250*, 9008.
41. Lewis, R.V., Roberts, M.F., Dennis, E.A., and Allison, W.S. (1977) *Biochemistry 16*, 5650.
42. Hazlett, T. L., and Dennis, E. A. (1985) *Biochemistry 24*, 6152.
43. Hazlett, T. L., and Dennis, E. A. (1988) *Biochim. Biophys Acta. 958*, 172.
44. Hazlett, T. L., and Dennis, E. A. (1988) *Biochim. Biophys. Acta. 961*, 22.
45. Lombardo, D., and Dennis, E. A. (1985) *J. Biol. Chem. 260*, 16114.
46. Van Eijk, J.H., Verheij, H.M., Dijkman, R., and de Haas, G.H. (1983) *Eur. J. Biochem. 132*, 183.

47. Mal'tsev, V.G., Zimina, T.M., Kurenbin, O.I., Belen'kii, B.G., Aleksandrov, S.L., Paulova, N.P., Dyakov, V.L., and Antonov, V.K. (1979) *Bioorgh. Khim.* 5, 1710.

48. Zhelkovskii, A.M., Dyakov, V.L., and Antonov, V.K. (1978) *Bioorgh. Khim.* 4, 1665.

49. Wells, M. (1972) *Biochemistry 11*, 1030.

50. Harmon, J.T., Nielsen, R.B., and Kempner, E.S. (1985) *Methods in Enzymology 117*, 65.

51. Kempner, E.S. (1988) *Advances in Enzymology 61*, 107.

52. Volwerk, J.J., Pieterson, W.A., and De Haas, G.H. (1974) *Biochemistry 13*, 1446.

53. Roberts, M.F., Deems, R.A., Mincey, T.C., and Dennis, E.A. (1977) *J. Biol. Chem. 252*, 2405.

54. Bonsen, P.P.M., de Haas, G.H., Pieterson, W.A., and van Deenen, L.L.M. (1972) *Biochim. Biophys. Acta. 270*, 364.

55. Meyer, H., Verhoef, H., Hendriks, F.F.A., Slotboom, A.J., and De Haas, G.H. (1979) *Biochemistry 18*, 3582.

56. Meyer, H., Puijk, W.C., Dijkman, R., Foda-Van der Hoorn, M.M.E.L., Pattus, F., Slotboom, A.J., and De Haas, G.H. (1979) *Biochemistry 18*, 3589.

57. Yang, C.C., Huang, C.S., and Lee, H.J. (1985) *J. Prot. Chem. 4*, 645.

58. Yang, C.C., King, K., and Sun, T.P. (1985) *Toxicon 19*, 645.

59. Yang, C.C., King, K., and Sun, T.P. (1981) *Toxicon 19*, 783.

60. Lombardo, D., and Dennis, E. A. (1985) *J. Biol. Chem. 260*, 7234.

61. Deems, R. A., Lombardo, D. Morgan, B. P., Mihelich, E. D., and Dennis, E. A. (1987) *Biochim. Biophys. Acta 917*, 258.

62. Reynolds, L. J., Morgan, B. P., Hite, G. A., Mihelich, E. D., and Dennis, E. A. (1988) *J. Am. Chem. Soc. 110*, 5172.

63. van Eijk, J.H., Verheij, H.M., and de Haas, G.H. (1983) *Eur. J. Biochem. 132*, 177.

64. Apsalom, U.R., Shainborant, O.G., and Miroshnikov, A.I. (1977) *Bioorg. Khim. 3*, 1553.

65. Viljoen, C.C., Visser, L., and Botes, D.P. (1977) *Biochim. Biophys. Acta 483*, 107.

66. Verheij, H.M., Egmond, M.R., and De Haas, G.H. (1981) *Biochemistry 20*, 94.

67. Slotboom, A.J., and de Haas, G.H. (1975) *Biochemistry 14*, 5394.

68. Haruki, H., Teshima, K., Samejima, Y., Kawauchi, S., and Ikeda, K. (1986) *J. Biochem. 99*, 99.

69. van Scharrenburg, G.J.M., Puijk, W.C., Egmond, M.R., Van der Schaft, P.H., de Haas, G.H., and Slotboom, A.J. (1982) *Biochemistry 21*, 1345.

70. Dijkstra, B.W., Kalk, K.H., Drenth, J., de Haas, G.H., Egmond, M.R., and Slotboom, A.J. (1984) *Biochemistry 23*, 2759.

71. Roholt, O.A., and Schlamowitz, M. (1961) *Arch. Biochim. Biophys. 94*, 364.

72. Brockman, H.L., Law, J.H., and Kezdy, F.J. (1973) *J. Biol. Chem. 248*, 4965.

73. Borgström, B., Erlamson-Albertsson, C., and Wieloch, T. (1979) *J. Lipid Res. 21*, 805.

74. Roberts, M. F., and Dennis, E. A. (1977) *J. Am. Chem. Soc. 99*, 6142.

75. Roberts, M. F., and Dennis, E. A. (1978) in *Biomolecular Structure and Function* (P.F. Agris, R.N. Loeppky, and B.D. Sykes, Edit.), Academic Press, New York, pp.71-78

76. Roberts, M. F., Bothner-By, A. A., and Dennis, E. A. (1978) *Biochemistry 17*, 935.

77. DeBony, J., and Dennis, E. A. (1981) *Biochemistry 20*, 5256.

78. Barlow, P. N., Lister, M. D., Sigler, P. B., and Dennis, E. A. (1988) *J. Biol. Chem. 263*, 12954.

79. Lin, G., Noel, J., Loffredo, W., Stable, H. Z., and Tsai, M. (1988) *J. Biol. Chem. 263*, 13208.

80. Fanni, T., Deems, R. A., and Dennis, E. A. (1989) *Biochim. Biophys. Acta.* 1004, 134.

81. Lombardo, D., Fanni, T., Plückthun A., and Dennis, E. A. (1986) *J. Biol. Chem 261*, 11663.

82. Plückthun, A., and Dennis, E.A. (1982) *Biochemistry 21*, 1750.

83. Plückthun, A., Rohlfs, R., Davidson, F. F., and Dennis, E. A. (1985) *Biochemistry 24*, 4201.

84. Dennis, E. A., and Plückthun, A. (1986) in *Enzymes of Lipid Metabolism II,* NATO ASI Series A: Life Sciences Vol. 116, (Freysz, L., Dreyfus, H., Massarelli, R. and Gatt S., Edit.), Plenum Press, New York, 121-132

PROBING THE MECHANISM OF PANCREATIC PHOSPHOLIPASE A2 WITH THE AID OF RECOMBINANT DNA TECHNIQUES

O.P. Kuipers, C.J. van den Bergh, H.M. Verheij and G.H. de Haas

Department of Biochemistry
State University of Utrecht
CBLE, University Center De Uithof
Padualaan 8
3584 CH Utrecht
The Netherlands

Introduction

Phospholipase A2 (E.C. 3.1.1.4) attacks the acyl ester bond at position 2 of 3-*sn*-phosphoglycerides (van Deenen en de Haas, 1963). The *in-vivo* importance of phospholipase A2 (PLA) is reflected by the fact that this enzyme occurs ubiquitously in nature and that PLA activity has been detected in a large number of cell types and cell organelles. In general their physiological role can be either a digestive or a regulatory one. The extracellular PLAS from mammalian pancreas and snake venom, which are responsible for the hydrolysis of dietary phospholipid, predominantly belong to the first class of PLAs, the digestive enzymes. To the cellular PLAs have been assigned regulatory functions (for a review see Waite, 1987), although certain cellular PLAs also exhibit a digestive role in the break-down of phagocytized materials (Elsbach and Weiss, 1988).

The extracellular, PLAs are relatively small proteins (14 kD; ~120 amino acids) and display a very high stability to denaturing conditions, like high temperature and low pH. The high stability probably stems from the high content of disulfide bridges (6 or 7), which is a common feature of all PLAs. The primary structure has been elucidated for about 55 extracellular PLAs from different origin and from a comparison of these sequences(van den Bergh et al.,1989a) a high degree of homology becomes evident. Crystallographic data from bovine and porcine pancreatic PLAs, and from the venom of *Crotalus atrox* also show striking similarities in three-dimensional structure (Dijkstra et al., 1981a; Dijkstra et al.,1983; Brunie et al., 1985). The large number of conserved and semi-conserved amino acid residues and the structural resemblances suggest a general mode of action for the extracellular PLAs. For pancreatic PLAs, chemical modification and X-ray crystallography studies (Verheij et al., 1980; Fleer et al., 1981; Dijkstra et al., 1983) have indicated the importance of His-48, Asp-99 and Asp-49 for PLA-action [for numbering of amino acid residues see Waite (1987)].

Biochemistry, Molecular Biology, and Physiology of Phospholipase A₁ and Its Regulatory Factors   **65**
Edited by A. B. Mukherjee, Plenum Press, New York, 1990

From these data a general catalytic mechanism has been proposed (Figure 1), in which an Asp-His couple serves as a proton relay system, resembling that of the serine proteases. Contrary to these esterases, PLA lacks a serine in the active center and a water molecule is held responsible for the nucleophilic attack. The role of the calcium ion in the active site, is presumed to be the polarisation of the carbonyl oxygen of the scissile ester bond and the stabilisation of the tetrahedral intermediate formed. The proposed interaction of Ca^{2+} with the phosphate group of the substrate molecule could help to fix the substrate productively in the active site. The carboxylate of Asp-49, the carbonyl oxygens from the residues that constitute the calcium-binding loop (Tyr-28, Gly-30 and Gly-32) and some water molecules have been proposed as ligands of the calcium ion (Verheij et al., 1980; Dijkstra et al., 1983). Although the carboxylate of Asp-49 is conserved in all active PLA species, its role as a calcium ligand was questioned after the finding of PLA homologs that contain a lysyl residue at this position (Maraganore et al., 1984; Maraganore and Heinrikson, 1986).

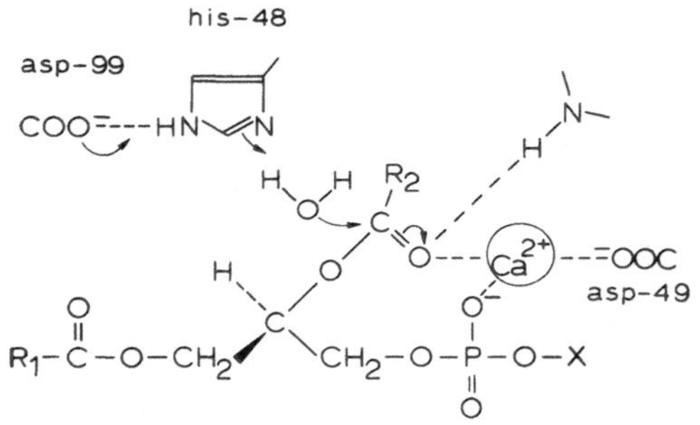

Fig. 1. Schematic representation of a proposed catalytic mechanism for PLA (Verheij et al., 1980)

In addition to the catalytically important calcium ion, porcine and equine PLAs are known to possess a second calcium binding site (van Dam-Mieras et al., 1975). Binding of a second calcium ion increases, at alkaline pH, the affinity of the latter PLA species for lipid-water interfaces and increases their activity. Ba^{2+}-ions are competitive inhibitors for this metal binding site, but the barium concentrations needed for the mediation of interface recognition are much higher than the required calcium concentrations. The side chains of Glu-71 and Glu-92 and the carbonyl oxygen of Ser-72 have been proposed as the ligands in the second metal binding site (Donné-Op den Kelder et al., 1983; Dijkstra et al., 1983).

Although porcine pancreatic PLA, is able to degrade monomers of short-chain phospholipids, low rates are observed below the critical micellar concentration (CMC). Upon passing the CMC the activity increases drastically, reaching a maximal velocity a factor 1000 higher as compared to the activity on monomers (Pieterson et al., 1974). A similar behaviour is observed for all extracellular PLAs since they exhibit maximal enzymatic activity toward aggregates of substrate molecules. Several explanations have been proposed for the activation of the enzyme by a lipid-water interface and these have previously been reviewed by Volwerk and de Haas (1982). The so-called interface-recognition site hypothesis (Verger et al., 1973) states that a surface region of the PLA-molecule, topographically distinct from the active site, is responsible for the interaction of the enzyme with organized lipid-water interfaces. Formation of the enzyme- lipid complex would induce, *via* the specific recognition site, a conformational change of the active site that results in an increase of the hydrolysis rate. The interface recognition site consists mainly of hydrophobic and positively charged residues that surround the entrance to the active site, as can be deduced from crystallographic data from bovine and porcine pancreatic PLAs Dijkstra et al., 1981b and 1983). Although these amino acids are not invariant among the different PLA species, the overall properties of the amino acid side chains are conserved. For porcine pancreatic PLA, X-ray analyses and semi-synthesis studies (van Scharrenburg, 1984) have indicated several residues in the amino-terminal α-helix (e.g. Leu-2, Trp-3, Arg-6 and Lys-10) and various amino acids in the rest of the enzyme (e.g. His-17, Leu-19, Met-20, Leu-31, Leu-64, Tyr-69, Thr-70, Lys-116, Asn-117 and Lys-121) to be part of the interface recognition site.

In summary, during the last 25 years a wealth of information on the action of extracellular PLAs has been obtained using various "classical" chemical techniques, like chemical modification, semi-synthesis, protein sequencing and X-ray crystallography. Moreover, the development of several spectroscopic methods and kinetic assays with a large variety of synthetic substrates have supplied the means to study in detail the PLA action. These experiments have culminated at an early stage in the proposition of a catalytic mechanism that is still used as a working hypothesis, although several questions remain to be answered. Without a crystal stucture of an enzyme-calcium-inhibitor complex it remains difficult to predict the orientation of substrate molecules in the active site. Hence the role of the cofactor calcium and of the water molecule which is presumed to be the nucleophile remain speculative. Even further away is an answer to the central question in lipolysis: what causes the efficient hydrolysis of aggregated substrates compared to that on monodisperse solutions?

Advances in microbiology and molecular genetics in the last decade have led to the development of recombinant DNA techniques which have become a powerful tool in the study of proteins, permitting the replacement of any given residue by any other amino acid. Also the development of molecular modeling techniques has opened the way to the design of proteins and enzymes with novel properties. The cloning and expression (de Geus et al., 1987a) of porcine pancreatic PLA made it possible to verify the proposed mechanism of catalysis and to probe the residues involved in the interaction of calcium as well as substrate molecules with this enzyme. These techniques have also

opened the way to the introduction of specific labels in PLA.
The combined use of these proteins, competitive inhibitors and
high-resolution NMR may shed light on the structure of the
enzyme-micelle complex. In the present paper the results
obtained sofar will be summarized.

Cloning of Prophospholipase A2 encoding cDNA

From a cDNA bank of porcine pancreas in *E. coli* PLA
specific clones were identified and one of them was sequenced.
It contained a 560 base pairs long cDNA insert, covering the
complete coding region for the preproPLA (Figure 2). The
predicted amino acid sequence is in full agreement with the one
known for porcine proPLA (Puijk et al., 1977). As expected for a
secreted protein, a 15 amino acids N-terminal extension,
matching the criteria for a signal peptide (Perlman and
Halvorson, 1983) was found. This signal peptide is not only
homologous to the ones found in cloned PLA cDNA's from rat and
dog pancreas and human lung (Ohara et al., 1986; Seilhamer et
al., 1986) but also with the ones of the mammalian
non-pancreatic (Seilhamer et al.,1989; Kramer et al., 1989) and
snake venom (Ducancel et al., 1988) PLAs. Expression experiments
in eukaryotic cell lines, confirmed the functional integrity of
this cDNA sequence, since the proPLA was accurately processed
into the culture medium (de Geus et al., 1987b).

Expression of (pro)phospholipase A2

Although several eukariotic proteins have been cloned,
their expression at levels that permit characterizations using

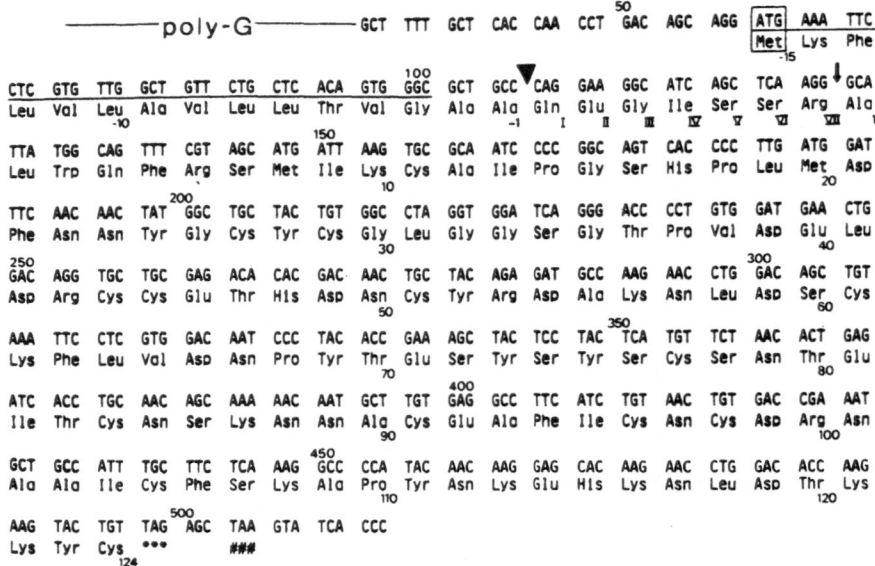

FIG. 2. Nucleotide sequence and amino acid sequence of the cDNA
encoding porcine pancreatic preproPLA. The solid
triangle represents the substrate site for signal
peptidase.Roman numerals denote residues of the
activation peptide, which is cleaved off by trypsin in
the duodenum (arrow).

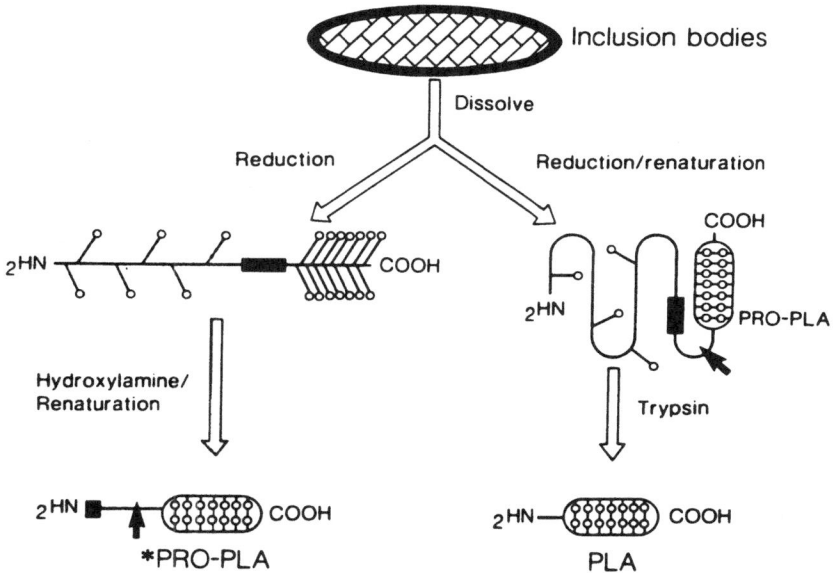

Fig. 3. Schematic representation of the processing of fusion
protein into either proPLA by hydroxylamine cleavage,
or PLA by trysin cleavage (arrow)

physicochemical methods which require substantial amounts of
proteins can be the limiting factor. A recent review on the
expression of eukaryotic proteins in *E. coli* (Marston, 1986)
lists but 5 enzymes out of a total of 33 eukaryotic proteins
expressed in the bacterial cytoplasm. Numerous other expression
systems, both prokaryotic and eukaryotic, have been developed
for the expression of biologically active proteins. Most of
these are designed to secrete the protein in order to obtain
correct N-terminal residues, glycosylation and/or disulphide
bonds, which appear to be the major difficulties associated with
intracellular expression in *E. coli* (Bebbington and Hentschel,
1985; Marston, 1986; Nicaud et al., 1986; Abrahmsen et al.,
1986). Some of these systems have been used for protein
engineering (Craik et al., 1985), but none of them seems to
equal *E. coli* yet in its potential to produce high amounts of
intracellular recombinant polypeptide per liter of culture
medium in an easy and very rapid way (Varadarajan et al., 1985;
Marston, 1986).

As it became evident that PLA could not be expressed to any
significant levels, either directly in the cytoplasm, or after
processing into the periplasmic space, we developed a strategy
for the expression of PLA as a fusion protein. This approach has
led in several cases to the successful expression of eukaryotic
proteins in *E. coli*, but inevitably requires a cleavage step in
the end. The developed approach combines several published
methods: (i) purification of insoluble fusion proteins from the
cytoplasm, (ii) S-sulphonation and subsequent reoxidation and
renaturation of recombinant proteins, and (iii) site specific
cleavage of fusion proteins with hydroxylamine or enzymes, for
the liberation of the protein of interest. The order of

individual steps and protein folding stages, is schematically outlined in Figure 3. For details see de Geus et al. (1987a).

Hydroxylamine cleavage of the peptide bond between Asn and Gly residues, a sequence which is absent in PLA, was optimized by: (i) insertion of an (Asn/Gly)3 linker sequence between proPLA and the bacterial leader fragment of the fusion protein and (ii) by deleting as much as possible of the leader sequence, without affecting the final yield of fusion protein too much. The resulting 59 kD fusion protein could be effectively cleaved with hydroxylamine. The peptide material thus obtained was renatured and the the active proPLA was purified on CM-cellulose. The yield of proPLA was about 25% based on fusion protein.

The tryptic cleavage at the "natural" cleavage site *i.e.* the Arg^{VII}-Ala^1 (see Figure 2) bond which is cleaved upon conversion of proPLA into PLA is even more efficient, approaching quantitative conversion. After S-sulphonation and renaturation in the presence of 2M Urea the renatured fusion protein can be cleaved either directly or after dialysis. In the presence of 1% of trypsin relative to fusion protein, cleavage is complete after 1-2 hours at room temperature. Two subsequent CM-cellulose columns and a DEAE-cellulose column yield pure PLA in 50-80% yield. Thus a typical 10 liter culture permits the isolation of about 15 mg of proPLA or 45 mg of PLA.

Expression of PLA in *Escherichia coli* does not lead automatically to the production of active (pro)PLA because no or incorrect formation of disulphide bonds occurs in the cytoplasm. In contrast to *E. coli* the yeast *Saccharomyces cerevisiae* is capable of both *in vivo* disulphide bridge formation and of efficient secretion of heterologous proteins. Therefore this yeast has been used as a host system for the expression and secretion of several mammalian enzymes (Mellor et al., 1983; Gardell et al., 1985; Lemontt et al., 1985; Hallewell et al., 1987). Expression and secretion of proPLA in *S. cerevisiae* was obtained after fusing the proPLA to the prepro-sequence of the yeast α-mating factor (van den Bergh et al., 1987). Upon secretion, the fusion protein was cleaved by the *KEX2* protease yielding a 140 amino acid zymogen-like form of the phospholipase A2. This protein was purified in high yield by ion-exchange chromatography. Yeast-proPLA shows the same specific activity on monomeric substrate as native proPLA, although it has an amino-terminal extension of nine amino acids. Apparently these extra amino acids do not hinder the entrance of substrate to the active site; they are probably fully exposed to the solvent and do not form contacts with any other part of the enzyme. This is also evident from the fact that yeast-proPLA is as rapidly cleaved by trypsin as is native proPLA. The resulting yeast-PLA is indistinguishable from authentic pancreatic PLA showing that a protein with a disulphide bridge content as high as 7 per 124 amino acid residues, can be correctly processed by the yeast secretory apparatus. The results of this study show that *S.cerevisiae* is a suitable host for the expression of proPLA, and probably of other small eukaryotic proteins which are rich in disulphide bonds. The level of expression that was obtained under control of the α-mating factor promotor was about 0.6 mg per liter culture. When this constitutive promotor was replaced by the inducible GAL-7 promotor (St. John and Davis, 1981) the level of expression was raised five to ten fold thereby making it comparable to that obtained in *E. coli*. (C. J. van den Bergh, unpublished results).

Ligands involved in calcium binding

In order to probe the role of Asp-49 in the active site of porcine pancreatic PLA two mutant proteins were constructed (van den Bergh et al., 1989a) containing Glu or Lys at position 49. Their enzymatic activities and their affinities for substrate and for Ca^{2+}-ions were examined in comparison with the native enzyme. Enzymatic characterization indicated that the presence of Asp-49 is essential for effective hydrolysis of phospholipids. Conversion of Asp-49 to Glu reduces the affinity for Ca^{2+}-ions more than fourty fold, whereas introduction of a Lys at this position essentially abolishes calcium binding (Figure 4). The affinity of both mutant PLAs for monomeric and aggregated substrate analogues is hardly affected relative to the native enzyme. The mutant D49E was about 1000 times less active than native PLA. This low but reproducible activity may

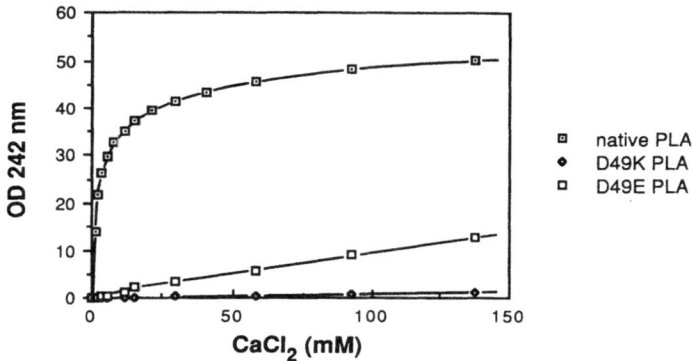

Fig. 4. Binding of calcium ions to native and mutant porcine pancreatic PLAs at pH 6 as determined by ultra violet difference spectroscopy

be caused by the fact that the Ca^{2+}-ion which is bound to Glu instead of Asp is too far away from the susceptible ester bond of the substrate to effectively polarize the carbonyl of this ester bond and stabilize the tetrahedral intermediate, which has been suggested to be crucial for catalysis (Figure 1). The D49K mutant shows an activity which is at least 25000 times lower than native PLA even in the presence of Ca^{2+} concentrations as high as 250 mM. The D49K-PLA is unaffected in its binding to monomeric and to micellar substrate analogues. Therefore in kinetic assays this mutant protein can still be saturated with substrate and hence the loss of activity must be due to the loss of affinity of the mutant protein for calcium ions.

Because in the D49E mutant the negative charge is retained, albeit shifted to a somewhat different position, the possibility existed that ions other than calcium could activate. Several di- and tri-valent kations were tested but none of them did activate the mutant D49E. To our surprise, however, considerable activity (about 15% relative to calcium) was found with native PLA and strontium ions at pH 6 both on monodisperse and on micellar lecithins. At pH 8 the activity on monomers remained 15%, whereas the activity on micelles dropped to values below 2%. In the presence of barium ions no hydrolysis was observable neither at pH 6 nor at pH 8. This finding extends the notion by Pieterson et al. (1974) who used porcine PLA and dioctanoyl lecithin micelles and concluded that at pH 8 both barium and strontium are competitive inhibitors of calcium. We now conclude that strontium can substitute calcium effectively in the active site, but that at elevated pH values other effects dominate (see also below).

In addition to the catalytically essential calcium, it has been demonstrated that porcine and equine PLAs contain a second calcium binding site. Compared with the calcium ion at the active site, the second ion binds with almost ten-fold lower affinity. The importance of this site for enzyme kinetics became clear from the observation that binding of a second metal ion is essential for effective interaction of PLA with organized lipid-water interfaces at alkaline pH (van Dam-Mieras et al., 1975; Slotboom et al., 1978). Different pancreatic PLA

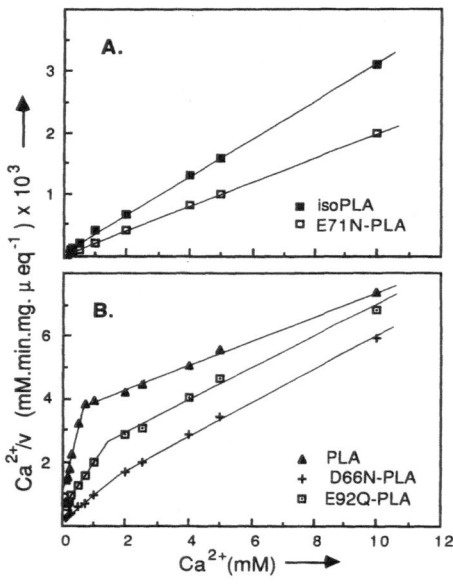

Fig. 5. Hanes plot of the hydrolysis at pH 8.5 of micellar 1,2 dioctanoyl lecithin by native and mutant porcine pancreactic PLAs as a function of calcium concentration.

Table 1. Kinetic parameters of the hydrolysis of 1,2 dioctanoyl
lecithin by native and mutant porcine pancreatic PLAs.

enzyme	pH 6.0		pH 8.5			
			first site		second site	
	K_{Ca}^{app}	v_{max}	K_{Ca}^{app}	v_{max}	K_{Ca}^{app}	v_{max}
	[mM]	[u/mg]	[mM]	[u/mg]	[mM]	[u/mg]
PLA	0.41	1088	0.23	238	8.1	2430
isoPLA	0.44	1280	0.09	3144		
D66N-PLA	0.49	1003	0.24	1268	1.8	2074
E71N-PLA	0.35	1644	0.05	5043		
E92Q-PLA	0.40	1187	0.30	650	2.7	1742

species either possessing a second calcium binding site (porcine
and equine PLA) or lacking this site (bovine and human PLA;
porcine isoPLA) have been studied by proton titration
experiments (Donné-Op den Kelder et al., 1983) and it was
concluded that Glu-71 in porcine PLA is the ligand of the second
calcium ion. In contrast crystallographic data indicated that
the second calcium ion was liganded by Glu-92 rather than to
Glu-71 (Dijkstra et al.,1983). Comparison of the 3-D structure
of bovine and porcine PLAs shows that in the ox enzyme a third
carboxylate (of Asp-66) is in the direct vicinity of both Asn-71
and Glu-92. Since in porcine pancreatic PLA Asp-66 is more
distant from Glu-71 and Glu-92, a role for this residue in the
second calcium binding process has never been considered.
      To study the identity of the acidic amino acid residues
involved in liganding the second calcium ion in detail,three
mutant PLA species were constructed, which lacked one of the
potentially important carboxylates: Asp-66 , Glu-71 and Glu-92,

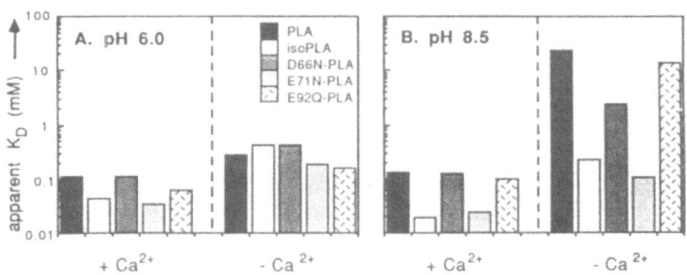

Fig. 6. The influence of calcium ions on the affinity of
various PLA species for micelles of n-hexadecyl
phosphocholine

respectively (van den Bergh et al 1989b). The Gln-92 mutant-PLA displayed the same properties as native PLA indicating that Glu-92 is not important for binding the second metal ion (Figure 5). However, Glu-71 and, to a lesser extent, Asp-66 are both directly involved in the low-affinity calcium binding, indicating the vicinity of Asp-66 and Glu-71 to the second calcium binding site in porcine PLA. From this it seems conceivable that electrostatic forces are important in determining the configuration of the 62-72 surface loop which adopts a different structure in porcine and in bovine PLA. At basic pH, repulsion of the clustered acidic amino acids (id est Asp-66, Glu-71 and perhaps Glu-92) might distort the 3-D structure of this part of the molecule. Binding of a second calcium ion and (partial) neutralisation of the negative charges might, subsequently, induce an altered loop configuration that brings back Asp-66 in the vicinity of Glu-71. As a result of this stabilization of the loop both the affinity of the porcine PLA for lipid aggregates (Figure 6) and the activity (Table 1) is increased. The results with the strontium ions suggest that this ion cannot efficiently substitute calcium at the second calcium binding site.

Deletion of the Surface Loop 62-66

The extracellular PLAs from mammalian pancreas and from snake venom (Waite , 1987), and also the mammalian intracellular PLAs (Seilhamer et al., 1989), exhibit a high degree of sequence homology. An interesting difference between snake venom PLAs and the pancreatic enzymes is that the former ones in general have higher turnover numbers and a greater affinity for phospholipid molecules aggregated in e.g. micelles than the pancreatic ones (van Eijk et al., 1983; Verheij et al., 1981). The X-ray analyses of several PLAs from pig, ox and *Crotalus atrox* venom have been published and it has been pointed out that these enzymes are structurally very similar (Renetseder et al., 1985). A comparison of the structures of active bovine PLA and that of inactive bovine precursor (Dijkstra et al., 1984) shows that in the active enzyme the N-terminal helix and loop 62-72 are well defined, whereas in the precursor this loop and the first three residues of the N-terminal α-helix are mobile. Since proPLA, contrary to active PLA, does not bind to aggregates of zwitterionic phospholipids, it has been suggested that a low mobility of the N-terminal helix and the surface loop are required for efficient binding (Dijkstra et al., 1984). In particular the residues 65, 67, 70 and 72, which are part of the 62-72 surface loop mentioned above, are of interest, since these residues presumably participate in the binding of phospholipid aggregates (Dijkstra et al., 1981b). Snake venom PLAs and also the mammalian intracellular rat platelet PLA, lack part of the 62-72 surface loop. In fact, the deletion of 5 amino acid residues is the most conspicious difference between pancreatic and elapid venom PLAs. In an attempt to improve the catalytic properties of pancreatic PLAs, Kuipers et al. (1989a) have studied the effect of deleting residues 62-66 from porcine pancreatic PLA. In redesigning the primary sequence other adjustments were made to maintain maximal sequence homology with elapid venom PLA in the region of the loop (Figure 7). Thus the mutations D59S, S60G and N67Y were introduced simultaneously. The desired mutant (Δ62-66 PLA) was expressed and isolated as described previously for the wild-type enzyme.

To analyse the three-dimensional structure of the mutant at
the atomic level, the crystal structure of the mutant PLA was
determined and a comparison of the crystal structures of
wild-type and mutant porcine PLA at 2.5 Å resolution was made.
The folding of both molecules is very similar, with the largest
differences occurring at the site of the deletion. The structure
of the 62-72 loop appears to be intermediate between those in
porcine and *C. atrox* PLAs. The *C. atrox* PLA lacks 3 more
residues in this region than the mutant and *Naja melanoleuca*
PLA (Figure 7). Moreover the Tyr-69 side chain has moved outward
towards the molecule's surface. In contrast to the residues in
or near the deleted loop, the region around the $\alpha-NH_3^+$-group of
residue Ala^1 and the N-terminal α-helix are virtually the same.
The residues in the active site are also, within error, in the
same position as in the wild-type structure.

Kinetic parameters of the mutant enzyme were determined
both on monomeric and on micellar substrates; these data are
presented in Table 2. To allow comparison, the values of the
native pancreatic and of a venom PLA are also included. The data
obtained with monomeric substrate show that Δ62-66 PLA has a
fully functional active site. The dissociation constant for
Ca^{2+}-ions at pH 6.0 decreased from 1.8 mM for the wild-type to
0.8 mM for the mutant enzyme. These changes might indicate an
improved positioning of the amino acid side chains in the active
site. Such conformational changes could remain undetected by
X-ray crystallography at 2.5 Å resolution. The most striking
observation is that the mutant Δ62-66 PLA has much higher
activities than wild-type PLA, although not quite as high as
Naja melanoleuca PLA. Furthermore, it is clear that the
difference between Δ 62-66 PLA and native PLA increases with
decreasing acyl chain length in the substrate, the ratio of
these activities being as high as 16 for dihexanoyllecithin. On
the negatively charged substrate diheptanoylglycerosulphate,
Δ62-66 PLA has a much lower activity than the native enzyme,
despite the removal of one net negative charge in the mutant. So
deletion of the loop has induced a considerable change in the
enzyme's preference for zwitterionic and negatively charged
phospholipids, and has conferred properties to the mutant enzyme
similar to those found in e.g. *Naja melanoleuca* (van Eijk et
al., 1983) and *Crotalus adamanteus* (van Deenen en de Haas, 1963;

	50					55					60					65					70				
a)	H	D	N	C	Y	R	D	A	K	N	L	D	S	C	K	F	L	V	D	N	P	Y	T	E	S	Y
b)	*	*	*	*	*	K	Q	*	*	K	*	*	*	*	*	V	*	*	*	*	*	*	N	N	*
c)	*	*	*	*	*	G	E	*	E	K	I	S	G	*	-	-	-	-	-	W	*	*	I	K	T	*
d)	*	*	C	*	*	G	K	*	T	-	-	*	-	*	-	-	-	-	-	*	*	K	*	V	*	*
e)	*	*	*	*	*	*	*	*	*	*	*	S	G	*	-	-	-	-	-	Y	*	*	*	*	*	*

Fig. 7.Comparison of part of the sequences of a mutated and of
four native PLAs. a) porcine pancreas; b) bovine
pancreas; c) *Naja melanoleuca*, DE-III; d) *Crotalus atrox*
and e) mutant Δ62-66 PLA. The full sequences of the
native PLAs have been compared before (van den Bergh et
al., 1989a). An asterisk denotes homology with the upper
sequence.Lacking residues are indicated by a dash.

Table 2. Kinetic properties of different PLAs:
PC, phospatidylcholine; GS, glycerosulfate.* Monomers
of zwitterionic substrates induce lipid-protein
aggregation.+ Due to lipid-protein aggregation below
the CMC, no K_m values could be determined.

	Monomers (diC6dithioPC)			Micelles diC6PC		diC7PC		diC8PC		diC7GS	
Enzyme	k_{cat} (s^{-1})	K_m (mM)	k_{cat}/K_m $(s^{-1}\cdot M^{-1})$	k_{cat} (s^{-1})	K_m (mM)	k_{cat} (s^{-1})	K_m (mM)	k_{cat} (s^{-1})	K_m (mM)	k_{cat} (s^{-1})	K_m (mM)
Wild-type	0.62	0.7	890	5	14	25	3.7	410	3.2	45	+
Δ62-66	0.90	0.5	1790	80	8	240	1.9	980	1.9	10	+
N. mel.	*	*	*	830	+	980	+	3490	+	1	+

Wells, 1974) PLAs. The question remains how the increased
activity on micelles can be explained. Answering this question
is seriously hampered by the lack of structural knowledge of the
enzyme-micelle complex. What is clear, however, is that the
mutations and deletions in the loop 62-72 have affected part of
the putative binding site for phospholipid aggregates: Asn-67
has been replaced by Tyr, Val-65 has been deleted and the X-ray
analysis shows that Tyr-69 in the mutant enzyme has a position
and orientation which is different from that in the wild-type
enzyme. All these changes could cause a different orientation of
the active site of the mutant with respect to the lipid
aggregate, allowing a more efficient interaction with the
individual phospholipid molecules.

The affinity of the mutant for n-hexadecyl phosphocholine
micelles is slightly (1.5 times) lower than that of native

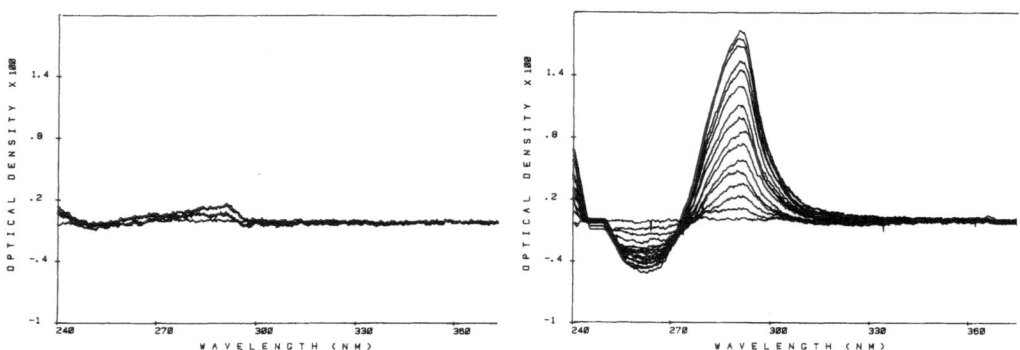

Fig. 8. Ultraviolet difference spectra resulting from the
addition of micelles of n-hexadecyl phosphocholine to
native (left) and to Δ62-66 (right) proPLA at pH 6 in the
presence of 10 mM calcium.

Table 3. The affinity of various native and mutant (pro)PLAs for micelles of n-hexadecyl phosphocholine at pH 6 in the presence of 10 mM calcium.

Enzyme	$N \cdot K_d$ (µM)	activity
Naja mel.PLA	20	++++
Pig Δ62-66 PLA	~ 400	+++
Pig PLA	~ 250	++
α-keto Naja mel.PLA	~ 400	−
Ox PLA	~ 4000	+
α-keto Δ62-66 PLA	~ 4000	−
pro Δ62-66 PLA	~ 4000	−
Pig pro PLA	no detectable binding	−
Pig α-keto PLA	no detectable binding	−

porcine pancreatic PLA. Thus the removal of the sequence Lys.Phe.Leu.Val apparently destabilizes the mutant-micelle complex. Surprisingly the precursors of native and Δ62-66 PLA behave differently upon addition of micelles. Addition of *n*-hexadecyl phosphocholine to the native precursor caused no spectral changes (Figure 8, left), whereas the Δ62-66 precursor yielded a strong perturbation spectrum (Figure 8, right). From these spectra an apparent dissociation constant (N.Kd) of 4 mM was calculated (Table 3). Similar spectra and an identical N.Kd value were obtained with α-keto Δ62-66 PLA. Although these proteins have about ten fold lower affinity for micelles than the active Δ62-66 PLA they can still be saturated with the aggregates.In fact their affinity for micelles equals that of native bovine pancreatic PLA. Despite their potential to bind to interfaces both proteins with a blocked or removed α-amino group are inactive on dioctanoyl lecithin micelles, but are fully active on monomeric lecithins. A similar loss in activity and a ten fold reduction in affinity for micelles was observed by Verheij et al.(1981) after transamination of *Naja melanoleuca* venom PLA. Thus whereas deletion of the surface loop 62-66 reduces the affinity of *active* phospholipases for micelles, deletion increases the affinity · of the precursor from immeasurably small to values permitting significant binding. A comparison of the structures of active bovine PLA and that of inactive bovine precursor or α-keto PLA shows that in the active enzyme the N-terminal helix and loop 62-72 are well defined, whereas in the precursor and α-keto PLA this latter loop and the first three residues of the N-terminal α-helix are mobile (Dijkstra et al., 1984) . This finding suggests that a low mobility of the N-terminal helix and the surface loop are required for efficient binding. Removal of the 62-66 surface loop might stabilize and immobilize the N-terminal helix of the Δ62-66 precursor thereby inducing an increased affinity for lipid aggregates. In the active Δ62-66 PLA where the N-terminal helix is already immobilized the negative effect of removing Lys.Phe.Leu Val apparently dominates any positive effect of a further immobilization of the N-terminal helix.

Residues involved in Substrate Binding

From an analysis of the three-dimensional structure of PLA and from chemical modification studies on porcine and bovine PLA it was proposed that the binding site for lipid aggregates consists of a surface region, formed by the side-chains of hydrophobic and basic residues (Dijkstra et al.,1981b; Volwerk and de Haas, 1982). Which residues in particular are important for the interaction with monomeric substrate remains speculative, since no enzyme-substrate complexes have been crystallized yet. Two residues seem, however, likely candidates since they are located at the edge of the entrance to the active site: Leu-31 and Tyr-69 which are found in all pancreatic PLAs. Using mononitrated PLA2 from porcine, bovine and equine pancreas Meyer et al. (1979) concluded that a non-ionized Tyr-69 hydroxyl group is important for the binding of substrate monomers. From an analysis of the 3-D structure of bovine PLA inhibited by p-bromo phenacyl bromide (Renetseder et al., 1988) it appears conceivable that the Tyr-69 sidechain will interact with the phosphate moiety of the substrate.

To gain a better understanding of the role of Tyr-69 in porcine pancreatic PLA Kuipers et al. (1989b) substituted Phe for Tyr-69 and in later experiments a lysine was introduced(O.P. Kuipers, manucript in preparation). Kinetic characterization revealed that the Phe-69 (Y69F) and the Lys-69 (Y69K) mutants, like the native enzyme, have retained considerable activity on monomeric and on aggregated substrates. Whereas the native PLA and the Y69K mutant are only able to catalyze the degradation of *sn*-3 phospholipids, the Y69F mutant hydrolyses both the *sn*-3 isomers and, at a low (1-2%) rate, the *sn*-1 isomers (Figure 9). Although the stereospecificity of the mutant phospholipase was changed, Y69F PLA still requires Ca2+ ions as a cofactor and also retains its specificity for the *sn*-2 ester bond. These data suggest that in native porcine PLA the hydroxyl group of Tyr-69 serves to fix and orient the phosphate group of phospholipid monomers by hydrogen bonding and that in the Y69K mutant the ε-amino group can play a similar role. The Y69F mutant looses (part of) its stereospecificity because no such interaction can occur between the Phe-69 side chain and the phosphate moiety of the substrate monomer.

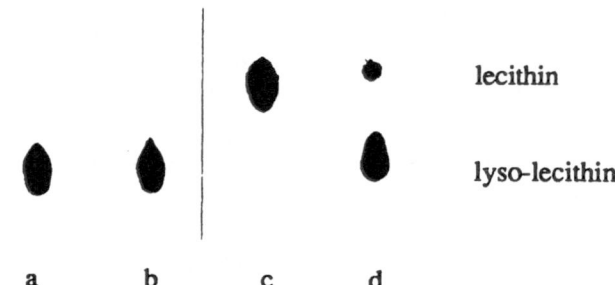

Fig. 9. Thin-layer chromatogram showing the degradation of *sn*-1 and *sn*-3 didodecanoyl lecithins by native and Y69F PLA in the presence of deoxycholate and 5 mM calcium at pH 8. To the *sn*-3 lecithins were added 5 µg native (lane a) or 20 µg mutant (lane b) PLA; to the *sn*-1 isomer were added 25 µg native (lane c) and 100 µg mutant (lane d) PLA. Incubations were for 8 h at 37°C.

Table 4. The activities of native, Y69K and Y69F PLAs on
1,2-didodecanoyl-sn-3-lecithin and on the
corresponding two isomers of the thion analog in the
presence of tauro dexycholate

Enzyme	activity (μmol·min^{-1}·mg^{-1})		
	diC12PC	diC12thionPC	
		R_p-isomer	S_p-isomer
wild-type	55	13	< 0.1
Lys-69	11	2	< 0.02
Phe-69	14	30	1.2

The interaction between the side chains of residue 69 in
native and in both mutant PLAs was further investigated by
measuring the activities on substrates with a modified phosphate
moiety. The Rp isomer of thion lecithins (Figure 10) has been
shown to be a sustrate for PLAs from various sources (Tsai et
al., 1985). The activities of native and of mutant PLAs was
tested on di-dodecanoyllecithin and on its Rp and Sp thion
analogs. The activity of the native PLA on the Rp-thion PC is
lower than on the normal PC (Table 4). More striking is the
observation that both mutant PLAs respond quite differently to
the introduction of a sulfur in the phosphate moiety. The Y69K
mutant has a reduced activity, the Y69F mutant has an increased
activity on the thion PC compared to normal PC. Also with the Sp

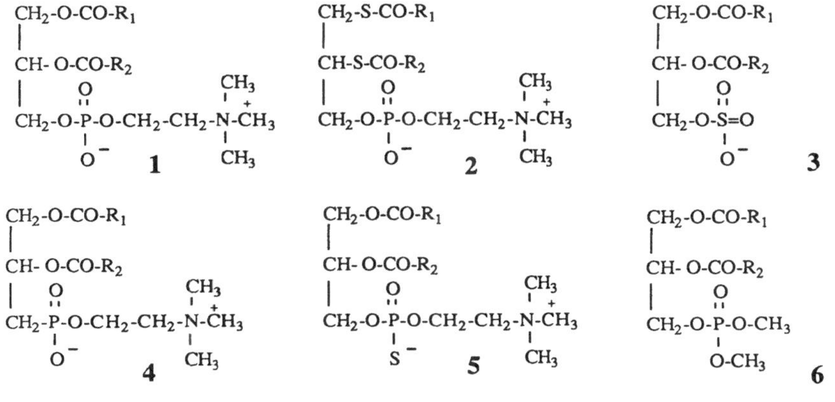

Fig. 10. Schematic stuctures of: lecithin, 1; diacyl dithiolecithin, 2; diacyl glycero sulfate, 3; phosphono
lecithin, 4; thion lecithin, 5; and dimethyl
phosphatidic acid, 6.

Table 5. The activity of native, Y69K and Y69F PLA on dimethyl phosphatidic acid in the presence of 10 mM *n*-hexadecyl phosphocholine, 10 mM calcium at pH 8.

Enzyme	activity (μmol·min^{-1}·mg^{-1})	
	diC12PC	diC12dimethylPA
wild-type	12	1.4
Lys-69	4.5	0.2
Phe-69	7	7

isomer there is a clear distinction: both native PLA and the Y69K mutant show very low activities on this compound whereas the Y69F mutant has about 4% activity on the Sp compared to the Rp isomer. Thus the latter mutant not only has a changed specificity at *sn*-2 but likewise the stereochemical preference at phosphorus is affected. Other modifications that were made in the polar headgroup of the substrate molecule were the removal of the negative charge of the phosphate (dimethylester of didodecanoyl phosphatidic acid in mixed micelles) and a change of the distance between the phosphate and the *sn*-2 ester bond (phosphono phospholipids). On the dimethyl ester native and Y69K PLA have lower activities than on the corresponding lecithin.The Y69F mutant, however, has lost this preference for substrates with a free ionization on phosphorus since it hydrolyses the dimethyl ester and the lecithin equally well (Table 5). In the phosphonolipids the distance between the negative charge and the susceptible ester bond is reduced by approximately 2 Å. Despite this change in the structure 1-dodecanoyl-2-tetradecanoyl-*sn*-3-phosphono lecithin is a rather good substrate for native and both mutant PLAs and hydrolysis occurs with retention of positional specificity. Surprisingly the *sn*-1 isomer is hydrolysed not only by the Y69F mutant but also by native porcine pancreatic PLA at 1-2% rate compared to the *sn*-3 isomer. Clearly the interaction between the side chain of residue 69 can be relieved by taking out the phenolic hydroxyl or by increasing the distance between the hydrogen bond acceptor and donor and in both cases a loss of stereospecificity is the result.

The other residue flanking the entrace of the active site, *id est* Leu-31 is conserved in all pancreatic PLAs, but in more than 85 percent of the venom PLAs either Trp, Ala or Arg is found at this position. To study the role of Leu-31 this residue has been replaced by Trp, Ala, Arg, Gly, Ser and Thr (O.P. Kuipers, manuscript in preparation). The introduction of a smaller or more polar residue than Leu markedly reduced kcat for the hydrolysis of monomeric substates without affecting Km significantly. With dioctanoyl lecithin micelles as substrate the apparent Km is hardly changed, but Vmax has been reduced to values between 9% (Arg) and 2% (Gly). Measured on the negatively charged substrate diheptanoyl glycerosulfate the activity of all mutants is within a close range of about 20% of the native PLA. This could mean that for the interaction of PLA with negative surfaces the contribution of lysines in the lipid binding domain overrules the contribution of Leu-31. With neutral surfaces the contribution of Leu-31 is relatively more important. Of the six mutants the L31W mutant was the only one having improved catalytic properties. The specificity constant for monomeric substrates was increased 1.5 fold, mainly due to a five fold increased affinity for the substrate 1,2-dihexanoyl dithiolecithin. With micellar dioctanoyl lecithin as the substrate the apparent affinity for the substrate had increased with a factor ten, although the Vmax was reduced 60%. It seems thus apparent that in native PLA Leu-31 is important for the interaction with neutral interfaces, but that it has few contacts with the monomeric substrate. Thus replacing it by a smaller residue does not lower the affinity for monomeric phospholipids. The introduction of the larger and hydrophobic tryptophan side chain has profound effects on the interaction both with monomers and micelles. Apparently the side chain is large enough to penetrate into the hydrophobic core of the micelle and to stabilize the complex. It is conceivable that with monodisperse substrates in water there is considerable hydrophobic interaction of (a part) of one of the fatty acyl chains with the Trp side chain. Whether this is the acyl chain esterified to *sn*-1 or to *sn*-2 position remains to be solved.

One possible approach to study the interaction of substrate molecules and calcium with PLA is the use of time resoved fluorescence (Beecham and Brand, 1985). Such studies are greatly facilitated if proteins are being used which contain a single tryptophan residue. Native porcine pancreatic PLA meets this requirement since it contains a single tryptophan at position 3 which is involved in binding to lipid aggregates (van Dam-Mieras et al., 1975). Substitution of phenylalanine for tryptophan yielded an enzyme with reduced lipid binding properties that was used as the parent molecule to create PLAs with unique tryptophan residues at various positions (O. P. Kuipers, manuscript in preparation). Two of these mutants, W3F,L31W PLA and W3F,L31W,Δ62-66 PLA, are of particular interest since they exhibit a five to ten fold increased affinity for monomers of *n*-dodecyl phosphocholine than the parent enzymes, whereas the Km values for dihexanoyl dithiolecithin was reduced about three fold. Due to this higher affinity for monomeric product analogs, the active site of both enzymes is about 90% saturated at the CMC of *n*-dodecyl phosphocholine. Time resolved fluorescence anisotropy studies of such complexes are in progress.

Concluding Remarks

The use of recombinant DNA techniques in the study of porcine pancreatic PLA has added significant information to our knowledge of this enzyme. Many interesting questions pertaining to the physiological and pharmacological properties, like neurotoxicity, anti-coagulant activities of extracellular and the activation of cellular PLAs, remain to be solved yet. On a molecular level the role of many of the residues which are (nearly) completely conserved in PLA has not been elucidated. A major challenge remains the elucidation of the 3-D structure of the enzyme in the presence of substrates or inhibitors and a comparison with the structure of the apo-enzyme. Since it is unlikely that PLA-micelle complexes can be crystallized, structural information of such structures probably has to come from NMR studies. Recombinant DNA techniques may facilitate the identification of signals in NMR spectra since specifically labelled amino acids can be introduced in PLA. Crystallization of PLA in the presence of monomeric phospholipid analogs may be a realistic goal. The mutant W3F,L31W,Δ62-66 PLA combines good crystallization properties with high affinity for phospholipids. In combination with amide phospholipids which are potent inhibitors of porcine pancreatic PLA (de Haas et al., 1989) and calcium this mutant PLA indeed yields crystals of good quality (B.W. Dijkstra, personal communication). Solving this stucture might unravel the contacts between the inhibitor and individual amino acid side chains.

References

Abrahmsen, L., Moks, T., Nilsson, B. and Uhlen, M., 1986, Nucleic Acids Res. 14, 7487-7500

Bebbington, C. and Hentschel, L., 1985, Trends Biotech. 3, 314-317

Beechem, J.M. and Brand, L., 1985, Ann. Rev. Biochem. 54, 43-71

Brunie, S., Bolin, J., Gewirth, D. and Sigler, P.B., 1985, J. Biol. Chem. 260, 9742-9749

Craik, C.S., Largman, C., Fletcher, T., Roczniak, S., Barr, P.J., Fletterick, R., and Rutter, W.J., 1985, Science 228, 291-297

de Geus, P., van den Bergh, C.J., Kuipers,O., Verheij, H.M., Hoekstra, W.P.M. and de Haas, G.H., 1987a, Nucleic Acids Res. 15, 3743-3759

de Geus, P., Kuipers, O.P., van den Heuvel, M., Verheij, H.M. and de Haas, G.H., 1987b, Chimicaoggi Ottore, 73-77

de Haas, G.H., van Oort, m.G., Dijkman, R. and Verger, R., 1989, Biochemical Society Transactions 17, 274-276

Dijkstra, B.W., Kalk, K.H., Hol, W.G.J. and Drenth, J., 1981a, J. Mol. Biol. 147, 97-123

Dijkstra, B. W., Drenth, J. and Kalk, K.H., 1981b, Nature 289, 604-606

Dijkstra, B.W., Renetseder, R., Kalk, K.H., Hol, W.G.J. and Drenth, J., 1983, J. Mol. Biol. 168, 163-179

Dijkstra, B.W., Kalk, K.H., Drenth, J., de Haas, G.H., Egmond, M.R. and Slotboom, A.J., 1984, Biochemistry 23, 2759-2766

Donné-Op den Kelder, G.M., de Haas, G.H. and Egmond, M.R., 1983, Biochemistry 22, 2470-2478

Ducancel, F., Guignery-Frelat, G., Bouchier, C., Menez, A. and Boulain, J.C., 1988, Nucleic Acids Res. 16, 9049

Elsbach, P. and Weiss, J., 1988, Biochim. Biophys. Acta 947, 29-52

Fleer, E.A.M., Verheij, H.M. and de Haas, G.H., 1981, Eur. J. Biochem. 113, 183-188

Gardell, S.J., Craik, C.S., Hilvert, D., Urdea, M.S. and Rutter, W.J., 1985, Nature 317, 551-555

Hallewell, R.A., Mills, R., Tekamp-Olson, P., Blacher, R., Rosenberg, S., Ötting, F., Masiarz, F.R. and Scandella, C.J., 1987, Bio/Technology 5, 363-366

Kramer, R.M., Hession, C., Johansen, B., Hayes, G., McGray, P., Chow, E.P., Tizzard, R. and Pepinsky, R.B., 1989, J. Biol. Chem. 264, 5768-5775

Kuipers, O.P., Thunnissen, M.M.G.M., de Geus, P., Dijkstra, B., Drenth, J., Verheij, H.M. and de Haas, G.H., 1989a, Science, 244 82-85

Kuipers, O.P., Dijkman, R., Pals, C.E.G.M., Verheij, H.M. and de Haas, G.H., 1989b, Protein Engineering 2, 467-471

Lemontt, J.F., Wei, C.-M. and Dackowski, W.R., 1985, DNA 4, 419-428

Maraganore, J.M., Merutka, G., Cho, W., Welches, W., Kezdy, F.J. and Heinrikson, R.L., 1984, J. Biol. Chem. 259, 13839-13843

Maraganore, J.M. and Heinrikson, R.L., 1986, J. Biol. Chem. 261, 4797-4804

Marston, F.A.O., Biochem. J., 1986, 240, 1-12

Mellor, J., Dobson, M.J., Roberts, N.A., Tuite, N.F., Emtage, J.C., White, S., Lowe, P.A., Patel, J., Kingsman, A.J. and Kingsman, S.M., 1983, Gene 24, 1-14

Meyer, H., Puijk, W.C., Dijkman, R., Foda-van der Hoorn, M.M.E.L., Pattus, F., Slotboom, A.J. and de Haas, G.H., 1979, Biochemistry, 16, 3589-3597

Nicaud, J.M., Mackman, N. and Holland, I.B., 1986, J. Biotechmol. 3, 255-270

Ohara, O., Tamaki, M., Nakamura, E., Tsuruta, Y., Fujii, Y., Shin, M., Teraoka, H. and Okamoto, M., 1986, J. Biochem. 99, 733-739

Perlman, D. and Halvorson, H.O., 1983, J. Mol. Biol. 167, 391-409

Pieterson, W.A., Vidal, J.C., Volwerk, J.J. and de Haas, G.H., 1974, Biochemistry 13, 1455-1459

Puijk, W.C., Verheij, H.M., de Haas, G.H., 1977, Biochim. Biophys. Acta 492, 254-259

Renetseder, R., Brunie, S., Dijkstra, B.W., Drenth, J. and Sigler, P.B., 1985, J. Biol. Chem. 260, 11627-11634

Renetseder, R, Dijkstra, B.W., Huizinga, K., Kalk, K.H. and Drenth, J., 1988, J. Mol. Biol., 200, 181-188

Seilhamer, J.J., Randall, T.L., Miles, Y. and Johnson, L.K., 1986, DNA 5, 519-527

Seilhamer, J.J.,Vadas, P., Plant, S., Millar, J.A., Kloss, J.,Pruzanski, W. and Johnson, L.K. ,1989, J. Biol. Chem. 264, 5335-5338

Slotboom, A.J., Jansen, E.H.J.M., Vlijm, H., Pattus, F., de Araujo, P.S. and de Haas, G.H., 1978, Biochemistry 17, 4593-4600

St. John, T.P. and Davis, R.W., 1981, J. Mol. Biol. 152, 285-315

Tsai, T.-C., Hart, J., Jiang, R.-T., Bruzik, K., and Tsai, M.-D., 1985, Biochemistry 24, 3180-3188

van Dam-Mieras, M.C.E., Slotboom, A.J., Pieterson, W.A. and de Haas, G.H., 1975, Biochemistry 14, 5387-5394

van Deenen, L.L.M. and de Haas, G.H., 1963, Biochem. Biophys. Acta 70, 538-553

van den Bergh, C.J., Bekkers, A.C.A.P.A., de Geus, P., Verheij, H.M. and de Haas, G.H., 1987, Eur. J. Biochemistry 140, 241-246

van den Bergh, C.J., Slotboom, A.J., Verheij, H.M. and de Haas, G.H., 1989a, J. Cellular Biochem. 39, 379-390

van den Bergh, C.J., Verheij, H.M. and de Haas, G.H., 1989b, Eur. J. Biochem. , 182 307-313

van Eijk, J.H., Verheij, H. M., Dijkman, R. and de Haas, G.H., 1983, Eur. J. Biochem. 132, 183-188

van Scharrenburg, G.J.M., 1984, Ph. D. Thesis, University of Utrecht, Utrecht, The Netherlands.

Varadarajan, R., Szabo, A. and Boxes, S.G., 1985, Proc. Natl. Acad. Sci. U.S.A. 82, 5681-5684

Verger, R., Mieras, M.C.E. and de Haas, G.H., 1973, J. Biol. Chem. 248, 4023-4034

Verheij, H.M., Volwerk, J.J., Jansen, E.H.J.M., Puijk,W.C., Dijkstra, B.W., Drenth, J. and de Haas, G.H., 1980, Biochemistry 19, 743-750

Verheij, H.M., Egmond, M.R. and de Haas, G.H., 1981, Biochemistry 20,94-99

Volwerk, J.J., Dedieu, A.G.R., Verheij, H.M., Dijkman, R. and de Haas, G.H., 1979, Recl. Trav. Chim. Pays-Bas 98, 214-220

Volwerk, J.J. and de Haas, G.H.,1982, in Molecular Biology of Lipid-protein Interactions (Griffith, O.H. and Jost, P.C., eds.) pp. 69-149, Wiley, New York.

Waite, M., 1987, in Handbook of Lipid Research (Hanahan, D.J., ed.) vol. 5, Plenum Press, New York.

Wells, M.A., 1972, Biochemistry 11, 1030-1041

Wells, M.A., 1974, Biochemistry 13, 2248-2257

MOLECULAR ASPECTS OF

PHOSPHOLIPASE A$_2$ ACTIVATION

Rodney L. Biltonen, Thomas R. Heimburg, Brian K. Lathrop, and
John D. Bell

Departments of Biochemistry and Pharmacology
University of Virginia
Charlottesville, VA

INTRODUCTION

A striking characteristic of soluble phospholipases A$_2$ is their tendency
to become activated at a lipid–water interface.[1] The activity of these
enzymes is much greater with aggregated phospholipid substrates than with
monomeric substrates. Furthermore, the activity is heavily influenced by the
state of the aggregated lipids. Our interests in phospholipase A$_2$ are
focussed on the mechanism of the activation process and the physical basis of
the role of lipid structure and/or dynamics in that process.

Early studies is our laboratory involved the various interactions of
porcine pancreatic phospholipase A$_2$ and either small sonicated unilamellar
vesicles of dipalmitoylphosphatidylcholine (DPPC) or large fused unilamellar
vesicles (LUV) of DPPC.[2,3] It was found that rates of hydrolysis were maximum
at initial time with small unilamellar vesicles at temperatures well below the
thermotropic transition to the liquid crystalline phase of the phospholipid
vesicles (T$_m$ = ~37°C). At temperatures above about 36°C, lag phases prior to
rapid hydrolysis are observed, and these lag times become longer at
temperatures in the liquid crystalline phase.[2] Somewhat different results
were obtained with DPPC LUV. First, lag periods in the hydrolysis time
courses were seen at all temperatures. Second, the length of these lag phases
is inversely proportional to temperature below T$_m$, minimal at T$_m$ (~41.5° in DPPC
LUV) and increases as a function of temperature in the liquid crystalline
phase.[3] Analysis cf temperature, calcium, and substrate and enzyme concentration dependence data from those two studies[2,3] led to three conclusions:
1) enzyme activation and substrate hydrolysis depend on the structure of the
vesicles; 2) binding of the enzyme to the vesicles is stonger in the gel state
and does not require calcium; 3) activation involves aggregation of the enzyme
on the surface of the vesicles.

Subsequent investigations focussed on a quantitative analysis of time
courses describing pancreatic phospholipase A$_2$ hydrolysis of DPPC LUV at 38°C.
Under such conditions, the initial 10–20% of the time course was directly
proportional to At^2 and could be analyzed in terms of a simple activation
model.[4] Analysis of the second–order coefficient (A) demonstrated that the
rate of activation was proportional to the square of the enzyme concentration
and that the rate of activation was an inverse function of the vesicle
concentration. Implicit in the analysis was the assumption that activation

was irreversible within the time frame of the experiment. The conclusion of this work was that the mechanism of activation of phospholipase A_2 involved the formation of enzyme dimers on the surface of the vesicles.[4] Recently, it has been reported that the pancreatic and other soluble phospholipases A_2 become acylated during phospholipid hydrolysis, that the acylated enzyme is a stable dimer and that it is more active than the virgin monomer.[5,6] This acylation reaction, then, could be the basis of the irreversible dimer activation mechanism derived from the kinetic data.

Any simple activation model in which the rate of enzyme activation is constant with time will yield a time course in which the product formed is a second-order function of time. In all other cases, the hydrolysis time course will be a more complex function of time. The complete time courses of phospholipase A_2 hydrolysis of LUV all exhibit such complex behavior, even in cases where simple behavior was observed up to 20% hydrolysis. The deviation from At^2 behavior generally indicates that a second process which increases the rate of activation occurs as hydrolysis proceeds. This second process is most apparent at temperatures above the lipid phase transition temperature for the pancreatic enzyme and at all temperatures for the monomeric aspartate-49 phospholipase A_2 from Agkistrodon piscivorus piscivorus (AppD49).[7]

Unlike the pancreatic enzyme, AppD49 exhibits a large fluorescence change upon interaction with DPPC. Thus, time-dependent changes in the enzymatic activity, the intrinsic enzyme fluorescence and changes in the physical properties of the lipid bilayer can be measured simultaneously in this system.[7] Correlation of these observables has allowed us to begin an examination of the temporal sequence of events in phospholipase A_2 activation and address various models proposed to describe the mechanism of that activation. This paper constitutes a first report of our analysis of experimental hydrolysis data using computer simulations of selected models.

TEMPORAL SEQUENCE

The time course of AppD49-catalyzed hydrolysis of DPPC LUV is complex both as a function of time and temperature.[7] At low temperatures (below 30°C) initial hydrolysis is very slow and is almost undetectable at 25°C. In contrast, hydrolysis is readily detectable at temperatures in the vicinity of the main thermotropic phase transition of the lipid (41.3°C for DPPC LUV). The time courses of DPPC LUV hydrolysis by AppD49 phospholipase A_2 at 39, 41 and 45°C are shown in Fig. 1. Hydrolysis initially proceeds at a relatively slow rate. After a lag time of several hundred seconds, the activity suddenly increases by two to three orders of magnitude within 5 to 10 s. The length of the lag period is dependent on temperature, enzyme concentration and substrate concentration. The lag period is a minimum at approximately the phase transition temperature and becomes dramatically longer as temperature is further increased. The length of this initial phase of the time course increases with increasing substrate concentration and is inversely proportional to enzyme concentration. These observations are consistent with the observations that led to proposal of the dimer activation model,[4] but the complexity of the time courses (Fig. 1) preclude analysis of the data with that model in its simplest form.

Phospholipase A_2 Fluorescence

The interaction of many phospholipases A_2 with lipid substrate can be detected by changes in the intrinsic tryptophan fluorescence of the protein upon mixing with the lipid substrate.[8-10] Generally, the fluorescence intensity increases and the maximum of the emission spectrum shifts some 5 to 10 nm toward lower wavelength. Such spectral changes have been traditionally interpreted as indicating a change in the tryptophan environment to lower polarity.

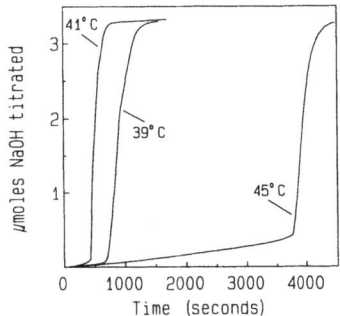

Fig. 1. Time courses of hydrolysis of DPPC LUV by AppD49
phospholipase A_2. Reactions were initiated by
the addition of enzyme (final concentration =
310 nM) to 2.5-ml samples containing 35 mM KCl,
10 mM $CaCl_2$ and 1.5 mM DPPC LUV at 39, 41 or 45°C
and monitored at pH 8.0 with a pH stat mounted
in the sample chamber of an SLM 8000C spectro-
fluorometer. Used by permission from Ref. 7.

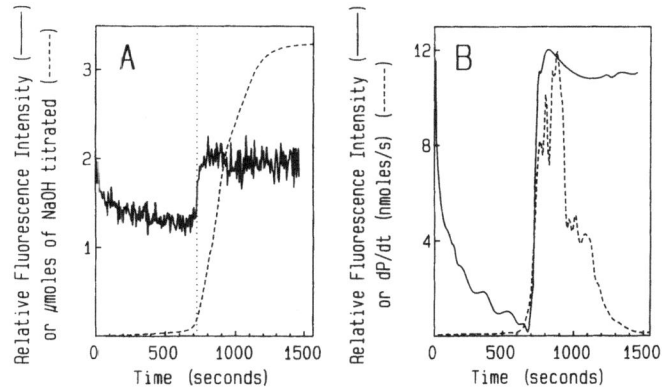

Fig. 2. Correlation of AppD49 fluorescence with the time courses of vesicle
hydrolysis (dashed line) and enzyme fluorescence (excitation = 280
nm, emission = 340 nm, solid line) from the 39°C experiment described
in Fig. 1. Panel B, the derivative of the hydrolysis time course
(dP/dt) was calculated from the data in A. The enzyme fluorescence
was rescaled and smoothed to demonstrate more clearly the temporal
correlation of the fluorescence change with dP/dt. The decrease in
enzyme fluorescence during the first 500 s is independent of the presence of vesicles and is due to binding of some of the enzyme to the
walls of the cuvette. Used by permission from Ref. 7.

Upon mixing with DPPC small unilamellar vesicles, the AppD49 phospho-lipase A_2 fluorescence increases 70%, and the emission spectrum shifts 6 nm to lower wavelength. This fluorescence change does not require calcium but is sensitive to temperature. In the absence of vesicle hydrolysis, it only occurs in the gel state of the lipid.[10] These results substantiate the conclusion derived from studies with the pancreatic phospholipase that binding does not require calcium but is stronger to the gel state of the lipid.[3]

The addition of AppD49 phospholipase A_2 to DPPC LUV results in no measurable fluorescence change at time zero. However, after an initial latency, the fluorescence intensity suddenly increases some 70%.[7] This increase in intensity is accompanied by a shift in the maximum of the emission spectrum from 348 nm to about 340 nm. By combining the pH stat instrument used to measure the hydrolysis reaction with the fluorometer cell,[7] we are able to temporally correlate the fluorescence and hydrolysis time courses. As shown in Fig. 2, the abrupt increase in enzyme fluorescence intensity superimposes on the derivative of the hydrolysis time course (Fig. 2B). This result has been found to be reproducible at all temperatures and concentrations of enzyme or substrate and strongly suggests that enhanced fluorescence intensity is directly correlated to formation of the active state of the enzyme-lipid complex.

Membrane Structure

Two observations are of substantial importance in an effort to understand the mechanisms of these unusual time courses. First, the onset of rapid hydrolysis can be an extremely abrupt phenomenon. Second, the mole fraction of hydrolysis products formed at the time of the abrupt increase in activity is constant at a given temperature regardless of enzyme or substrate concentration (.071 ± standard deviation of .016 at 39°C).[7] These ob-servations suggest that a cooperative change in membrane structure occurs within a specific concentration range of the reaction products and that this change promotes formation of the active enzyme-lipid complex. We have employed two spectroscopic observations that can be monitored simultaneously with the hydrolysis or fluorescence time courses looking for evidence of such a phenomenon. The first is the intensity of the light scattered by the vesicles and the second is the fluorescence of the membrane probe trimethylammonium diphenylhexatriene (TMA-DPH).

Changes in light scattering are indicative of changes in vesicle size which could be related to the abrupt increase in activity. As shown in curve b of Fig. 3, a sudden change in the intensity of scattered light at 290 nm occurs near the onset of the enzyme fluorescence increase (curve c) and the onset of rapid hydrolysis of the substrate (dotted line in Fig. 3). The light scattering first decreases and then increases as a function of time. The ultimate increase in both the intensity and noise of the light scattering is due to the formation of a precipitate of the calcium and palmitic acid released in the late phase of vesicle hydrolysis. At low calcium concen-trations (≤100 μM), no visible precipitate is formed, and the increases in intensity and noise of the light scattering are not observed. The magnitude of the initial decrease in light scattering is less than or equal to that produced by sonication of the DPPC LUV to form small unilamellar vesicle (corresponding to a reduction in vesicle diameter from 900 to 200 Å). The most important aspect of the light scattering change is that it consistently occurs after the onset of the fluorescence change and rapid hydrolysis. In over 50 experiments, the onset of increased protein fluorescence has occurred from 0-60 s prior to the onset of the light scattering change depending on the experimental conditions. We have never found any evidence that the light scattering change occurs prior to the enzyme fluorescence increase. Therefore, changes in the vesicles during the hydrolysis time course which are reflected by the change in light scattering are not responsible for the sudden increase in AppD49 phospholipase A_2 fluorescence or hydrolytic activity.

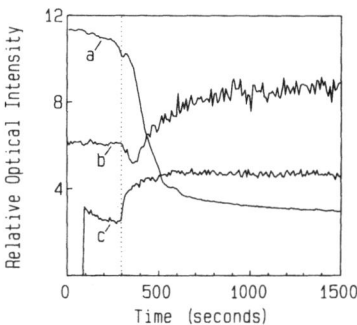

Fig. 3. Correlation of time courses of AppD49 fluorescence, vesicle light scattering and TMA-DPH
fluorescence during the hydrolysis of DPPC LUV.
LUV (0.4 mM DPPC) were equilibrated with 0.4 μM
TMA-DPH in 35 mM KCl, 10 mM $CaCl_2$ and 10 mM
sodium borate at pH 8. TMA-DPH fluorescence
(excitation = 360 nm, emission = 430 nm, curve
a), light scattering at 290 nm (curve b) and
enzyme fluorescence (curve c) were simultaneously recorded as described in Figs. 1 and 2. The
enzyme (140 nM final) was added to the sample
at 90 s into the time course. For explanation
of the initial time dependence of the enzyme
fluorescence, see legend to Fig. 2. Used by
permission from Ref. 7.

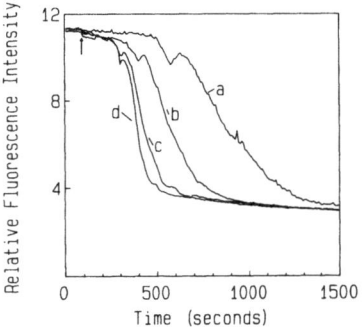

Fig. 4. Enzyme concentration dependence of TMA-DPH fluorescence during hydrolysis of DPPC LUV by AppD49
phospholipase A_2. LUV (0.4 mM DPPC) equilibrated
with TMA-DPH were mixed with 36 nM (curve a),
71 nM (curve b), 140 nM (curve c) or 570 nM
(curve d) phospholipase A_2 at 39°C, and the TMA-
DPH fluorescence was monitored as described in
Fig. 3. The enzyme was added to the sample at
90 s as indicated by the arrow. Used by permission from Ref. 7.

The fluorescence of TMA-DPH is sensitive to changes in membrane structure such as those that occur at the thermotropic phase transition.[11] Fig. 3 (curve a) shows a typical time course of TMA-DPH fluorescence during hydrolysis of DPPC LUV. During the initial slow phase of hydrolysis, the TMA-DPH fluorescence gradually decreases as a function of time. Just prior to the increase in AppD49 phospholipase A_2 fluorescence, the TMA-DPH fluorescence increases slightly and then rapidly decreases. The small temporary increase in TMA-DPH fluorescence is a highly reproducible phenomenon and is most apparent at low enzyme concentrations when the hydrolysis time course is expanded in time (Fig. 4). In all experiments in which TMA-DPH and phospholipase A_2 fluorescence have been monitored concurrently, the time of onset of the sudden increase in intensity for both fluorophors has been identical. The possibility that the coincidence is due to energy transfer from tryptophan to TMA-DPH or other optical artifacts has been ruled out. We thus conclude that the TMA-DPH data indicates a change in the internal bilayer structure that is presumably coupled to the concentration of reaction products in the bilayer and promotes the rapid activation of the phospholipase A_2. Jain et al. have recently reported that the autoquenching of the fluorescent probe NK-529 increases during hydrolysis of vesicles of dimyristoyl-phosphatidylcholine.[12] Their study also suggested a temporal coincidence between the fluorescence changes of this probe and a rapid increase in hydrolytic activity of the porcine pancreatic phospholipase A_2.

The putative lipid structural change induced by reaction products is probably not an isolated phenomenon produced only by palmitic acid and lysophospholipid. Rather, we think that membrane changes that promote enzyme activation can be induced by other perturbations. Several observations are consistent with such a notion. First, it has been found with vesicles of egg phosphatidylcholine mixed with cholate that the rate of activation of porcine pancreatic phospholipase A_2 is directly proportional to the mole fraction of cholate in the vesicle.[13] This dependence is smooth until about 0.22 mole fraction where a further increase of .02 to .04 results in a 15 to 30-fold increase in the rate of enzyme activation. Second, osmotic shock of DPPC LUV causes apparently instantaneous activation of pancreatic phospholipase A_2 under conditions where lag periods would normally be observed.[3,4] Third, Cunningham and co-workers have reported that critical concentrations of diacylglycerol increase pancreatic phospholipase A_2 activity several-fold.[14] An additional interesting observation that may also be related is that a threshold electric field can induce large increases in phopholipase A_2 activity toward monolayer substrate.[15]

MODELS FOR REACTION PRODUCT COUPLING

Important questions are how might a cooperative lipid structural change be induced by products and how might such a process be coupled to activation of the enzyme. Two-component systems in which each molecule occupies a site of a two dimensional lattice are topologically described in terms of the distribution of the two different molecules among the sites. If the two components demonstrate no preferential association (i.e. nearest neighbors), then the distribution will be random. These distributions will include small clusters of connected molecules of the minor component until a critical value of the mole fraction, called the percolation point, is reached. At that point, all molecules become essentially connected in a single, very large cluster, analogous to phase-separation. The percolation point for a random system is precisely defined by the coordination geometry (i.e. the number of nearest neighbors) of the lattice. For a hexagonal lattice this is when the mole fraction of the minor component equals 0.5.[16]

The details of percolation become more complex when preferential interaction exists between molecules. If the interactions between the components are very strong, complexes may be formed and percolation can be acheived at concentrations well below the "random" percolation point. Such appears to be the case for cholesterol-DPPC vesicles where 2:1 lipid-cholesterol complexes form and the percolation point is deduced to be about 0.2 from Monte Carlo calculations and calorimetric data.[17,18] In the case of two immisible lipids, the components will always be "phase-separated" and the minor component will always exist in a single distinct cluster. Thus, in an hexagonal lattice, the sudden change from a large number of small clusters to a small number of large clusters can occur at any mole fraction of the minor component, dependent only on the magnitude of its nearest neighbor interactions.

It is this clustering phenomenon that may be coupled to activation of the phospholipase A_2. Three current pieces of evidence substantiate this hypothesis: 1) Calorimetric and fluorescence polarization experiments indicate that the critical mole fraction for the compositional transition from mixtures of phosphatidylcholine and diacylglycerol to formation of "specific complexes or preferred packing assays" occurs at 0.25 mole fraction diacylglycerol.[14] The activity of pancreatic phospholipase A_2 increases several-fold abruptly as the mole fraction of diacylglycerol is raised from 0.20 to 0.25 mole fraction. 2) The apparent rate of activation of porcine pancreatic phospholipase A_2 increases up to 30-fold between 0.22 and 0.26 mole fraction cholate in egg phosphatidylcholine vesicles.[13] The structural similarity between cholate and cholesterol suggests that the two may have similar percolation points in mixtures with phosphatidylcholine. As stated above, the percolation point for cholesterol is estimated to be 0.20.[17,18] 3) The fluorescent probe NK-529 has been found to detect apparent sudden clustering of reaction products at the time of the burst in activity of pancreatic phospholipase A_2 during the time course of phosphatidylcholine vesicle hydrolysis.[12]

The structural change induced by reaction products during vesicle hydrolysis can hypothetically be described by a two-state model. At very low concentrations of reaction products, the bilayer exists in state A. At high concentrations, it exists in state B. The critical mole fraction of reaction products is defined as the concentration at which the ratio of A to B equals 1.0. The observed activation of the phospholipase A_2 could be coupled to this compositional transition in at least two ways. The first is that either the equilibrium amount of active enzyme or the rate of activation is directly proportional to the amount of lipid in state B. One can mathematically describe the proportion of lipid in state B with the following relationship:

$$f_B = \frac{1}{e^{n(X_c - X_{(t)})} + 1} \tag{1}$$

where f_B is the fraction of lipid in state B and the exponential term is the statistical weight of lipid in state A. X_c is the critical mole fraction of reaction products and $X_{(t)}$ is the mole fraction existing in the bilayer as a function of time. The coefficient n is the cooperative unit size or the number of molecules participating as a cluster in the transition. At small values for n, the transition from state A to B will occur gradually as the concentration of reaction products increases. Large values produce a sharp transition. To simulate a given model for the mechanism of phospholipase A_2 activation and its coupling to the compositional transition, one would simply set the appropriate model parameter (i.e. rate constant or equilibrium constant) proportional to f_B.

A second possibility for the coupling of enzyme activation to the compositional transition is that activation depends not on the presence of state A or B of the lipid; but rather, it depends on dynamic changes or fluctuations in structure that occur during the transition from state A to B. If this possibility were true, activation would be a maximum at the critical mole fraction of reaction products and would decrease at higher or lower concentrations of reaction products. This suggestion is analogous to a previous proposal that the effect of the thermotropic phase transition to enhance phospholipase A_2 activation is coupled to fluctuations of clusters of gel or liquid crystalline lipid at that temperature.[3] This type of model for the effect of reaction product accumulation on phospholipase A_2 activation assumes that it is the <u>rate</u> of enzyme activation that is coupled to the compositional transition and that the activation is essentially irreversible for the duration of the time course. Formulation of this coupling to the dynamics of the compositional transition is analogous to the formulation described above for coupling to the fraction of lipid in state B except that the rate of activation is proportional to $f_T = f_B(1-f_B)$ where f_T is the magnitude of the structural fluctuations as a function of the mole fraction of reaction product.

MODELS FOR PHOSPHOLIPASE A_2 ACTIVATION

As stated in the introduction, the simple time dependence of the model for activation that led to the conclusion that activation of the pancreatic phospholipase A_2 involves dimerization is not adequate to describe the complex time courses shown in Fig. 1. Furthermore, the possible effect of reaction products on the activation process were not explicitly addressed by the model.[4] However, the inclusion of a product-dependent activation step could be readily incorporated as will be described. Other models which will be considered are based on a variety of experimental evidence and suggestions in the literature including activation being the result of enhanced binding of the enzyme to the bilayer surface,[19] penetration of the enzyme into the bilayer,[1] increased catalytic efficiency due to reorientation of the phospholipid molecules[20] and conformational changes of the enzyme.[21,22] In simple terms, three general models will be used to describe these various hypotheses: 1) binding activation, 2) equilibrium activation, and 3) kinetic activation.

Theory

For the purpose of this discussion, we will defer several of the proposed details of specific activation mechanisms such as whether the activation step is an enzyme conformation change, a substrate conformation change or enzyme dimerization. Some of these details can be incorporated into the various models but are beyond the scope of this manuscript.

The binding activation model assumes that increased activity is solely the result of the products increasing the amount of enzyme bound to the substrate surface and is described by the following equilibrium:

$$
E + S \underset{}{\overset{K_B}{\rightleftharpoons}} \begin{array}{c} S_m \\ + \\ E_B \\ \updownarrow K_m \\ E_B \cdot S_m \\ \downarrow k_{cat} \\ E_B + P_n \end{array}
$$

where E and S are the enzyme and lipid, S_m is the lipid mole fraction in the bilayer and P_n are the hydrolysis products (H^+, fatty acid, lysophospholipid). K_B is the apparent association constant of the enzyme to the surface of the membranes, K_m is the Michaelis constant for the bound enzyme interacting with phospholipid monomers in the bilayer and k_{cat} is the catalytic rate constant or turnover number. K_m is a thermodynamic quantity relating to the conditional probability that a substrate molecule is bound to the enzyme's active site given that the enzyme is bound to the vesicle surface. The increase in activity occurs because the enzyme binds better to state B of the lipid. This is reflected in the value of K_B which is related to f_B in the following way

$$K_B = K_{B0} + f_B(K_{B1} - K_{B0}) \tag{2}$$

where K_{B0} is the initial value of K_B and K_{B1} is the maximum value. The rate of hydrolysis is

$$\frac{dP}{dt} = \frac{k_{cat}E_B S_m}{S_m + K_m} \tag{3}$$

where

$$E_B = \frac{E_T K_B S}{1 + K_B S} \tag{4}$$

with

$$S = \frac{\sqrt{b^2 + 4K_B S_T} - b}{2K_B} \tag{5a}$$

$$b = 1 + K_B(NE_T - S_T) \tag{5b}$$

E_T and S_T are the total concentrations of enzyme and phospholipid and N is the number of phospholipid molecules that define the surface binding site of one enzyme molecule. Thus, it is assumed that the first enzyme to bind to the vesicles has an equal probability of binding to a number of sites equal to the total phospholipid concentration. Each enzyme that binds reduces the probability of the next enzyme binding by removing N binding sites. Note that S_m and K_m are both expressed in mole fraction units since they represent phospholipid concentrations within the bilayer. The depletion of substrate by hydrolysis is included by setting S_m equal to $1-X_{(t)}$ where $X_{(t)} = P_{(t)}/S_T$. The possibility of product inhibition[23] has not been incorporated into these models.

The equilibrium activation model is one step more complex than the binding activation. It states that the enzyme exists as an equilibrium between active and inactive species both in solution and on the vesicle surface and that the equilibrium on the membrane surface is altered by the presence of products as described by the following scheme.

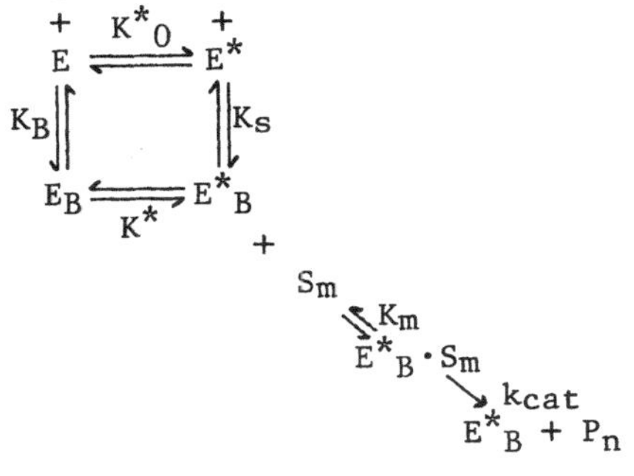

93

The alteration of the position of the equilibrium is described explicitly by assuming that the apparent equilibrium constant K^* equals K^*_0 when the lipid is in state A and K^*_1 when the lipid is in state B.

$$K^* = K^*_0 + f_B(K^*_1 - K^*_0) \qquad (6)$$

The equations describing the hydrolysis reaction for this equilibrium model, then, are:

$$\frac{dP}{dt} = \frac{k_{cat}E^*_B S_m}{S_m + K_m} \qquad (7)$$

$$E^*_B = \frac{E_T K^* K_B S}{1 + K^*_0 + K_B S + K^* K_B S} \qquad (8)$$

where

$$S = \frac{\sqrt{b^2 + 4ac} - b}{2a} \qquad (9a)$$

$$a = K_b(1 + K^*) \qquad (9b)$$

$$b = 1 + K^*_0 + K_B(1 + K^*)(NE_T - S_T) \qquad (9c)$$

$$c = S_T(1 + K^*_0) \qquad (9d)$$

The third model to be discussed is a kinetic activation model.

$$E + S \underset{K_B}{\rightleftharpoons} E_B \xrightarrow{k_a} E^*_B \underset{K_S}{\rightleftharpoons} E^* + S$$
$$+$$
$$S_m$$
$$\updownarrow K_m$$
$$E^*_B \cdot S_m$$
$$\downarrow k_{cat}$$
$$E^*_B + P_n$$

In this model, it is the rate of activation, k_a, that is enhanced by reaction product accumulation. The activation step on the membrane surface has been written as irreversible for simplicity, but it could also be written as a slowly reversible reaction. The most important feature of this model is that $E \to E^*$ is a reaction that does not occur in the absence of phospholipid. Conformation changes of the protein–lipid complex such as penetration into the bilayer[1] or covalent modification of the enzyme[5,6] are examples of such a phenomenon. This model requires two differential equations:

$$\frac{dP}{dt} = \frac{k_{cat}E^*_B S_m}{S_m + K_m} \qquad (7)$$

where

$$E^*_B = \frac{E^*_T K_S S}{1 + K_S S} \qquad (10)$$

and

$$\frac{dE^*_T}{dt} = k_a E_B \qquad (11)$$

We assign two rates of activation k_{a0} and k_{a1} depending on whether the lipid exists in state A or B as defined by equation 1. Thus,

$$k_a = k_{a0} + f_B(k_{a1} - k_{a0}) \qquad (12)$$

where k_{a0} and k_{a1} are the initial and maximum values of k_a. The other relationships necessary to integrate equations 7 and 11 for this model are:

$$E_B = \frac{(E_T - E^*_T)K_B S}{1 + K_B S} \qquad (13)$$

and equations 5a and 5b from the binding activation model using a constant value for K_B. The calculation of S using equations 5a and 5b for this model does ignore the fact that E^*_B will become very large after the latency period and the value of S will therefore be overestimated by equations 5a and 5b. However, this only produces inaccuracies in the calculations at low substrate concentrations ($S < K_B^{-1}$) after the lag phase in the time course. Since most of our analysis of these time courses involves the early phase (see below), this is not a problem.

Simulations

Numerical integration of the differential equations describing the three models yields theoretical time courses of phospholipid hydrolysis (Fig. 5) reminiscent of the experimental data shown in Fig. 1. These simulated results suggest that the models are distinguishable by the shapes of the time course. However, this requires extremely precise and reproducible data which are difficult to obtain. For the time being, we have adopted the strategy of finding as many measurable parameters of the time courses as possible and then comparing relative changes in the values of these parameters as a function of experimental perturbations explicitly contained within the models.

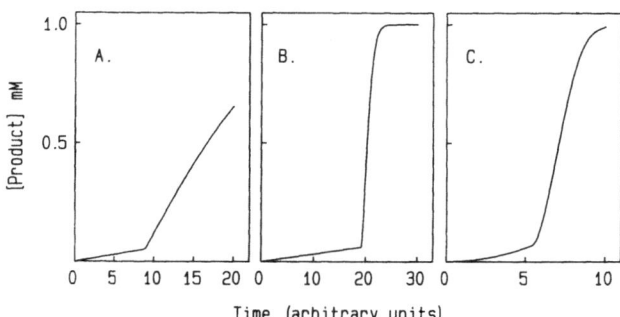

Fig. 5. Time courses of DPPC LUV hydrolysis simulated by the A, binding activation; B, equilibrium activation; or C, kinetic activation model. The time courses were calculated by numerical integration of the differential equations described in the text pertaining to each model using the Runge-Kutta fourth order algorithm. The integration interval was 0.01 arbitrary time units. Parameter values: $E_T = 10^{-7}$ M, $S_T = 10^{-3}$ M, $K_m = 0.5$, $n = 10^3$, $X_C = 0.07$, $N = 50$; Panel A: $k_{cat} = 10^3$ units time^{-1}, $K_{B0} = 100$ M^{-1}, $K_{B1} = 10^9$ M^{-1}; Panel B: $k_{cat} = 10^4$ units time^{-1}, $K_B = 10^3$ M^{-1}, $K^*_0 = 0.01$, $K^*_1 = 10$; Panel C: $k_{cat} = 10^4$ units time^{-1}, $K_B = 3 \times 10^3$ M^{-1}, $K_S = 3 \times 10^4$ M^{-1}, $k_{a0} = 0.01$, $k_{a1} = 1.0$.

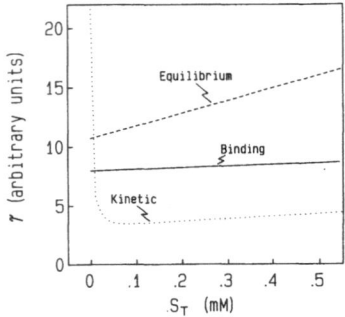

Fig. 6. Simulated dependence of τ on S_T for each of the
three models. Time courses were simulated as
in Fig. 5 using the same parameters at the values
of S_T indicated. τ was calculated as the time
at which $P/S_T = X_C$. Solid curve: binding activation model, dashed curve: equilibrium activation model, dotted curve: kinetic activation
model.

We have initiated such a protocol using time course data of DPPC LUV
hydrolysis by the AppD49 phospholipase A_2. One promising experimental
observable is to measure the time at which the activity of the enzyme rapidly
increases (τ) as a function of substrate concentration. The fact that the
enzyme fluorescence change correlates well with enzyme activation (Fig. 2),
allows us to measure τ at concentrations of substrate lower than we use in
the pH stat enzyme activity assay. Fig. 6 demonstrates the τ versus substrate
concentration dependence predicted by each of the three models. Both the
binding (solid curve) and equilibrium (dashed curve) activation models predict
that τ will be independent of substrate concentration at low concentrations
and will be an increasing function at high concentrations. This behavior is
always true of these two models, and in the case of the equilibrium model,
does not matter whether the equilibrium is slow or fast. In contradistinction, the kinetic model (dotted curve) predicts a completely
different dependence of τ on substrate concentration. The value of τ is first
a decreasing function of substrate concentration, reaches a minimum and then
increases with further increases in substrate concentration. The concentration at which the minimum occurs depends on the values of K_B and K_s.

The experimentally determined behavior of τ as a function of substrate
concentration is shown in Fig. 7. The value τ is a decreasing function at
low substrate concentration, an increasing function at high substrate
concentrations and reaches a minimum at ~0.1 mM DPPC. This kind of behavior
is highly reproducible and has been found without failure in 10 experiments,
although the exact position of the minimum varies between 0.03 mM and 0.2 mM
DPPC. Similar behavior has been seen in preliminary experiments with the
phospholipase A_2 from <u>Crotalus</u> <u>atrox</u> venom. Obviously, these results are not
consistent with either the binding or equilibrium models as defined here.
They are completely consistent with the kinetic model.

The state of our interpretations at this point regarding the mechanism
of activation of the pancreatic and AppD49 phospholipase A_2 is as follows.
1) Both enzymes can bind to DPPC vesicles in the absence of calcium and this
binding is stronger when the lipid is in the gel phase. 2) Both enzymes are
activated on the surface of DPPC LUV. 3) This activation is, at least in
part, coupled to a membrane structural change which occurs as reaction
products accumulate in the bilayer. 4) The kinetics of hydrolysis for both
enzymes are consistent with the occurrence of a quasi−irreversible step in the
activation mechanism. 5) The enzyme and substrate concentration dependencies

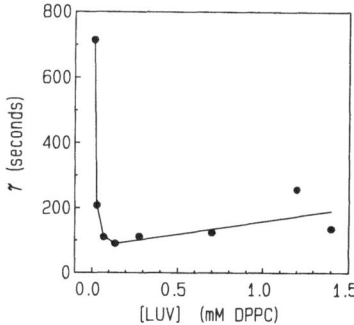

Fig. 7. Experimental dependence of τ on the concentration of DPPC LUV. The intrinsic AppD49 enzyme
fluorescence (290 nM) was monitored during hydrolysis of LUV at the indicated DPPC concentrations at 39°C using the protocol described in
Fig. 3 (no TMA–DPH). τ was measured as the time
of the fluorescence increase.

of the rate of activation of the pancreatic enzyme at 38°C indicate that the
activation mechanism involves dimerization on the membrane surface. Both a
dimer and monomer kinetic activation model predict that τ versus S_t will
exhibit a minimum as shown in Fig. 7 and distinction between those two models
for the AppD49 enzyme will require a more sophisticated analysis which is
currently underway. It is worth noting, however, that both the pancreatic and
AppD49 enzymes have been reported to become acylated and stabilized as a dimer
during the activation process.[5,6]

While the binding activation model cannot account for the results shown
in Fig. 7, the binding of the enzyme to the vesicles must indeed improve upon
activation. Thermodynamically, K_s must be greater than K_B if enzyme activation
is to occur on the surface of the vesicles. We therefore do not dispute
previous data reporting that the enzyme binds better in the presence of the
appropriate concentration of reaction products;[19] rather, we argue that
improved binding is not the activation step _per se_ but is a thermodynamic
consequence of the fact that activation occurs on the vesicle surface.

CALCIUM

The calcium requirement for the activity of most phospholipases A_2 has
generally been reported to be in the millimolar concentration range.[1]
However, the role of calcium in the temporal sequence of events in
phospholipase A_2 activation is not clear. Specifically, is calcium required
for enzyme binding to the bilayer, for activation and/or for catalysis? It
has frequently been reported that calcium is required for "catalytically
effective" phospholipase A_2 binding to lipid substrate.[19] However,
calorimetric[3] and fluorescence,[10] studies have demonstrated that calcium is not
required for enzyme–lipid interaction. We have also obtained similar results
with the _Crotalus atrox_ enzyme.

Fig. 8 shows time courses of DPPC LUV hydrolysis by AppD49 phospholipase
A_2 at three concentrations of calcium. These preliminary results show that
the lag time is sensitive to calcium in the low millimolar to high micromolar
ranges; but the maximum rate of hydrolysis is not. The increase in the
intrinsic enzyme fluorescence at time τ, simultaneously measured for each of
these time courses, was found to be identical for all three calcium
concentrations. Therefore, we conclude that the maximum degree of enzyme
activation can be acheived at very low calcium.

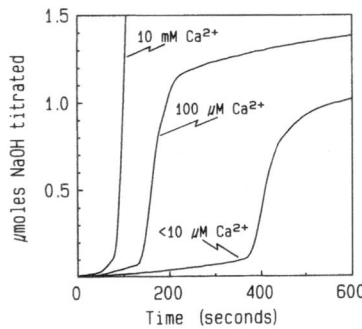

Fig. 8. Time courses of DPPC LUV hydrolysis by AppD49
phospholipase A_2 as a function of $CaCl_2$ concentration. Hydrolysis reactions were monitored at
41.5°C with 380 nM enzyme and 1.5 mM DPPC as
described in Fig. 1. $CaCl_2$ concentrations were
10 mM, 0.1 mM and <0.01 mM (50 mM KCl instead
of $CaCl_2$).

Preliminary studies on the effect of calcium on the hydrolysis of small
unilamellar vesicles sheds some additional light on the problem. As
mentioned, the maximal rate of hydrolysis of small unilamellar vesicles is
observed at time zero. At 10 mM calcium, the hydrolysis time—course is
essentially a true hyperbolic function of time. However, at lower calcium
concenrtions (in 50 mM KCl) the time course is non—hyperbolic and of such a
form as to suggest severe product inhibition as the reaction progresses. Such
inhibition has previously been reported during the hydrolysis of vesicular
substrate by Crotalus atrox PLA_2.[23] Furthermore, the concentration of calcium
required to acheive half maximal initial activity is on the order of 0.5 to
1 mM. This can be seen in Fig. 9 where the estimated initial velocity versus
[calcium] is displayed (circles). The apparent inhibition by product can be
relieved by 10 mM magnesium which, in the absence of calcium, is incapable
of supporting catalysis.[1] This relief of apparent product inhibition is
reflected by an increase in the estimated initial velocity. In Fig. 9, the
initial velocity as a function of calcium in the presence of 10 mM magnesium
is displayed and the estimated calcium dissociation constant under these
conditions is approximately 2×10^{-5} M (triangles). These results clearly show
that the maximal catalysis can be achieved at calcium concentrations much
lower than previously reported and at calcium concentrations much lower than
the dissociation constant for the isolated enzyme.

These results suggest the following. Since calcium does not appear to
affect the binding of phospholipase A_2 to zwitterionic vesicles and since the
apparent calcium binding constant for catalysis is much stronger than that of
the free enzyme, the structure of the active enzymes on the membrane surface
must be different than the inactive solution form or the form which is
initially bound to the lipid. This conclusion, if correct, means that
increased activity of the protein cannot be solely due to increased binding,
as has been deduced from comparison of computer simulations of such a model
and the experimental data from r versus S_T (Figs. 6 and 7). These calcium
results reported here do not directly address the role of calcium in the
activation process per se, but are consistent with calcium being required as
previously suggested.[2,3] A detailed reevaluation of the role of calcium in the
overall process of phospholipase activation is currently underway.

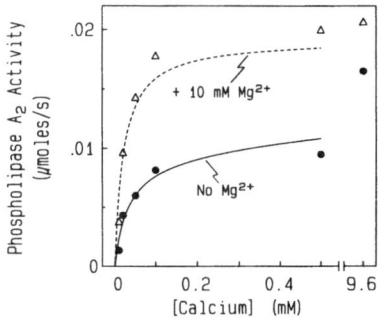

Fig. 9. CaCl₂ dependence of AppD49 phospholipase A₂ activity without (solid line) or with (dashed line) MgCl₂. Enzyme (90 nM final) was added to a 2-ml reaction cocktail containing small unilamellar vesicles (1 mM DPPC), the indicated concentrations of CaCl₂, ± 10 mM MgCl₂ and 50 mM KCl at 25°C. The reaction was monitored at pH 8.0 with a pH stat. The activity was determined from the maximum initial slope of the time courses. The curves have no theoretical significance.

SIGNAL TRANSDUCTION

What might be learned from these soluble phospholipases A_2 in terms of the regulation of the enzyme in a living cell? A cell must accomplish four tasks in regulating phospholipase A_2 for signal transduction. 1) It must protect itself from extensive hydrolysis which could lead to solubilization of membrane components and lysis. 2) It must be capable of reversibly activating the enzyme. 3) It must be capable of limiting the number of phospholipids that will be hydrolyzed by the active enzyme. 4) The activation process must be specific.

A plethora of phospholipase A_2 have now been purified from a variety of mammalian cells and tissues. One interesting consistency is that the enzymes are generally active upon purification requiring only calcium and the appropriate pH and substrate. One could argue that the enzymes responsible for signal transduction have not yet been isolated and would behave differently. Nevertheless, the cell must have some way of protecting itself from all these phospholipases. An obvious answer is that the enzyme requires calcium which is normally very low intracellularly. Calcium may indeed be one level of regulation of phospholipase A_2 activity, but it probably does not suffice. Certainly, it would lack specificity in the sense that all the phospholipase A_2 in the cell would be activated every time there was a calcium flux. Such could be disastrous in muscle cells and neurons. In addition, some studies have suggested that hormonal-stimulated calcium flux is insufficient in itself to induce activation of phospholipase A_2.[24,25]

It would seem, then, that cellular regulation of phospholipase A_2 must involve more than modulation of intracellular calcium concentration. A variety of other mechanisms have also been suggested in the literature which could presumably act as additional levels of regulation. One suggestion is that hormonally-regulated phospholipase is activated by phosphorylation via protein kinase C.[26] Other reports suggest a role of intracellular pH[27] and the cytoskeleton.[28] Substrate specificity is probably also important.[29-32] Finally, a variety of reports have proposed that phospholipase A_2 is hormonally regulated by one or more GTP-binding proteins analogous to those that regulate adenylate cyclase.[33]

Notwithstanding these several possible mechanisms, one might ask whether membrane structure could not play a significant role in the regulation of cellular phospholipase A_2 as it apparently does for the soluble enzymes from snake venom or pancreas. Initially, the cellular membrane would exist in a state which is either not susceptible to hydrolysis or incapable of activating the phospholipase. This would confer protection to the membrane. The activation of the enzyme could then be initiated by applying a reversible perturbation to the membrane structure which renders a small number of lipids susceptible to hydrolysis or which activates a small number of enzymes. Upon removal of the perturbation, the enzyme could presumably relax to the inactive state. The specificity could be achieved in the following ways. First, the extreme sensitivity of phospholipase A_2 to membrane perturbation could mean that a perturbation of small magnitude that would not affect the majority of cellular membrane-bound proteins could have a larger effect on the phospholipase A_2. Second, a typical theme in cellular homeostasis is to regulate an important process at several levels to make the system fail-safe and finely controllable. One can imagine that phospholipase A_2 activation might require a calcium influx plus interaction with a transducer protein plus an appropriate membrane perturbation. Such a multi-level mechanism would be especially important if the GTP-binding transducer protein component responsible for phospholipase A_2 regulation turns out to be the beta-gamma subunit as has been proposed.[33] The same or a similar beta-gamma subunit exists in all GTP-binding transducer proteins which couple to a large number of hormone receptors.[34] Thus, since a number of receptors on a given cell would cause release of the beta-gamma subunit upon stimulation, additional levels of phospholipase A_2 regulation would be absolutely required to maintain specificity.

A variety of experimental evidence supports the possibility that the membrane structure and perturbations of the same play a role in cellular phospholipase A_2 regulation. Like the soluble enzymes, the activity of many cellular phospholipases A_2 depends on the structure of the bilayer upon which they act.[35-39] Perturbations by molecules such as glycerol[37] or diacylglycerol[38-40] have also been reported to activate cellular phospholipase A_2. A phospholipase A_2 activating protein structurally similar to mellitin which probably activates phospholipase A_2 by effects on the membrane[20] has also been described.[41] In fact, it has even been speculated that the conformational change of hormone receptors or transducer proteins upon binding of ligands could induce the necessary membrane perturbations.[20] Clearly, increased physical characterization of the role of membrane structure on the temporal sequence of events in the activation of phospholipase A_2 will aid in the understanding of the regulation of this and other membrane-bound cellular proteins.

ACKNOWLEDGEMENT

This work was made possible by funding from the National Institute of General Medical Sciences (Grants GM37658 and GM11838) and from the Office of Naval Research (N00014-88-K-0326). We also thank Drs. Guillermo Romero and Dov Lichtenberg for their many helpful discussions.

REFERENCES

1. H. M. Verheij, A. J. Slotboom, and G. H. De Haas, Structure and function of phospholipase A_2. Rev. Physiol. Biochem. Pharmacol. 91:91 (1981).
2. M. Menashe, G. Romero, R. L. Biltonen, and D. Lichtenberg, Hydrolysis of dipalmitoylphosphatidylcholine small unilamellar vesicles by porcine pancreatic phospholipase A_2, J. Biol. Chem. 261:5328 (1986).
3. D. Lichtenberg, G. Romero, M. Menashe, and R. L. Biltonen, Hydrolysis

of dipalmitoylphosphatidylcholine large unilamellar vesicles by porcine pancreatic phospholipase A_2, J. Biol. Chem. 261:5334 (1986).

4. G. Romero, K. Thompson, and R. L. Biltonen, The activation of porcine pancreatic phospholipase A_2 by dipalmitoylphosphatidylcholine large unilamellar vesicles: analysis of the state of aggregation of the activated enzyme, J. Biol. Chem. 262:13476 (1987).

5. W. Cho, A. G. Tomasseli, R. L. Heinrikson and F. J. Kézdy, The chemical basis for interfacial activation of monomeric phospholipases A_2: autocatalytic derivatization of the enzyme by acyl transfer from substrate, J. Biol. Chem. 263:11237 (1988).

6. A. G. Tomasselli, J. Hui, J. Fisher, H. Zürcher-Neely, I. M. Reardon, E. Oriaku, F. J. Kézdy, and R. L. Heinrickson, Dimerization and activation of porcine pancreatic phospholipase A_2 via substrate level acylation of lysine 56, J. Biol. Chem. 264:10041 (1989).

7. J. D. Bell, and R. L. Biltonen, The temporal sequence of events in the activation of phospholipase A_2 by lipid vesicles: studies with the monomeric enzyme from Agkistrodon piscivorus piscivorus, J. Biol. Chem. 264:12194 (1989).

8. M. C. E. van Dam-Mieras, A. J. Slotboom, W. A. Pieterson, and G. H. de Haas, The interaction of phospholipase A_2 with micellar interfaces. The role of the N-terminal region, Biochemistry 14:5387 (1975).

9. M. K. Jain, J. Rogers, J. F. Marecek, F. Ramirez, and H. Eibl, Effect of the structure of phospholipid on the kinetics of intravesicle scooting of phospholipase A_2, Biochim. Biophys. Acta 860:462 (1986).

10. J. D. Bell, and R. L. Biltonen, Thermodynamic and kinetic studies of the interaction of vesicular dipalmitoylphosphatidylcholine with Agkistrodon piscivorus piscivorus phospholipase A_2, J. Biol. Chem. 264:225 (1989).

11. F. G. Prendergast, R. P. Haugland, and P. J. Callahan, 1-[4-(Trimethylamino)phenyl]-6-phenylhexa-1,3,5-triene: synthesis, fluorescence properties, and use as a fluorescence probe of lipid bilayers, Biochemistry 20:7333 (1981).

12. M. K. Jain, B.-Z. Yu, A. Kozubek, Binding of phospholipase A_2 to zwitterionic bilayers is promoted by lateral segregation of anionic amphiphiles, Biochim. Biophys. Acta 980:23 (1989).

13. N. Gheriani-Gruszka, S. Almog, R. L. Biltonen, and D. Lichtenberg, Hydrolysis of phosphatidylcholine in phosphatidylcholine-cholate mixtures by porcine pancreatic phospholipase A_2, J. Biol. Chem. 263:11808 (1988).

14. B. A. Cunningham, T. Tsujita, and H.L. Brockman, Enzymatic and physical characterization of diacylglycerol-phosphatidylcholine interactions in bilayers and monolayers, Biochemistry 28:32 (1989).

15. T. Thuren, A.-P. Tulkki, J. A. Virtanen, and P. K. J. Kinnunen, Triggering of the activity of phospholipase A_2 by an electric field, Biochemistry 26:4907 (1987).

16. J. M. Ziman in Models of Disorder, pp. 370-379, Cambridge University Press, Cambridge (1979).

17. J. C. Owicki, and H. M. McConnell, Lateral diffusion in inhomogeneous membranes: model membranes containing cholesterol, Biophys. J. 30:383 (1980).

18. B. Snyder, and E. Freire, Compositional domain structure in phosphatidylcholine-cholesterol and sphingomyelin-cholesterol bilayers, Proc. Natl. Acad. Sci. U.S.A. 77:4055 (1980).

19. M. K. Jain, and O. G. Berg, The kinetics of interfacial catalysis by phospholipase A_2 and regulation of interfacial activation: hopping versus scooting, Biochim. Biophys. Acta 1002:127 (1989).

20. A. Achari, D. Scott, P. Barlow, J. C. Vidal, Z. Otwinowski, S. Brunie, and P. B. Sigler, Facing up to membranes: structure/function relationships in phospholipases, Cold Spring Harbor Symposia on Quantitative Biology 52:441 (1987).

21. M. F. Roberts, R. A. Deems, and E. A. Dennis, Dual role of interfacial phospholipid in phospholipase A_2 catalysis, Proc. Natl. Acad. Sci. U.S.A. 74:1950 (1977).

22. D. O. Tinker, and J. Wei, Heterogeneous catalysis by phosphlipase A_2: formulation of a kinetic description of surface effects, Can. J. Biochem. 57:97 (1979).

23. J. P. Kupferberg, S. Yokoyama, and F. J. Kézdy, The kinetics of the phospholipase A_2-catalyzed hydrolysis of egg phosphatidylcholine in unilamellar vesicles: product inhibition and its relief by serum albumin, J. Biol. Chem. 256:6274 (1981).

24. M. F. Crouch, and E. G. Lapetina, No direct correlation between Ca^{2+} mobilization and dissociation of G_i during platelet phospholipase A_2 activation, Biochem. Biophys. Res. Commun. 153:21 (1988).

25. R. Bicknell and B. L. Vallee, Angiogenin stimulates endothelial cell prostacyclin secretion by activation of phospholipase A_2, Proc. Natl. Acad. Sci. U.S.A. 86:1573 (1989).

26. J. H. Gronich, J. V. Bonventre, and R. A. Nemenofff, Identification and characterization of a hormonally regulated form of phospholipase A_2 in rat renal mesangial cells, J. Biol. Chem. 263:16645 (1988).

27. J. D. Sweatt, T. M. Connolly, E. J. Cragoe, and L. E. Limbird, Evidence that Na^+/H^+ exchange regulates receptor—mediated phospholipase A_2 activation in human platelets, J. Biol. Chem. 261:8667 (1986).

28. T. Nakano, K. Hanasaki, and H. Arita, Possible involvement of cytoskeleton in collagen—stimulated activation of phospholipases in human platelets, J. Biol. Chem. 264:5400 (1989).

29. C. C. Leslie, D. R. Voelker, J. Y. Channon, M. M. Wall, and P. T. Zelarney, Properties and purfication of an arachidonoyl—hydrolyzing phospholipase A_2 from a macrophage cell line, RAW 264.7, Biochim. Biophys. Acta 963:476 (1988).

30. D. K. Kim, I. Kudo, and K. Inoue, Detection in human platelets of phospholipase A_2 activity which preferentially hydrolyzes an arachidonoyl residue, J. Biochem. 104:492 (1988).

31. R. A. Wolf and R. W. Gross, Identification of neutral active phospholipase C which hydrolyzes choline glycerophospholipids and plasmalogen selective phospholipase A_2 in canine myocardium, J. Biol. Chem. 260:7295 (1985).

32. J. Balsinde, E. Diez, A. Schüller, and F. Mollinedo, Phospholipase A_2 activity in resting and activated human neutrophils: substrate specificity, pH dependence, and subcellular localization, J. Biol. Chem. 263:1929 (1988).

33. J. Axelrod, R. M. Burch, and C. L. Jelsema, Receptor—mediated activation of phospholipase A_2 via GTP—binding proteins: arachidonic acid and its metabolites as second messengers, Trends in Neurosciences 11:117 (1988).

34. A. G. Gilman, G—Proteins: transducers of receptor-generated signals Ann. Rev. Biochem. 56:615 (1987).

35. R. Kannagi, K. Koizumi, Effect of different physical states of phospholipid substrates on partially purified platelet phospholipae A_2 activity, Biochim. Biophys. Acta 556:423 (1979).

36. D. H. Petkova, A. B. Monchilova—Pankova, and K. S. Koumanov, Effect of liver plasma membrane fluidity on endogenous phospholipase A_2 activity, Biochimie 69:1251 (1987).

37. R. J. Ulevitch, Y. Watanabe, M. Sano, M. D. Lister, R. A. Deems, and E. A. Dennis, Solubilization, purification, and characterization of a membrane—bound phospholipase A_2 from the $P388D_1$ macrophage—like cell line, J. Biol. Chem. 263:3079 (1988).

38. R. M. C. Dawson, R. F. Irvine, J. Bray, and P. J. Quinn, Long—chain diacylglycerols cause a perturbation in the structure of phospholipid bilayers rendering them susceptible to phospholipase attack, Biochem. Biophys. Res. Commun. 125:836 (1984).

39. R. M. Kramer, G. C. Checani, and D. Deykin, Stimulation of Ca^{2+}—activated human platelet phospholipase A_2 by diacylglycerol, Biochem. J. 248:779 (1987).

40. M. M. Billah and M. I. Siegel, Phospholipase A_2 activation in chemotactic peptide—stimulated HL60 granulocytes: synergism between diacylglycerol

and Ca^{2+} in a protein kinase C-independent mechanism, <u>Biochem</u>. <u>Biophys</u>. <u>Res</u>. <u>Commun</u>. 144:683 (1987).

41. M. A. Clark, T. M. Conway, R. G. L. Shorr, and S. T. Crooke, Identification and isolation of a mammalian protein which is antigenically and functionally related to the phospholipase A_2 stimulatory peptide melittin, <u>J</u>. <u>Biol</u>. <u>Chem.</u> 262:4402 (1987).

STIMULATION OF PHOSPHOLIPASES A$_2$ BY TRANSGLUTAMINASES

Eleonora Cordella-Miele, Lucio Miele, Simone Beninati and Anil B. Mukherjee

Section on Developmental Genetics
Human Genetics Branch, NICHD, NIH
Bethesda, Maryland 20892

INTRODUCTION

Among the enzymes catalyzing post-translational modifications of proteins, transglutaminases (TG; EC 2.3.2.13) have been extensively characterized from the enzymological point of view (1-4). Nevertheless, the physiological role(s) of these enzymes, particularly the intracellular TGs, are still poorly understood. TGs are a class of enzymes which catalyze a Ca^{++}-dependent acyl-transfer reaction in which the γ-carboxamide group of a peptide-bound glutamine residue is the acyl-donor (1-4). Primary amino groups of many low-molecular weight amines may act as acyl-acceptors with the formation of mono-substituted γ-carboxamides of peptide-bound glutamic acid. In the absence of small molecular weight amines, TGs catalyze the formation of an ϵ-(γ-glutamyl)-lysine isopeptide bond between endo-γ-glutaminyl and endo-ϵ-lysyl residues in polypeptides (1-4). The latter reaction results in the formation of inter or intramolecular covalent crosslinks. These enzymes have been detected both intra and extracellularly in higher animals including man. The best characterized extracellular TG is variously known as fibrin-stabilizing factor, Laki-Lorand factor or coagulation Factor XIII. The active form of this enzyme (Factor XIIIa) is generated by the action of thrombin and Ca^{++} on the catalytically inactive zymogen. Following coagulation of blood, Factor XIIIa cross-links fibrin molecules via the formation of interchain ϵ-(γ-glutamyl)-lysine isopeptide bonds, thereby stabilizing the clot by preventing its hydrolysis by proteases (5-7). Besides Factor XIII, several extracellular and intracellular TGs have been described (1-4). One of the most thoroughly studied intracellular TGs is derived from guinea pig liver (3). TGs apparently identical to the liver enzyme are present intracellularly in many mammalian tissues and organs. These enzymes are commonly designated "tissue TG". TG-catalyzed reactions have been proposed to be involved in several biological phenomena besides fibrin clot stabilization. These include the formation of seminal plugs in rodents (8), cataract formation (9), aging of the erythrocyte membrane (10) and neuronal aging (11). Additionally, TG-mediated reactions have been suggested to be involved in masking the immunogenicity of rabbit spermatozoa in the female genital tract following coitus (12) and in

Biochemistry, Molecular Biology, and Physiology of Phospholipase A$_2$ and Its Regulatory Factors
Edited by A. B. Mukherjee, Plenum Press, New York, 1990

105

protecting implanting rabbit embryos from maternal immunological assault (13). These results have been confirmed in the rat (14). Additionally, it has been reported that the maturation of human spermatozoa may involve TG-mediated processes (15). Recently, TG has been suggested to be involved in the terminal differentiation of keratinocytes (16). Moreover, it has been proposed that TG participates in the biochemical events leading to programmed cell death (apoptosis) of hepatocytes (17).

Phospholipases A_2 (PLA$_2$) (EC 3.1.1.4) are a family of lipolytic enzymes which specifically catalyze the hydrolysis of the 2-acyl ester bond in 3-*sn*-phosphoglycerides, generally in the presence of Ca^{++} (18). The most abundant sources of these enzymes are the mammalian pancreas and snake venoms (18). Mammalian pancreatic PLA$_2$ is secreted as a zymogen which is converted to the active enzyme in the intestinal lumen after proteolysis and liberation of the N-terminal heptapeptide. Many of these enzymes have been thoroughly studied from the structural point of view, and complete amino acid sequences of more than 35 phospholipases are known (18). Characterization of a large number of PLA$_2$s from different phyla has revealed that the members of the PLA$_2$ family are closely related to each other. Several important features have been conserved in these enzymes through evolution. These include: specificity for hydrolysis of the *sn*-2 position acyl group of 3-*sn* phosphoglycerides, requirement for Ca^{++}, a minimal molecular mass of 14 kDa, differential enzymatic reactivity towards different substrates, a large number of intrachain disulfide bridges and a very high degree of amino acid sequence homology (18). Extensive conservation of amino acid sequence has been discovered in more than 35 phospholipases from different sources, including those mammalian intracellular PLA$_2$s the sequences of which have been determined so far (19-22). The most conserved regions in the primary structure of PLA$_2$s are: the N-terminal amphiphilic helix, the Ca^{++} binding loop and the active site (18). In all the PLA$_2$ sequences that have been reported so far, except for the bee venom enzyme (18), the N-terminal sequences revealed the same α-helical structure and the same pattern of amino acid residues. Recently, the crystallographic analysis of a PLA$_2$ inhibitory protein, blastokinin (23) or uteroglobin (UG) (24) at 1.34 Å resolution has revealed a remarkable similarity between the calculated molecular surfaces of this protein and that of PLA$_2$ (25). UG inhibits PLA$_2$s from porcine pancreas and mouse macrophage RAW 264.7 cells *in vitro* (26). UG has also been shown to be an excellent substrate of TG *in vitro* (27). In addition, TG treatment dramatically potentiates some of the biological activities of UG (12, 13). On the basis of these data we decided to investigate if PLA$_2$ is a substrate of TG and if TG treatment could affect the catalytic properties of PLA$_2$ (28).

ACTIVATION OF PORCINE PANCREATIC PLA$_2$ BY GUINEA PIG LIVER TG AND FACTOR XIIIa

Our initial approach was to investigate the possible effects of preincubating PLA$_2$ with various TGs on the catalytic activity of the enzyme (28). The dose-response curves obtained by preincubating PLA$_2$ with several concentrations of guinea pig TG are shown in Fig. 1A, while the curves obtained with purified rabbit plasma Factor XIIIa and human plasma Factor XIIIa are shown in Figs 1B and 1C respectively. Under the experimental conditions used (see legend to Fig. 1) an increase in PLA$_2$ activity up to about 200% of the control was obtained with guinea

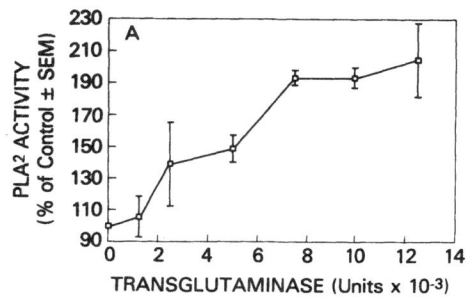

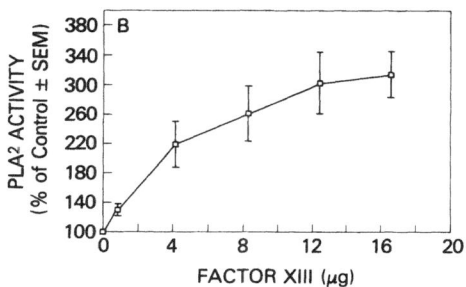

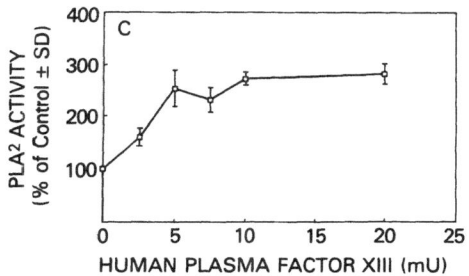

Fig. 1 **Dose response curve of PLA$_2$ stimulation by guinea pig TG (A), rabbit Factor XIIIa (B) and human Factor XIIIa (C).** PLA$_2$ was preincubated with the indicated amounts of TG or Factor XIIIa in a total volume of 80 μl. Controls were kept in which PLA$_2$ was preincubated with buffer only, or with thrombin-containing buffer in experiments with Factor XIIIa. Additional controls were run in which TG or Factor XIIIa or thrombin alone were preincubated with buffer. PLA$_2$ activity was assayed as described (28) using deoxycholate/phosphatidylcholine mixed micelles as the substrate. Each point represents the average of at least three determinations, each performed in duplicate, $\pm$ SEM. Reproduced with permission from ref. 28.

pig TG. A higher stimulation was obtained by pretreating PLA$_2$ with rabbit Factor XIIIa, which increased PLA$_2$ activity up to about 300% of control. Human plasma Factor XIIIa, used in concentrations comparable to those of guinea pig liver TG, also caused an increase of PLA$_2$ activity up to 300% of control. It should be noted that in all experiments where PLA$_2$ activity was measured immediately after preincubation, we excluded dithiothreitol (DTT) from the preincubation mixture. DTT, which is generally used to protect the reactive sulphydryl group in TG, was avoided in order to prevent any possible interference with the seven disulfide bridges of PLA$_2$. Results identical to those shown in Fig. 1 were obtained (unpublished data) using guinea pig liver TG (Sigma) on highly purified PLA$_2$ from porcine pancreas kindly provided by Dr. M. K. Jain (Department of Chemistry, Delaware University, Wilmington, Delaware). Negative controls with non-specific proteins such as ovalbumin, chicken egg lysozyme, myoglobin and hemoglobin showed no stimulation of PLA$_2$ catalytic activity in this assay. In another series of controls, we verified that the TGs used were devoid of PLA$_2$ activity (unpublished data). These results indicate that the observed increase of PLA$_2$ activity is unlikely to be due to non-specific artifacts or to the presence of contaminating PLA$_2$ in the preparations of TG and Factor XIIIa.

The time courses of PLA$_2$ stimulation by TG and Factor XIIIa are shown in Fig. 2. The maximum activity obtained with guinea pig TG reached 350% of control at 1 hr. In the absence of added Ca^{++}, the observed stimulation of PLA$_2$ was drastically reduced, although not completely abolished. In fact, after 50 min of preincubation some stimulation was observed also in the absence of added Ca^{++}. This is probably due to the presence of Ca^{++} bound to TG and/or to PLA$_2$. When TG was tested for enzymatic activity in a conventional assay, in the absence of added Ca^{++} at the concentration used in the experiments described in Fig. 2A, it retained about 25 % of the enzymatic activity observed with 1 mM Ca^{++} (unpublished data). We avoided the use of EGTA, EDTA or other chelating agents in the control preincubation mixture in order to avoid potential artifacts due to interferences with the concentration of Ca^{++}. As an additional control, heat-inactivated TG was used to perform an identical experiment in the presence of 1 mM Ca^{++}. In this case we observed no increase in PLA$_2$ activity (Fig. 2A). As in dose response experiments, human plasma Factor XIIIa produced a better stimulation of PLA$_2$ activity than guinea pig TG (Fig. 2B), reaching 600 % of the control in 1 hr. In the absence of added Ca^{++}, Factor XIIIa was completely inactive as a stimulator of PLA$_2$ activity. These results indicate that the effect of TG and Factor XIIIa on PLA$_2$ is time-dependent, Ca^{++}-dependent and, at least in the case of guinea pig liver TG, sensitive to thermal inactivation.

Fig. 3 shows the dependence of TG-mediated activation of PLA$_2$ upon the concentration of PLA$_2$ in the preincubation mixture. Under these conditions the increase of PLA$_2$ activity reached a maximum at about 125 nM PLA$_2$. The shape of the curve suggests that the PLA$_2$ activation process is saturable. This may be an indirect indication that the process of TG-induced PLA$_2$ activation is an enzymatic reaction. Assuming that we are dealing with a TG-catalyzed modification of PLA$_2$, the activity of the TG-treated samples, expressed in percent of the activity of control samples run in parallel, should be proportional to the velocity of TG-catalyzed modification. Therefore, a double-reciprocal plot of percent PLA$_2$ activity versus PLA$_2$ concentration should give us an approximate estimate of the apparent K$_m$ of TG for PLA$_2$.

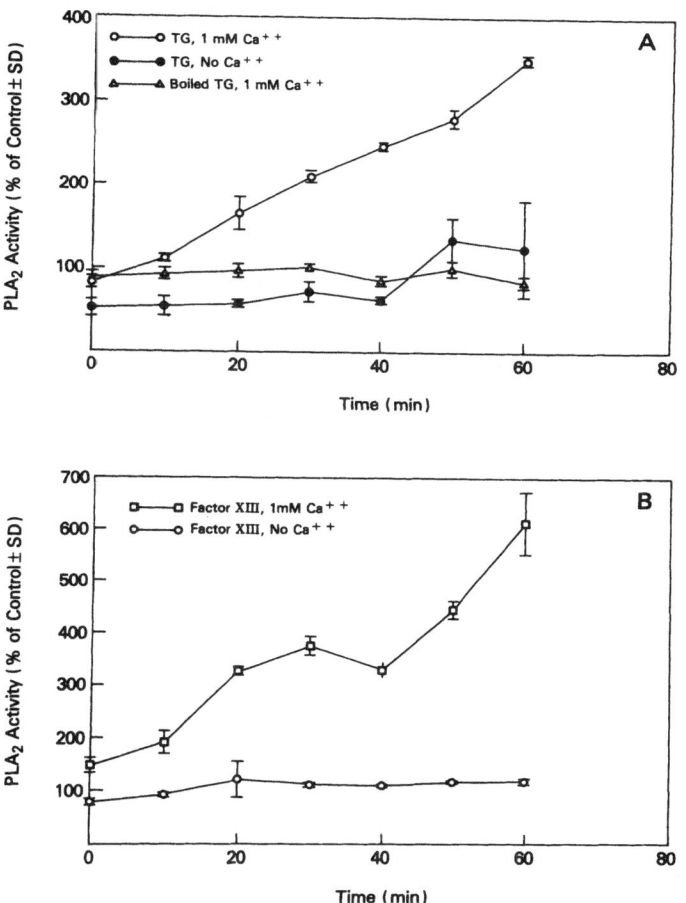

Fig. 2. **Time course of PLA$_2$ preincubation with tissue TG and human plasma Factor XIIIa with and without calcium.** A: PLA$_2$ (5 nM) was preincubated with 5 mU of Guinea pig liver TG in a total volume of 80 μl for the specified times, in 10 mM Tris HCl pH 8.0. *Open circles* indicate that the preincubation mixture contained 1 mM CaCl$_2$. *Solid circles* indicate that the preincubation mixture was devoid of Ca^{++}. *Open triangles* indicate that the preincubation mixture contained 1 mM CaCl$_2$ but TG was inactivated by boiling for 45 min. B: PLA$_2$ (5nM) was preincubated with thrombin activated Factor XIII (1 x 10^{-2} units) in 10 mM Tris HCl pH 8.0 in a total volume of 80 μl. *Open squares* indicate the presence of 1 mM CaCl$_2$ in the preincubation mixture . *Open circles* indicate the absence of Ca^{++} in the preincubation buffer. Each point represents the average of at least three determinations, each performed in duplicate, ± SEM. Reproduced with permission from ref. 28

As shown in Fig. 3 (inset), a double reciprocal plot of these data yielded an apparent K_m value of 0.99 nM.

The time-course of product formation by TG- and Factor XIIIa-treated and untreated PLA_2 is shown in Fig. 4. In both cases (Fig. 4a, 4b) no appreciable lag period was observed, and the rate of product accumulation is clearly much higher with TG- and Factor XIIIa-treated PLA_2 than in control experiments. It is noteworthy that the absolute values of percent hydrolysis obtained in the experiments with Factor XIIIa (Fig. 4b) are lower than those obtained in the experiments with guinea pig liver TG (Fig 4a). This difference was consistently observed in experiments involving thrombin or thrombin-activated Factor XIII. Since the preparation of human thrombin used in these experiments contained citrate, it is possible that this substance may have interfered with the concentration of free Ca^{++}. The shape of the curves is similar to that of time courses obtained with *Crotalus atrox* PLA_2 using egg phosphatidylcholine unilamellar vesicles (29). These curves show a biphasic behaviour, with two distinct components: an initial, rapid phase in which the rate of product accumulation is linear and a more prolonged phase of slower hydrolysis. This is clearly shown in plots of log[S] versus time (Fig. 4c, 4d). The curves show that the rate of product accumulation is increased by TG and Factor XIIIa in both the initial and the slower phase of the time course. The absence of a lag phase in PLA_2-catalyzed reactions, when mixed micellar substrates containing detergents are used, is a well-known phenomenon (30). In the case of *Crotalus atrox* PLA_2 the decrease of the reaction rate in the second phase of the reaction has been shown to be due to product inhibition by monomeric and micellar lysophosphatidylcholine (29). We have not determined if product inhibition is actually responsible for the decrease in reaction rate observed in our system. If this is the case, monomeric lysophosphatidylcholine alone is probably responsible for the phenomenon, since the concentrations of products obtained in our system are all far below the estimated CMC for egg lysophosphatidylcholine (29). Approximate estimates of the initial rates for both the components of the reaction were obtained from the slopes of the semilogarithmic plots (Fig. 4c and 4d). We designated these slopes as k_1 (for the fast component) and k_2 (for the slow component). From the data shown in Fig. 4c the calculated values for control reactions were: $k_1 = -1.96 \times 10^{-4}$ and $k_2 = -2.79 \times 10^{-5}$ while for TG-stimulated PLA_2 the values were: $k_1 = -4.99 \times 10^{-4}$ and $k_2 = -1.12 \times 10^{-4}$. From the data shown in Fig. 4d the values obtained were: $k_1 = -1.65 \times 10^{-4}$ and $k_2 = -2.29 \times 10^{-5}$ for the control curve; $k_1 = -3.04 \times 10^{-4}$ and $k_2 = -6.90 \times 10^{-5}$ for Factor XIIIa-treated PLA_2. Both rates appear to be increased by TG and by Factor XIIIa, with the increment of k_2 being slightly more pronounced than that of k_1 in both cases.

Since we routinely assay pancreatic PLA_2 activity in the presence of deoxycholate-phosphatidylcholine mixed micelles, we decided to investigate whether TG-induced stimulation of PLA_2 activity could also be observed in a different assay, with a detergent-free substrate. For these experiments we used a modification of the procedure described by Haigler et al. (31, 32), which uses autoclaved *E. coli* cells, metabolically labeled with [^{3}H]-oleic acid, as PLA_2 substrate. We studied the time course of PLA_2 reaction at two different temperatures and the results of this group of experiments are shown in Fig. 5. At 37° C, we observed no lag period in the hydrolysis of *E. coli* phospholipids (Fig. 5a). Again, the reaction catalyzed by TG-

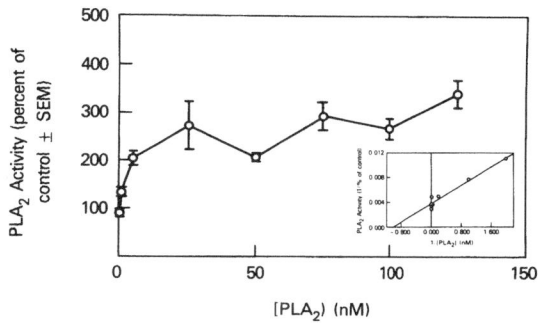

Fig. 3. **Effect of PLA$_2$ concentration on the TG-induced stimulation.** PLA$_2$ at the indicated concentrations was preincubated with 5mU of guinea pig liver TG in the presence of 10 mM Tris HCl pH 8.0, 1 mM CaCl$_2$ in a total volume of 80 μl. Preincubation was performed for 30 min at 37° C and was followed by a 1 min reaction at 37° C. After the preincubation each sample was diluted with prewarmed buffer to reach a PLA$_2$ concentration of 5 nM and 20 μl of the diluted samples were used to start PLA$_2$ reactions. Inset: double reciprocal plot obtained from these data. The apparent K$_m$ for PLA$_2$ is about 1 nM. Each point represents the average of at least three determinations, each performed in duplicate, $\pm$ SEM.

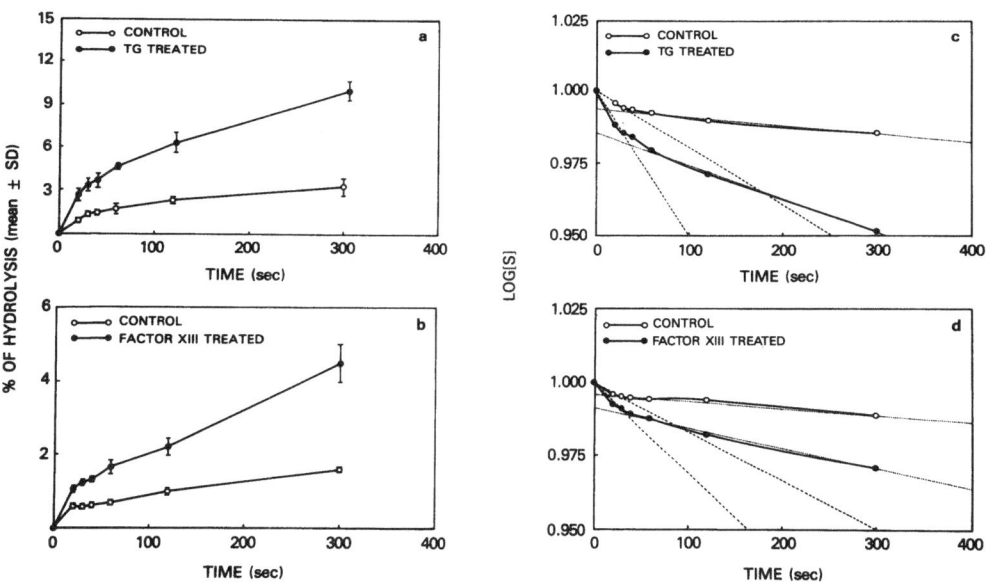

Fig. 4. **Time course of PLA$_2$ reaction after preincubation with TG or Factor XIIIa.**
A: PLA$_2$ (5 nM) was preincubated with guinea pig liver TG (5 mU) in a total volume of 80 μl for 30 min at 37° C. After preincubation PLA$_2$ reaction was started by addition of 20 μl of the preincubated enzyme to 30 μl of the reaction mixture and run for the indicated times. Controls (open circles) were preincubated with buffer only. B: The experiments were performed as described for panel A, but in place of TG, thrombin activated human Factor XIII (1 x 10^{-2} units) was used. Controls (open circles) were preincubated with thrombin only. Each point represents the average of at least three determinations, each performed in duplicate, $\pm$ SEM. Panels C and D are semilogarithmic plots of data from panels A and B respectively. Reproduced with permission from ref. 28.

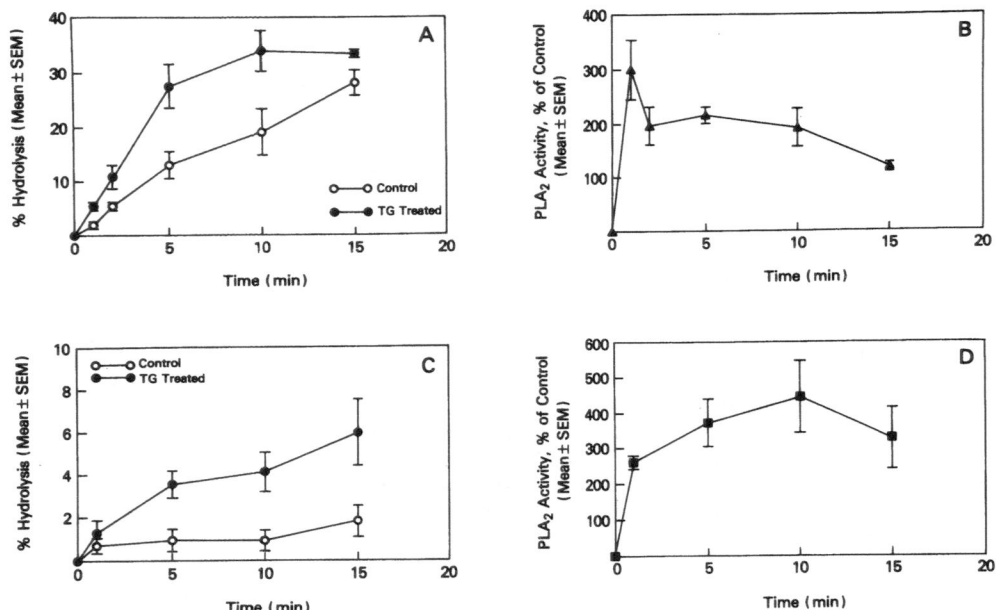

Fig. 5. **Time course of PLA$_2$ reaction after preincubation with TG in E. coli assay.** PLA$_2$ (60 nM) was preincubated with guinea pig liver TG (7.5 mU) in a total volume of 80 μl for 30 min at 37°C. After preincubation, PLA$_2$ reactions were started by adding 10 μl of the preincubated mixture to the reaction mixture. Assays were run using autoclaved *E. coli* metabolically labeled with ^{3}H-oleic acid as substrate (28). A: Reactions were run for the specified times at 37° C. Controls (*open circles*) were preincubated with buffer only and simultaneously samples containing only buffer were kept as blanks. B: Data from the experiments of panel A were replotted to show the PLA$_2$ activity of the TG-stimulated samples expressed as percent of control $\pm$ SEM. C: PLA$_2$ reactions were run at 12° C for the indicated times. Controls (*open circles*) were preincubated with buffer only. Blanks containing only buffer were kept for each data point. D: Data from the experiments shown in panel C were replotted to demonstrate the PLA$_2$ activity of the TG-stimulated samples expressed as percent of control, $\pm$ SEM. Reproduced with permission from ref. 28.

pretreated PLA$_2$ was much faster, reaching about 28% hydrolysis in 5 min, while the same value was reached in 15 min by TG-untreated PLA$_2$. Moreover, the time course of hydrolysis by TG-treated enzyme was linear from time 0, while with TG-untreated enzyme linearity was not reached until 10 min.

When the activity of TG-treated enzyme, expressed as percent of control was plotted versus the time of reaction (Fig. 5b), it was evident that the main difference between control and TG-treated enzyme was in the first 10 min of reaction. When the assays were performed at 12° C (Fig. 5c), the control curve showed a three-phase behavior, reminiscent of the one described by Apitz-Castro et al. (33) for the hydrolysis of dimiristoylphosphatidylcholine vesicles by porcine pancreatic PLA$_2$ at 17° C. After a small initial burst, a long latency phase was observed with virtually no product accumulation until 10 min, when the reaction started to progress again. Under the same conditions TG-pretreated PLA$_2$ showed no latency phase, although the rate of hydrolysis did increase after 10 min. Accordingly, the enzymatic activity of TG-treated enzyme, expressed as percent of control, increased from 0 to 10 min, and decreased slightly thereafter (Fig. 5d). These results indicated that TG-treated PLA_2 also has an enhanced enzymatic activity on detergent-free substrates. Moreover, these observations gave us some indirect information on the mechanism of stimulation of PLA_2 activity. The observed higher activity of TG-treated PLA_2 during the first 2 minutes of reaction at 37° C and particularly the absence of a true latency phase at 12° C, suggest that TG-treated enzyme had a much faster "penetration" and/or "interfacial activation" than untreated enzyme (34). Thus, these results may suggest that TG-catalyzed activation of PLA_2 could effectively bypass, and/or dramatically increase the speed of the activation step that takes place upon binding of PLA_2 to organized substrates (18, 34).

Michaelis-Menten kinetics of TG-treated and untreated PLA_2 were performed with the mixed micellar assay, since in such systems hyperbolically shaped Michaelis-Menten curves are usually obtained (18, 34-35). Fig. 6 shows double reciprocal plots obtained between 2.8 and 90 μM phosphatidylcholine, with a fixed concentration of 1 mM deoxycholate. Under these conditions, we obtained straight double reciprocal plots, with apparent K_m values of 42 and 53 μM for control and TG-treated PLA_2 respectively. Apparent V_{max} values were 0.32 μmoles x liter^{-1} x min^{-1} for PLA_2 alone and 0.8 μmoles x liter^{-1} x min^{-1} for TG-treated PLA_2. Thus, it appears that the kinetic parameter which is significantly altered by TG is the apparent V_{max}. This was confirmed by analyzing the same data by iterated weighted fit with the computer program ENZYME (36). With this program we obtained apparent K_m values of 27.7 $\pm$ 7.3 μM for PLA_2 and 23.6 $\pm$ 10.9 μM for TG-treated PLA_2, while V_{max} values were 0.26 $\pm$ 0.03 and 0.52 $\pm$ 0.07 μmoles x liter^{-1} x min^{-1} respectively. The effect of TG on the V_{max} of PLA_2 was found by ENZYME to be statistically significant (P < 0.05).

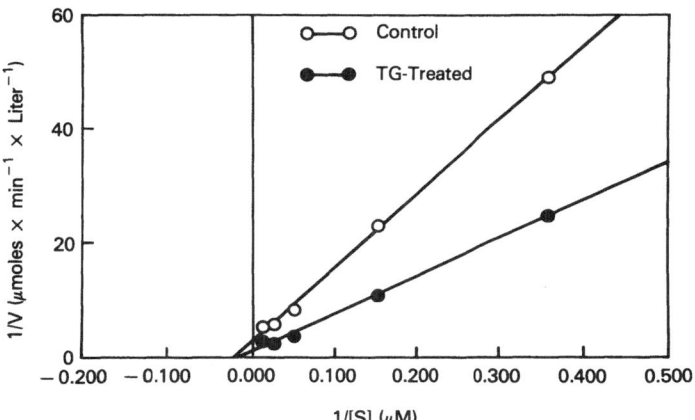

Fig. 6. **Kinetic study on TG-treated and untreated PLA$_2$ (double reciprocal plot).** PLA$_2$ (5 nM) was preincubated with guinea pig liver TG (5 mU) or with buffer (controls). The total volume of preincubation was 80 μl and the preincubation step was carried out for 30 min at 37° C. After preincubation, PLA$_2$ reactions were run for 1 min at 37° C. *Open circles* indicate PLA$_2$ controls which had been preincubated with buffer only. *Closed circles* indicate TG-treated PLA$_2$ reactions. Each point represents the average of at least three determinations, each performed in duplicate, $\pm$ SEM.

Although the data described above clearly indicated that the activation of PLA_2 was due to a TG-catalyzed reaction, several possible mechanisms of activation needed to be considered. These included: i) the formation of an intermolecular ϵ-(γ-glutamyl)-lysine isopeptide bond between molecules of PLA_2 or between one molecule of PLA_2 and TG; ii) a hydrolytic reaction with deamidation of Gln 4 in PLA_2 and iii) the formation of an intramolecular ϵ-(γ-glutamyl)-lysine cross-link within the same PLA_2 molecule. Our initial approach to this problem was to incubate PLA_2 with TG under appropriate conditions and then analyze the reaction mixture by size exclusion chromatography, assaying individual fractions for PLA_2 activity. In a preliminary experiment PLA_2 (5 nM) was incubated with 5 mU of TG for 1 hour in a volume of 720 μl. All other conditions were identical to those used for the experiments described in Fig. 2a. The incubation mixture was analyzed by size-exclusion chromatography on a Sephacryl-S200 column, and fractions were tested for PLA_2 activity. PLA_2 activity eluted in two poorly resolved peaks of approximately equal height, with apparent molecular masses of about 26 and 10 kDa respectively (data not shown). Since the molecular mass of porcine pancreatic PLA_2 is about 14.9 kDa, this observation suggested that PLA_2 may have been partially modified by TG into a higher molecular weight form. This experiment was then repeated with a higher concentration of PLA_2 (140 nM) and the reaction volume was doubled (Fig. 7a). After size exclusion chromatography PLA_2 activity was eluted in two poorly resolved peaks with apparent molecular masses of 26 kDa and 13 kDa respectively. When PLA_2 preincubated with buffer only was run through the same column, it eluted as a single peak at 14 kDa. In order to improve the resolution, we reduced the flow rate of the column from 6 to 4 ml/hour, and repeated the experiment. Fig. 7b shows that in this case both peaks of PLA_2 activity were retarded with respect to the profile shown in Fig. 7a, probably because of adsorption of this enzyme to the gel matrix due to the low ionic strength used. While peak 1 was only slightly retarded, peak 2 eluted with the total bed volume of the column. This resulted in a virtually complete separation of the two peaks. In order to test whether the appearance of peak 1 was due to a TG-catalyzed reaction, we repeated the experiment with heat-inactivated TG. In this case (Fig. 7b), peak 1 was virtually absent, except for a slight elevation of the baseline. Essentially, all PLA_2 activity eluted as a peak corresponding to peak 2. We tentatively interpreted these results as indirect evidence that the formation of a PLA_2 dimer was being catalyzed by TG. Therefore, we decided to confirm this hypothesis by testing peaks 1 and 2 for the presence of ϵ-(γ-glutamyl)-lysine. The results of this experiment are shown in Table I. This table shows that peak 1 contained 0.86 moles of ϵ-(γ-glutamyl)-lysine per mole of PLA_2 and that peak 2 contained a much lower amount of ϵ-(γ-glutamyl)-lysine. These data strongly supported the possibility that TG catalyzes the formation of an ϵ-(γ-glutamyl)-lysine isopeptide bond in PLA_2. However, under the assumption that an intermolecular isopeptide bond was formed by TG, the highest theoretical molar ratio of ϵ-(γ-glutamyl)-lysine to PLA_2 monomer should have been 0.5, and peak 2 should not have contained any ϵ-(γ-glutamyl)-lysine. A possible explanation may be that the ϵ-(γ-glutamyl)-lysine crosslink was intramolecular rather than intermolecular. In such a case, the dimerization could be a noncovalent process promoted by the conformational changes induced by the intramolecular crosslink. This hypothesis was also consistent with other

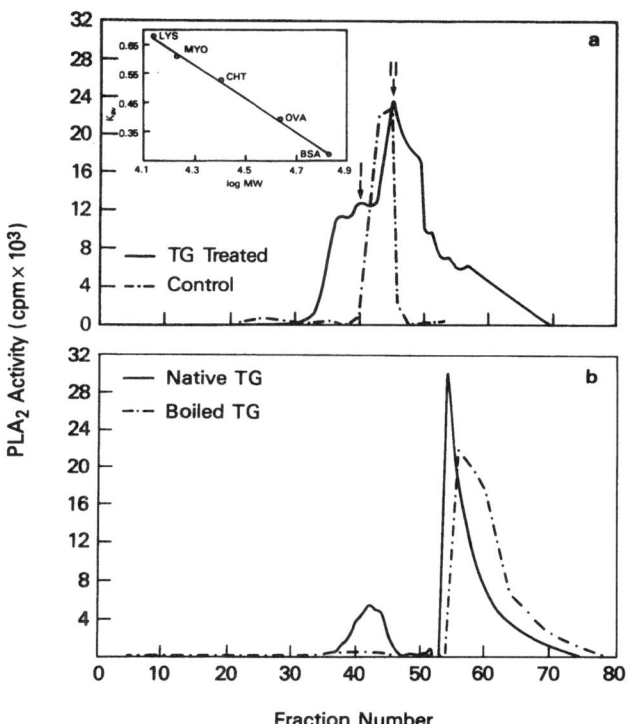

Fig. 7. **Size exclusion chromatography to estimate the molecular weight of TG-treated PLA$_2$.** Size exclusion chromatography was performed using a Sephacryl-S200 superfine column (see Materials and Methods). The column was calibrated as shown in the inset. A: The flow rate of the column was of 6 ml/hr. PLA$_2$ assay was performed on each fraction (20 μl) at 37°C for 10 min. Peak 1 (fractions 36-43) is in the 28 kDa m. mass range. Peak 2 (fractions 47-53) is in the 14 kDa range. B: The flow rate was reduced to 4 ml/hr. In this case the control was performed by preincubating PLA$_2$ with heat-inactivated TG. Abbreviations: LYS = lysozyme; MYO = myoglobin; CHT = chymotrypsinogen; OVA= ovalbumin; BSA= bovine serum albumin. Reproduced with permission from ref. 28.

Table I

Determination of ε–γ-glutamyl-lysine isopeptide
in peaks I and II from Sephacryl-S200 chromatography

Peak	PLA_2 (pmol)	ϵ-(γ-glutamyl) lysine (pmol)	ϵ-(γ-glutamyl) lysine/PLA_2
I	43.72	37.6	0.86
II	134.00	13.4	0.10

observations, such as the "skewed" aspect of the elution profile in Fig. 7a, and the noticeable decrease in the height of peak 1 after complete resolution from peak 2. Both these phenomena could be due to an alteration in the possible equilibrium between the monomer and dimer in solution as a result of the chromatographic process. In order to test this hypothesis, we decided to incubate TG with PLA_2 in a scaled-up amount and to attempt a semipreparative separation of peak 1, so that electrophoretic analysis under denaturing conditions could be performed using the material from this peak. After size-exclusion chromatography on Sephacryl-S200, peaks 1 and 2 were analyzed by SDS-PAGE. Both peaks showed a major band of apparent molecular mass of 15 kDa, but peak 1 contained a small amount of a protein with an apparent molecular mass of 35 kDa. However, this band was probably a TG fragment, since it did not react with the anti-PLA_2 antibody in a Western blot (see Fig. 9b). We further purified peak 1 by Superose 12 chromatography. In this column the PLA_2 activity in peak 1 eluted with an apparent molecular weight of about 25 kDa (Fig. 8). This value is slightly lower than the theoretical value of 29.6 kDa. However, the shape of the peak suggests the presence of a 30 kDa material which was in equilibrium with a lower molecular weight form during chromatography. The alteration of the monomer/dimer equilibrium associated with the chromatographic process might explain the slight shift of the apparent molecular weight as well as the characteristic aspect of the peak. When FPLC-purified PLA_2 from peak 1 was analyzed by SDS-PAGE under denaturing conditions, it was found to be apparently homogeneous and its molecular weight was indistinguishable from that of TG-untreated PLA_2 (Fig. 9). The results of electrophoretic analysis and Western blot are shown in Fig. 9. The material recovered after Superose 12 chromatography was found to contain ϵ-(γ-glutamyl)-lysine in a molar ratio of about 0.89 to 1 to PLA_2. These data strongly supported the hypothesis that TG catalyzed the formation of an intramolecular ϵ-(γ-glutamyl)-lysine isopeptide bond in PLA_2. The formation of this intramolecular crosslink appeared in turn to promote the formation of noncovalent PLA_2 dimers in solution. Dimerization of porcine pancreatic PLA_2 has been suggested to be the basis of "interfacial activation" of this enzyme (18, 30, 37-40). Thus, it is not unreasonable to speculate that this noncovalent dimerization of intramolecularly crosslinked PLA_2 may play a role in the observed TG-induced increase of PLA_2 activity. To test this hypothesis, we measured the specific activity of the putative PLA_2 dimer eluting in peak 1 after TG treatment and

Sephacryl S-200 chromatography. We treated PLA$_2$ (690 nM) with 45 mU of guinea pig liver TG in a total volume of 1.4 ml for 1 hour at 37°C and then separated peaks 1 and 2 by size exclusion chromatography on Sephacryl S200. When the specific activity of the two peaks was measured with the deoxycholate/phosphatidylcholine mixed micellar substrate, peak 1 was found to be about 10 fold more active than peak 2. Although this observation cannot be considered definitive evidence that the dimerization of crosslinked PLA$_2$ was the cause of its increased activity, it is consistent with such a hypothesis.

EFFECTS OF TG ON PLA$_2$s FROM DIFFERENT SOURCES

To determine if the observed TG-induced activation was specific for porcine pancreatic PLA$_2$ only, we investigated if PLA$_2$s from other species could also be stimulated by TG. The results of these experiments are summarized in Table II.

Table II

Stimulation of PLA$_2$ from different sources by guinea pig liver TG

PLA$_2$ source	Final concentration in the assay (nM)	PLA$_2$ activity (% of control ± SD)
Naja Naja	2	410 ± 66
	0.2	191 ± 26
	0.1	148 ± 13
Crotalus atrox	0.2	113 ± 11
	0.1	137 ± 10
Apis mellifera	0.2	167 ± 17
	0.1	104 ± 16

Reproduced with permission from ref. 28.

Interestingly, a considerable stimulation was observed with the *Naja naja* enzyme tested in three different concentrations, while no significant stimulation was observed with the *Crotalus atrox* enzyme. *Naja naja* PLA$_2$ exists as a monomer in solution below 4.5 μM (18). On the other hand, *Crotalus atrox* PLA$_2$ is a dimeric protein at concentrations as low as 2 nM (18). Thus, these observations support the hypothesis that the mechanism by which TG-mediated post-translational modification of PLA$_2$ increases PLA$_2$ activity is indeed by inducing a non-covalent dimerization of PLA$_2$. Further support for this hypothesis was provided by the fact that the observed stimulation of *Naja naja* PLA$_2$ is dependent on the final concentration of the enzyme in the reaction mixture (Table II). This is consistent with a concentration-dependent self association of PLA$_2$. Since some stimulation of *Naja naja* PLA$_2$ was observed at final concentrations as low as 0.1 nM, this might indicate that the TG-induced shift in the equilibrium of dimerization of this enzyme was quite dramatic. Interestingly, some degree of stimulation was also observed with bee venom PLA$_2$ at 0.2 nM. This enzyme could

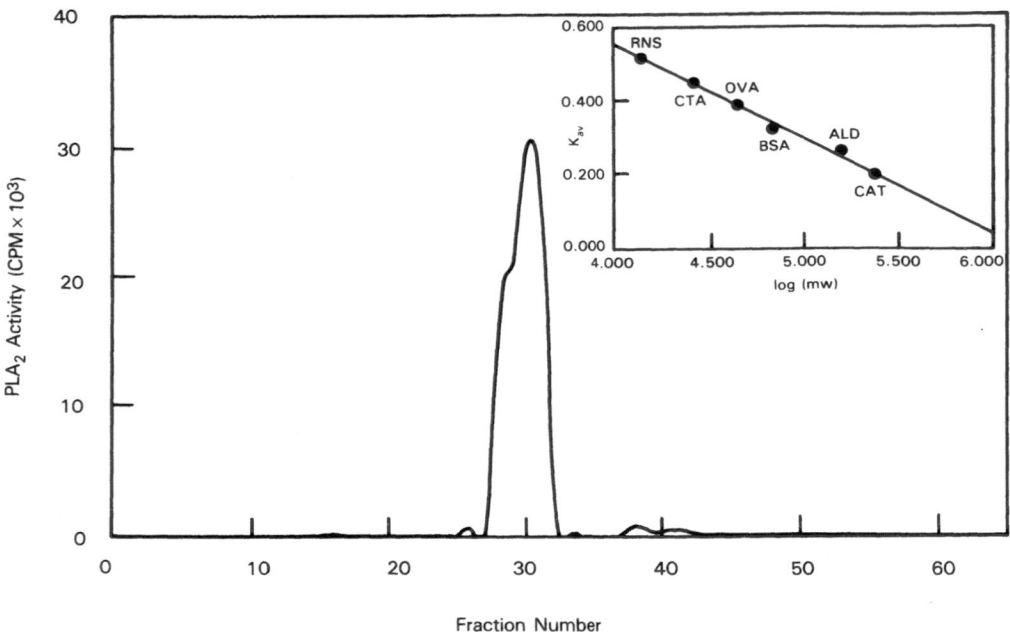

Fig. 8. **Purification of the putative dimer peak by FPLC.** The column (Superose 12) was calibrated as shown in the inset. The apparent molecular weight of the peak is 26 kDa. However, the shape of the peak indicates the presence of a 30 kDa component (shoulder) in equilibrium with a lower molecular weight form. The material was electrophoretically homogeneous and had an apparent molecular weight of 15 kDa by SDS-PAGE. Abbreviations: RNS = ribonuclease; CTA = chymotrypsinogen A; OVA = ovalbumin; BSA = bovine serum albumin; ALD = aldolase; CAT = catalase.

not be assayed at higher concentrations because its activity was too high, even in the absence of TG treatment. Bee venom PLA_2 lacks several amino acid residues present in the N-terminus of most pancreatic and snake venom enzymes (18, 41), including Gln 4. This residue is the only possible acyl donor in the pancreatic PLA_2 (see Discussion). It has been recently proposed, on the basis of computer-generated alignments and other considerations (41) that bee venom PLA_2 could be evolutionarily related to the pancreatic and snake venom enzymes. In this model, a region in the C-terminal part of bee venom PLA_2 was suggested to be structurally and functionally analogous to the N-terminal α-helix of the pancreatic and snake venom enzymes. This sequence includes Gln 121 which may be a putative homologue of Gln 4 (41). Should bee venom PLA_2 prove to be a substrate for TG, it would be interesting to investigate whether Gln 121 is indeed the acyl donor and functionally a counterpart of Gln 4 of *Naja naja* enzyme. All in all, our observations suggest that TG-catalyzed activation of PLA_2 is not restricted to porcine pancreatic PLA_2 only, but may be a more generalized phenomenon.

CONCLUSIONS

Our observations demonstrate that both Factor XIIIa and tissue TG can dramatically increase the enzymatic activity of porcine pancreatic PLA_2 in a time, concentration and Ca^{++}-dependent manner. In addition, we have shown that guinea-pig TG catalyzes the formation of an ϵ-(γ-glutamyl)-lysine isopeptide

A　　　　　　　　　　**B**

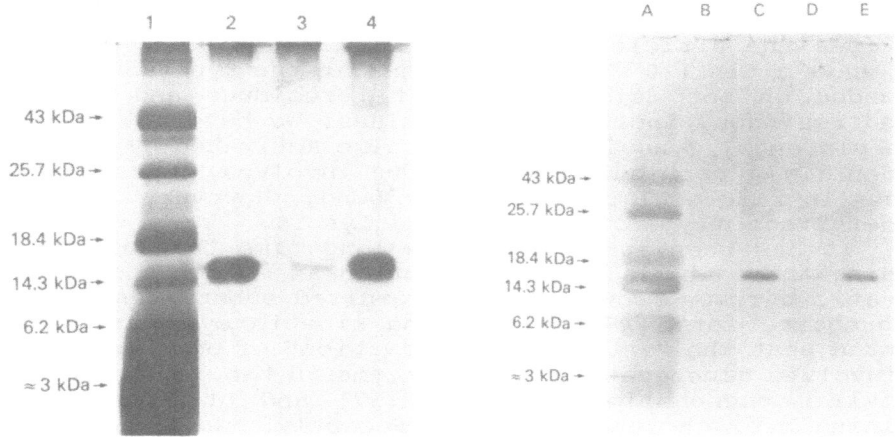

Fig. 9. **SDS-PAGE and Western Blot of TG-treated PLA$_2$.** A: SDS-PAGE. lane 1: prestained molecular weight standard (BRL), lane 2: pure PLA2 (2 μg), lane 3: 1 μg of dimer peak from the S-200 column; lane 4: 2 μg FPLC purified dimer.

B: WESTERN blot obtained using a monospecific anti-PLA2 antiserum from rabbit. lane A: molecular weight standard (prestained, BRL), lane B: pure PLA2 (1 μg), lane C: dimer peak from the S-200 (2 μg), lanes D and E were loaded with two concentrations of FPLC purified dimer (1 and 2 μg), respectively. Reproduced with permission from ref. 28.

bond when PLA$_2$ is used as a substrate. The chromatographic and electrophoretic data indicate that the isopeptide bond is intramolecular, and that the modified PLA$_2$ containing the ϵ-(γ-glutamyl)-lysine isopeptide has an increased tendency to form non-covalent dimers. This interpretation has been confirmed by isolating the putative dimeric PLA$_2$ and showing that the purified material: i) contained ϵ-(γ-glutamyl)-lysine in a molar ratio of 0.89 to 1 to PLA$_2$ protein; ii) when analyzed by size exclusion chromatography showed an apparent molecular mass of 25 to 30 kDa; iii) migrated as a monomer with an apparent molecular mass of 15 kDa when analyzed by SDS-PAGE under denaturing conditions and iv) reacted with an anti-PLA$_2$ antibody after electrophoresis and immunoblot. All in all, these observations strongly support the hypothesis that the TG-catalyzed post-translational modification of PLA$_2$ is the chemical basis for the observed increase in PLA$_2$ enzymatic activity.

Transglutaminases are highly selective enzymes in terms of the glutamine residues that can serve as acyl donors (2, 3, 42). In those proteins which are TG substrates usually only one specific glutamine residue is generally utilized as acyl donor even if more than one residue of this amino acid is present. It has been suggested that this extreme specificity for a particular glutamine residue indicates that TG-mediated post-translational modifications *in vitro* are likely to be functionally important *in vivo* (42). Since porcine pancreatic PLA$_2$ has only one glutamine residue, Gln 4, this residue is the only possible acyl donor for the TG-catalyzed reaction. Gln 4 is an almost invariant residue in the PLA$_2$ family, being conserved across the phyla. In particular, the type I enzymes of the "pancreatic" type, most snake venom enzymes (type I and type II), human lung PLA$_2$ (19) and human "non-pancreatic" PLA$_2$ (43) have a Gln residue in position 4. This could be an indication that TG-mediated post-translational modifications of PLA$_2$ might be of physiological importance. Interestingly, the rat and human placental/synovial PLA$_2$s (see the article by R. Crowl in this volume), which lack Gln 4, have a Gln 110 which, like Gln 4 in the type I enzymes, is surrounded on both sides by aromatic residues and positively charged residues, including two lysines. We have not determined if the placental/synovial PLA$_2$s are also activated by TGs. We did not identify a particular Lys residue involved in the formation of the intramolecular isopeptide bond. However, structural considerations suggest that it may be Lys 10.

It has been suspected for a long time that all PLA$_2$s, including those which exist in solution as monomers such as the pancreatic enzymes, could act on organized substrates in dimeric or aggregated form (18, 34). Strong kinetic evidence has been presented that the "interfacial activation" of porcine pancreatic PLA$_2$ involves dimerization of the enzyme in the presence of large unilamellar phospholipid vesicles (37) and of mixed vesicles containing deoxycholate (38, 39). Recently, the chemical basis for this activation has been shown to consist of a slow auto-catalytic acylation of Lys residues (Lys 56 in the case of porcine pancreatic PLA$_2$) (40). Autoacylation, in turn, promotes a non-covalent dimerization of the enzyme (40). Our data demonstrate the presence of a novel enzymatic mechanism by which the slow autoactivation step may be bypassed or accelerated *in vivo* through a TG-mediated post-translational modification. Like autoacylation, transglutamination promotes dimerization and activation of PLA$_2$. It is tempting to speculate that *in vivo* tissue TG or Factor XIIIa may convert a slowly auto-activating PLA$_2$ into a pre-activated or rapidly autoactivating form with enhanced enzymatic activity.

PLA$_2$ has been suggested to be involved in the onset and propagation of acute inflammation, and inflammatory cells such as the neutrophils release soluble PLA$_2$ upon activation (18, 44, 45). Macrophages also release arachidonate and PLA$_2$ upon phagocytosis (44). It is well known that inflammation and blood coagulation are intimately related. Since activation of the coagulation cascade may be triggered by inflammatory processes, it is not unreasonable to speculate that soluble PLA$_2$ released by phagocytes or by necrotic cells could be activated by Factor XIIIa. Moreover, activation of macrophages is accompanied by a striking increase in tissue TG activity (42, 46). TG-activated PLA$_2$ in the extracellular spaces during inflammation could participate in: i) releasing arachidonate from cell membranes, thereby increasing the production of proinflammatory eicosanoids; ii) killing microorganisms in the inflamed area and iii) lysing dead or dying cells in the exudate. Another possibility is that intramolecular crosslinking of PLA$_2$ may be a mechanism to induce a faster turnover of this activated enzyme *in vivo*. This may prevent the propagation of local inflammation to adjoining healthy tissues. As for the possible biological relevance of TG-catalyzed modification of intracellular PLA$_2$s, it has been suggested that intracellular TG activity is related to terminal differentiation and cell death (17, 42). It is entirely possible that one of the effects of this increased TG activity may be an activation of intracellular PLA$_2$(s). The irreversible activation of an intracellular PLA$_2$ could be part of biochemical events leading to programmed cell death and lysis. In this regard, it is worth noting that PLA$_2$s are active components of most animal poisons with cytolytic activity. Some poisons also contain peptides, such as melittin or mastoparan, which are well known activators of PLA_2 (18). Moreover, it has been recently demonstrated that cell death induced by the toxic peptide of vegetal origin *Pyrularia* thionin is mediated by Ca^{++} influx, followed by activation of an *intracellular* PLA_2 (47). It is not unreasonable to speculate that the final biochemical sequence of events leading to cell death might include an irreversible, TG-catalyzed activation of an intracellular PLA_2, converting the latter enzyme into a more cytolytic species. It should be emphasized that TG-mediated activation of PLA_2 is readily observed with PLA_2 concentrations as low as 5 nM (see Results), and the apparent K_m of TG for PLA_2 is about 1 nM. The fact that this reaction can take place at very low molar concentrations of PLA_2, suggests that it may be of physiological relevance. Moreover, in the case of a membrane-associated PLA_2 the actual concentration of this enzyme per unit of surface area could be much higher than the observed molar concentration. In this model an influx of Ca^{++} would trigger the activation of intracellular TG and PLA_2, since both are Ca^{++}-dependent enzymes. In the presence of Ca^{++}, TG would then irreversibly activate PLA_2 by forming an intramolecular crosslink. Although this model is at present entirely speculative, we feel that the possible role of TG-mediated activation of PLA_2 in inflammation and in programmed cell death deserves a thorough investigation.

There is compelling evidence to show that in inflammatory diseases, such as rheumatoid arthritis and juvenile rheumatoid arthritis, activation of PLA_2 may be pathophysiologically important (45). The mechanism of PLA_2 activation in these diseases is as yet unclear. Since inflammatory cells (e.g. neutrophils and macrophages) are known to secrete PLA_2 and PLA_2 may cause synovial tissue damage thereby releasing intracellular TG, further PLA_2 activation may occur as this process continues. On the other hand, activation of extracellular PLA_2 may be also catalyzed by Factor XIIIa present in the exudate. From the clinical standpoint it would be important to determine whether the synovial PLA_2 in rheumatoid and juvenile rheumatoid arthritis is a substrate of TG and if this enzyme can be found *in vivo* in tranglutaminated form.

REFERENCES

1. Lorand, L., Conrad, S. M. 1984. Mol. Cell. Biochem. 58:9-35.
2. Folk, J. E. 1980. Ann. Rev. Biochem. 49:517-531.
3. Folk, J. E., Chung, S. I. 1973. Adv. Enzymol. 38:109-191.
4. Chung, S. I. 1975. Multiple molecular forms of transglutaminases in human and guinea pig. In: "Isozymes", Markert, C. L. (ed.). Academic Press, New York 1:259-274.
5. Pisano, J. J., Finlayson, J. S., Peyton, M. P. 1968. Biochemistry 8:871-876.
6. Matacic, S., Loewy, A. G. 1968. Biochem. Biophys. Res. Commun. 30:356-362.
7. Lorand, L., Downey, J., Gotoh, T., Jacobsen, A., Tokura, S. 1968. Biochem. Biophys. Res. Commun. 31:222-230.
8. Williams-Ashman, H. G. 1984. Mol. Cell. Biochem. 58:51-61.
9. Lorand, L., Hsu, L. K. H., Siefring, G. E. Jr., Rafferty, N. S. 1981. Proc. Natl. Acad. Sci. USA 78:1356-1360.
10. Lorand, L., Siefring, G. E. Jr., Lowe-Krentz, L. 1978. J. Supramol. Struct. 9:427-440.
11. Selkoe, D. J., Abraham, C., Ihara, Y. 1982. Proc. Natl. Acad. Sci. USA 79:6070-6074.

12. Mukherjee, D. C., Agrawal, A. K., Manjunath, R., Mukherjee, A. B. 1983. Science 219:989-991.
13. Mukherjee, A. B., Ulane, R. E., Agrawal, A. K. 1982. Am. J. Reprod. Immunol. 2:135-141.
14. Paonessa, G., Metafora, S., Tajana, J., Abrescia, P., De Santis, A., Gentile, V., Porta, R. 1984. Science 226:852-855.
15. Porta, R., Esposito, C., De Santis, A., Fusco, A., Iannone, M., Metafora, S. 1986. Biol. Reprod. 35:965-970.
16. Yuspa, S. H., Ben, T., Hemings, H., Lichti, U. 1980. Biochem. Biophys. Res. Commun. 97:700-708.
17. Fesus, L., Thomazy, V., Falus, A. 1987. FEBS Lett. 224:104-108.
18. Waite, M. 1987. The phospholipases: handbook of lipid research, vol. 5. Plenum press. New York.
19. Seilhamer, J. J., Randall, T. L., Yamanaka, M., Johnson, L. K. 1986. DNA 5:519-527.
20. Forst, S., Weiss, J., Elsbach, P., Maranganore, J. M., Reardon, I. 1986. Biochemistry 25:8381-8385.
21. Hayakawa, M., Horigome, K., Kudo, I., Tomita, M., Nojima, S., Inoue, K. 1987. J. Biochem.(Tokyo) 101:1311-1314.
22. Kramer, R. M., Hession, C., Johanson, B., Hayes, G., McGray, P., Chow, E. P., Tizard, R., Pepinsky, R. B. 1989. J. Biol. Chem. 264:5768-5775.
23. Krishnan, R. S., Daniel, J. C. Jr. 1967. Science 158:490-492.
24. Beier, H. M. 1968. Biochim. Biophys. Acta 160:289-291.
25. Morize, I., Surcouf, E., Vaney, M. C., Epelboin, Y., Buehner, M., Fridlansky, F., Milgrom, E., Mornon, J. P. 1987. J. Mol. Biol. 194:725-739.
26. Levin, S. W., Butler, J. D., Wightman, P., Schumaker, U. K., Mukherjee, A. B. 1986. Life Sci. 38:1813-1819.
27. Manjunath, R., Chung, S. I., Mukherjee, A. B. 1984. Biochem. Biophys. Res. Commun. 121:400-407.
28. Cordella-Miele, E., Miele, L., Mukherjee, A. B. 1990. J. Biol. Chem. (in press)
29. Kupferberg, J. P., Yokoyama, S., Kezdy, F. 1981. J. Biol. Chem. 256:6274-6281.
30. Gheriani-Gruszka, N., Almog, S., Biltonen, R. L., Lichtenberg, D. 1988. J. Biol. Chem. 263:11808-11813.
31. Haigler, H. T., Schlaepfer, D. D., Burgess, W. H. 1987. J. Biol. Chem. 262:6921-6930.
32. Elsbach, P., Weiss, J., Franson, R. C., Beckerdite-Quagliata, S., Schneider, A., Harris, L. 1979. J. Biol. Chem. 254:11000-11009.
33. Apitz-Castro, R., Jain, M. K., de Haas, G. H. 1982. Biochim. Biophys. Acta 688:349-356.
34. Slotboom, A. J., Verheij, H. M., de Haas, G. H. 1982. In: Hawtorne, J. N., Ansell, G. B. (eds). Phospholipids. New comprehensive Biochemistry. Elsevier North Holland, Amsterdam. 4:359-434.
35. Deems, R. A., Eaton, B. R., Dennis, E. A. 1975. J. Biol. Chem. 250:9013-9020.
36. Lutz, R. A., Bull, C., Rodbard, D. 1986. Enzyme 36:197-206.
37. Romero, G., Thompson, K., Biltonen, R. L. 1987. J. Biol. Chem. 262:13476-13482.
38. Bell, J. D., Biltonen, R. L. 1989. J. Biol. Chem. 264:12194-12200.
39. Cho, W., Tomasselli, A. G., Heinrikson, R. L., Kezdy, F. J. 1988. J. Biol. Chem. 263:11237-11241.

40. Tomasselli, A. G., Hui, J., Fisher, J., Zurcher-Neely, H., Reardon, I. M., Oriaku, E., Kezdy, F. J., Heinrikson, R. L. 1989. J. Biol. Chem. 264:10041-10047.
41. Maraganore, J. M., Poorman, R. A., Heinrikson, R. L. 1987. J. Prot. Chem. 6:173-189.
42. Davies, P. J. A., Chiocca, E. A., Basilion, J. P., Poddar, S., Stein, J. P. 1988. In: Zappia, V. and Pegg, A. E. (eds.). Progress in Polyamine Research. Advances in Experimental Medicine and Biology. Plenum Press, N. Y. vol. 250:391-401.
43. Seilhamer, J. J., Randall, T. L., Johnson, L. K., Heinzman, C., Klisak, I., Sparkes, R. S., Lusis, A. J. 1989. J. Cell. Biochem. 39:327-337.
44. Vinegar, R., Truax, J. F., Selph, J. L., Johnston, P. R., Venable, A. L., McKenzie, K. K. 1987. Fedn. Proc. 46:118-126.
45. Vadas, P., Pruzansky, W. 1986. J. Lab. Invest. 55:391-404.
46. Leu, R. W., Herriott, M. J., Moore, P. E., Orr, G. R., Birckbichler, P. J. 1982. Exp. Cell Res. 141:191-199.
47. Evans, J., Wang, Y., Shaw, K., Vernon, L. P. 1989. Proc. Natl. Acad. Sci. USA 86:5849-5853.

FUNCTIONAL CONSEQUENCES OF PHOSPHOLIPASE A₂ ACTIVATION IN HUMAN

MONOCYTES

Thomas Hoffman, Clara Brando, Elaine F. Lizzio, Crystal Lee,
Michael Hanson, Karen Ting, Yoo Jin Kim, Tore Abrahamsen*,
Joseph Puri, and Ezio Bonvini

Laboratory of Cell Biology, Division of Blood and Blood
Products, Center for Biologics Evaluation and Research, Food
and Drug Administration and *Pediatric Oncology Branch
Division of Cancer Therapy, National Cancer Institute
Bethesda, MD 20892

ABSTRACT

Human monocytes release arachidonic acid upon stimulation with a
variety of soluble or particulate agents. These include: phorbol
esters (i.e., 12-0-tetradecanoate phorbol-13-acetate, TPA), calcium
ionophores (ionomycin), serum-treated zymosan (STZ) concanavalin A
(Con A), and, to a minor degree, lipopolysaccharides (LPS). Protein
Kinase C activation or increased intracellular Ca^{2+} are common
features of the actions of most, if not all, of these stimuli.
Prevention of PKC activation by the use of staurosporine or chelation
of extracellular calcium by EGTA selectively impaired AA release,
indicating that PLA₂ may be regulated by either pathway concurrently.
The generation of inositol phosphates and diacylglycerol by the
action of phospholipase C, notably upon interaction with opsonized
particles during phagocytosis, apparently constitutes the
physiological correlate of stimulation via these agents.

Release of arachidonic acid by the action of PLA₂ or other
phospholipid hydrolyzing enzymes leads directly to the formation of
cyclooxygenase products. In the presence of markedly elevated
calcium concentrations, 5-lipoxygenase (LO) is activated as well,
leading to the formation and release of leukotrienes.

Agents which stimulate AA release also initiate other monocyte
functions, including generation of reactive oxygen intermediates and
lymphokine release. This observation makes it tempting to implicate
PLA₂ activation in many aspects of monocyte physiology. However, no
correlation with PLA₂ activation and either superoxide or lymphokine
release was found when multiple stimuli, including TPA, ionomycin,
serum-treated zymosan, concanavalin A, or LPS, were compared
simultaneously. Instead, our results indicate that PLA₂ activation is
regulated by the same mechanisms, including PKC activation and
increased Ca^{2+}, as are other enzymes which determine expression of
monocyte function.

INTRODUCTION

Phospholipase A_2 (PLA_2) hydrolyzes fatty acid from the sn-2 position of a wide variety of phospholipids. Substrates for this (these) enzyme(s) include species which contain a variety of polar head groups (choline, serine, ethanolamine, etc.) and some phospholipids with ether linkages in sn-1. In many cell types, including human monocytes, phospholipase A_2 commonly acts on substrates containing arachidonic acid (AA). The liberation of free arachidonate is a first step in the metabolism of prostaglandins, hydroxyeicosatetraeinoic acids, (HETE'S), and leukotrienes (Lt's).

Monocytes and macrophages have been shown to be rich sources of arachidonate and its metabolites.[1] Some biologic properties of monocytes, notably their role as immunomodulating cells, have been attributed to eicosanoid production and release.[2] Accordingly, much of the interest regarding PLA_2 in human monocytes centers on this aspect of their function. In addition, PLA_2 has been implicated, directly or indirectly, in other aspects of monocyte-macrophage function. These include generation of reactive oxygen intermediates and phagocytosis.[3]

The physiologic regulation of PLA_2 in human monocytes[4] or rodent macrophages[5] has been partly elucidated only recently. In either cell type, increased intracellular calcium or activation of protein kinase C is associated with PLA_2 activation. In vitro treatment of monocytes with agents which result in PKC activation (e.g. phorbol esters), or others which cause release of Ca^{2+} from intracellular stores (i.e., serum-treated zymosan, STZ), or influx of Ca^{2+} from the extracellular compartment (ionophores), results in release of arachidonic acid. The additive actions of such agents has suggested a "dual" stimulation of PLA_2 by PKC and/or Ca^{2+}.[4] A common denominator for PKC and Ca^{2+} mobilization may be represented in a physiologic context by the activation of inositol phospholipid-specific phospholipase C (PLC). Cleavage of inositol phospholipids by PLC produces diacylglycerol (DAG), an endogenous PKC activator, and certain inositol phosphates, including inositol (1,4,5) trisphosphate (IP_3), that may act as Ca^{2+}-mobilizing agents.[6] When phospholipase C (PLC) is activated subsequent to receptor-agonist interaction, for example during zymosan phagocytosis[7], it would be expected to result in phospholipase A_2 activation. This was shown to be true even if the action of one arm of the pathway were blocked.[4]

With this fundamental understanding of the regulation of PLA_2 activation in monocytes come additional questions regarding the specifics of the response when monocytes are co-stimulated with additional agents, including cytokines, that do not necessarily directly modify PLA_2 activity. Furthermore, the precise roles of arachidonic metabolites in specific functions, particularly those related to normal immunity, remains to be fully comprehended. Considering that much of this research may so far only be realistically approached by pharmacological manipulation of these cells, we have spent considerable effort on characterizing arachidonate metabolism of human monocytes and the changes this pathway undergoes in response to manipulation in vitro. This report will review methods for evaluating PLA_2 activation and eicosanoid metabolism in the context of normal monocyte function and will describe specific inhibition of the generation of 5-lipoxygenase metabolites by a novel drug which acts on the translocation of the enzyme.

MATERIALS AND METHODS

Cells. Peripheral blood mononuclear cells (PBMNC) were obtained by ficoll-hypaque density gradient centrifugation[9] of buffy coats prepared from volunteer blood donors. Monocytes were isolated by countercurrent centrifugal elutriation as described previously.[9] Briefly, 500-800 x 10[6] mononuclear cells were introduced into a Beckman Instruments (Fullerton, CA) elutriator system comprising a JE6-B centrifuge and 4.2 ml elutriation chamber. The system was kept at 4°C, at a constant speed of 2000 rpm. Hank's balanced salt solution (HBSS) without Ca^{2+} and Mg^{2+} was passed through the system at increasing flow rates of 9 (fraction 1), 11 (fraction 2), 12.5 (fraction 3), 15 (fraction 4), 17 (fraction 5), and 20 (fraction 6) ml/minute. 100 ml fractions were collected at each flow rate. Cells from fractions 5-6 were used as a source of monocytes based on the presence of greater than 88% of cells with the morphologic features of monocytes revealed by Wright's staining, and reactive histochemistry for alpha naphthyl-acetate esterase (ANAE). Other contaminating cells in the monocyte fraction included: 5% large lymphocytes, 3% polymorphonuclear cells, 2% eosinophils and 1% unclassified cells. Platelets were not found in fractions eluting later than fraction 1. No rosetted platelets were seen surrounding monocytes. Alternatively, monocytes were isolated by a large scale elutriation technique devised at the NIH Clinical Center.[10] Monocytes obtained by either method were found to behave identically in the assays undertaken here.

Reagents. Arachidonic acid, concanavalin A, (Type IV) cytochrome C, zymosan, ethylene glycol bis (ß-aminoethyl ether) N,N,N'N'-tetraacetic acid (EGTA), essential fatty acid-free bovine serum albumin (BSA), superoxide dismutase (SOD), and lipopolysaccharide (LPS; serotype 111:B4) were purchased from Sigma Chemical Company (St. Louis, MO). Ionomycin and staurosporine were purchased from Calbiochem (San Diego, CA). Earle's balanced salt solution (EBSS) was obtained from GIBCO (Grand Island, NY). Iscove's medium (Isc) was from M.A. Bioproducts (Walkersville, MD). 12-O-tetradecanoate phorbol-13-acetate (TPA) was purchased from CCR, Inc. (Eden Prairie, MN). [5,6,8,9,11,12,14,15 - [3]H(N)] arachidonic acid (60-100 Ci/mmole), was bought from NEN, Inc. (Boston, MA). MK-886 was a generous gift from Dr. A. Ford-Hutchinson, Merck Frosst, Canada.

Arachidonate Release - Peripheral blood monocytes (2 x 10[6]/ml) were incubated for four hours in duplicate wells of a 96-well microplate or containing 200 µl (400,000 cells) or in a 24-well plate containing 2000 µl (4,000,000 cells) Iscove's' medium (BSA 1 mg/ml) with 1 µCi/ml of [^{3}H]-AA. At the end of the incubation, they were washed twice and the medium was replaced with 50 µl fresh medium (500 µl for 24-well plates) containing the indicated concentration of stimulus. After an additional two hours at 37°C, the medium was removed and its radioactivity determined by liquid scintillation counting in a LKB-Wallac 1218 Rackbeta counter calibrated for dpm measurement by external standard ratio.

Superoxide release - The microassay of Pick and Maizel[11] was used. Briefly, 4 x 10[5] elutriator-purified monocytes were incubated at 37°C with 200 µl EBSS containing 2 mg/ml cytochrome C and 1 mg/ml BSA with or without stimulating agents in 96-well flat-bottomed plates. Spectrophotometric readings at 550 nanometers were obtained in a VMax kinetic microplate reader (Molecular Devices, Menlo Park, CA) using cells incubated in SOD (300 U/ml) as a blank. Three replicates were used for each determination. Results were expressed as nmoles O_2^- per culture, using an extinction coefficient of 21×10^3 M^{-1} cm^{-1} corrected for the calculated length of the light path.

Radioimmunassay - Monocyte supernatants were prepared in 24-well plates as described above except for omission of labeled arachidonate. Assays were performed according to the manufacturer's instructions using standard curves constructed in Isc/BSA. PGE_2 assay was carried out using kits obtained from Seragen or NEN. LtB_4 was performed using kits obtained from NEN.

Enzyme-Linked immunoassay (EIA) - Interleukin 1 (IL-1ß) or tumor necrosis factor (TNFα) were assayed using kits obtained from Cistron (Pine Brook, NJ) according to the manufacturer's instructions.

RESULTS

Arachidonic Acid Release from Peripheral Blood Monocytes in Response to Different Stimuli. Monocytes labeled with ^3H-AA, took up 90% of the label within four hours. After thorough washing, monocytes released part of the incorporated label over a two-hour time course. during which maximal release was observed. Figure 1 shows release of label at optimal doses of agents known to stimulate a variety of monocyte functions. Also tested, but not shown in the figure, is the chemotactic peptide formyl-methiothyl-leucyl-phenylalanine (FMLP), which failed to induce significant arachidonate release at any dose between $10^{-6}M$ to $10^{-13}M$.

As shown previously, monocytes stimulated with TPA, ionomycin, or STZ release most of the incorporated label as unmetabolized arachidonate.[4] In order to assess whether other stimuli activated other pathways of AA metabolism, the supernatants were analyzed by HPLC. Chromatograms from Con A or LPS-stimulated monocytes, demonstrated the presence of cyclooxygenase products, including 12-L-hydroxy-5,8,10-heptadecatrienoic acid (HHT), thromboxane A_2, and 6-Keto-$PGF_{1\alpha}$ as the principal metabolites. 15-HETE was also detected, but 5-lipoxygenase products, including 5-HETE and leukotrienes were not seen unless ionomycin or STZ were used as stimuli (data not shown).

To verify the pattern of cyclooxygenase or lipoxygenase stimulation, representative products of each pathway were assessed by

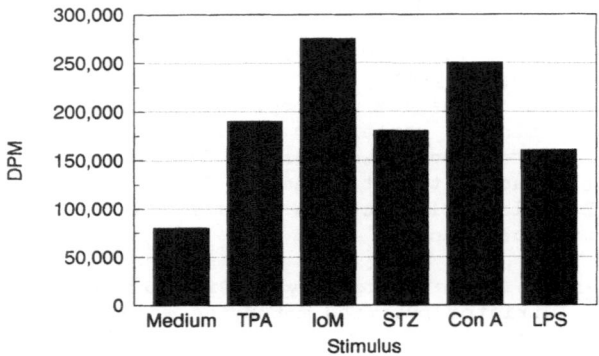

Fig. 1. Monocyte AA Release in Response to Stimulation In Vitro. Monocytes were labeled and stimulated as described in Methods. The stimuli employed were: TPA 20 nM, Ionomycin (IoM) 2 μM, STZ 300 μg/ml, Con A 30 μg/ml, LPS 10 μg/ml.

RIA. All stimuli which induced ^{3}H-AA release also induced PGE$_2$ release, but only ionomycin or STZ consistently stimulated formation and release of significant amounts of LtB$_4$ (Figure 2).

Role of extracellular Ca^{2+} and protein kinase C (PKC) activation in AA Release in Response to Con A stimulation. Con A could induce AA release and, as published previously[13], induced increased intracellular Ca^{2+}. Therefore, it was of interest to determine whether its mechanism of action conformed to the model of PLA$_2$ activation derived from observations on STZ-mediated release. Monocytes were therefore incubated in medium prepared without Ca^{2+} and Mg^{2+}. Advantage was taken from previous observations that the water used in media preparation contained sufficient calcium to support AA release in response to other stimuli, but could be easily chelated with EGTA at doses which were not toxic. Con A stimulation followed the same pattern seen with other stimuli, except TPA, which is known to act mainly independently of Ca^{2+} (Figure 3). Con A stimulation of AA release depended on the presence of extracellular calcium, but not on the presence of magnesium.

Con A has also been shown to induce the phosphorylation of certain cellular proteins in human monocytes, but not those involved in superoxide production, and to induce reverse PKC translocation, from the membrane to the cytosol.[13] It was therefore also of interest to test whether PKC activation played a role in Con A mediated AA release. Preliminary data showed that H-7, an inhibitor of protein kinases with preferential activity on PKC, could inhibit the Con A response at doses in the range of 100-300 uM (data not shown). In recent studies, we have employed a more selective inhibitor of PKC, staurosporine[14], which inhibits TPA-induced superoxide production by monocytes at much lower doses than H-7. In addition to inhibiting TPA and STZ-induced AA release, staurosporine treatment also inhibited the Con A response in a dose-dependent fashion (Figure 4). Evidence that the integrity of the monocytes was unimpaired by staurosporine treatment was obtained from trypan blue exclusion and the observation that ionomycin-induced AA release was not inhibited.

Lack of Correlation Between PLA$_2$ Activation and Superoxide Release Human Monocytes. A number of authors have concluded that AA release and superoxide are causally linked.[15] This conclusion is based on a number of observations: agents which release arachidonate have been shown to stimulate superoxide production, particularly in

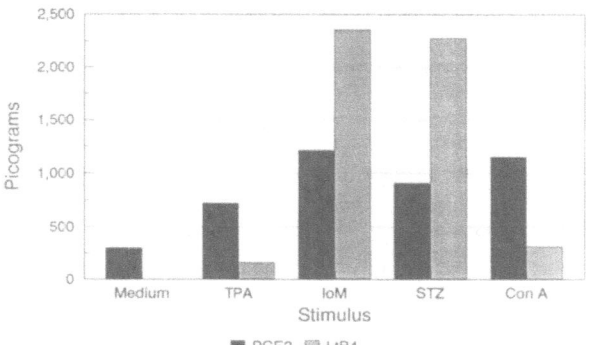

Fig. 2.   Eicosanoid Production by Human Monocytes. Monocytes were stimulated under the identical conditions as in Figure 1, except no labeled arachidonate was added.

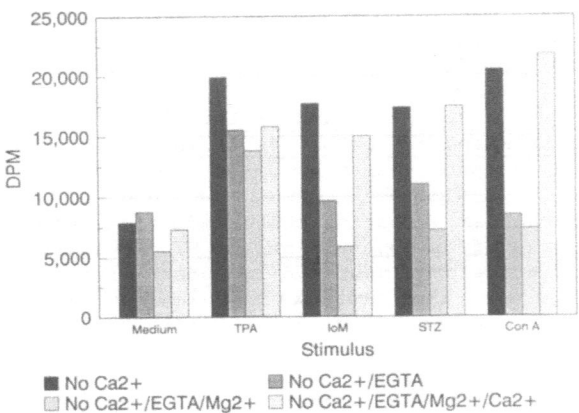

Fig. 3.   Dependence of AA Release on Extracellular Calcium.
Monocytes were incubated with ³H-AA-containing Iscove's
medium in microwells. After washing, the medium was
replaced with Ca²⁺-free EBSS with or without EGTA (0.5 mM).
Calcium or magnesium were added at 0.8 mM to the indicated
cultures. Stimulation was performed at the same
concentration of agents as in figures 1 and 2.

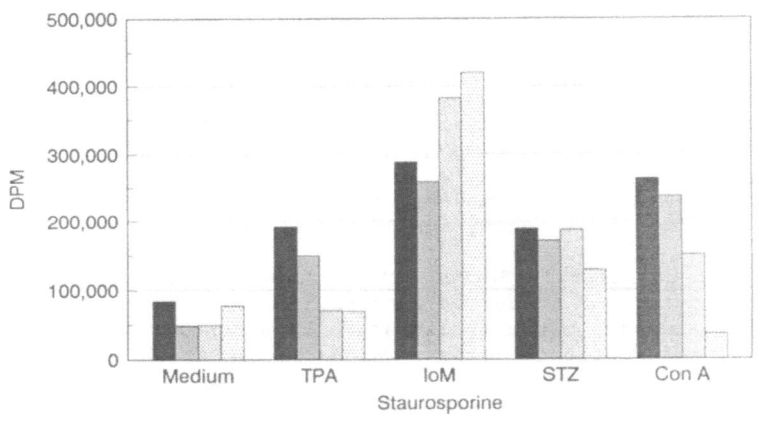

Fig. 4.   Effect of Staurosporine on Stimulated ³H-AA Release.
Monocytes were pre-incubated for five minutes at twice the
indicated concentration of Staurosporine. An equal volume
of medium containing the stimulant at 2x concentration was
added and the assay allowed to proceed without washing out
the drug.

neutrophils, and arachidonate and other unsaturated fatty acids may
stimulate NADPH-oxidase activity in vitro. Certain putative
inhibitors of PLA₂ have been shown to also inhibit superoxide
generation. To examine this point in monocytes, we measured
superoxide release in response to agents shown to stimulate AA
release (Figure 5). Only TPA and zymosan consistently induced
superoxide release from monocytes. Ionomycin and Con A, despite
being potent activators of PLA₂ in situ, failed to induce measurable

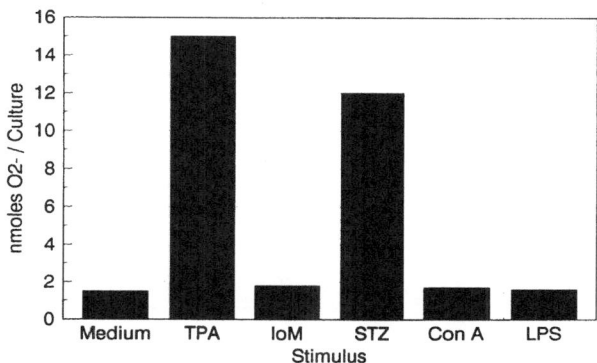

Fig. 5. Superoxide Release from Peripheral Blood Monocytes
Stimulated by Agents Which Induce AA Release. Monocytes
were assayed under identical conditions as described in
figure 2 except that the medium employed was EBSS containing
2 mg/ml cytochrome C. Superoxide was quantified by
monitoring cytochrome C reduction spectrophotometrically as
described in Methods. Release at two hours is depicted in
the graph.

superoxide release. LPS was also ineffective in mediating superoxide
generation.

Effect of Specific Inhibition of Cyclooxygenase or Lipoxygenase on
Cytokine Production and Release. Previous studies have reported a
variety of effects seen by the addition of exogenous AA metabolites
or inhibitors of AA metabolism, perhaps past PLA_2, on a variety of
cellular functions.[16] We were kindly given permission to study a new
agent, MK-886, which inhibits translocation of the 5-LO and has been
shown to inhibit leukotriene generation in vitro and in vivo in other
cell types, including granulocytes.[17] We compared the ability of
this agent to inhibit LtB_4 production by a panel of stimuli to their
ability to induce synthesis and release of two lymphokines which are
known to have potent inflammatory effects, IL-1 and TNF. As a
control, the effect of the putative inhibitor of 5-LO product
formation was assessed by RIA for PGE_2 or LtB_4 at the same time. MK-
886 treatment inhibited monocyte induced LtB4 release, as expected,
across a broad dose range. No effect on PLA_2 in terms of
arachidonate release was seen (data not shown) in parallel cultures
labeled with ^3H-AA. The drug also showed no effect on PGE_2 release at
any of the doses tested (Figure 6).

In other experiments, a dose of the drug which completely
inhibited LtB_4 release, alone or in combination with indomethacin,
had no effect on either Il-1 or TNF release by monocytes. This was
true regardless of the stimulus employed (Figure 7) even in response
to LPS, the most potent inducer. Some effect on STZ-induced TNF
release was observed, but this was not greater than that seen with
indomethacin alone or in combination.

DISCUSSION

Human monocytes can be induced to release arachidonic acid by a
wide variety of stimuli. Release of arachidonate is accompanied by
its metabolism to eicosanoids. Prostaglandins and HETE's are
generated concomitant with the liberation of free arachidonate, but
5-LO products are formed only under conditions of greatly enhanced
intracellular Ca^{2+} concentrations.

131

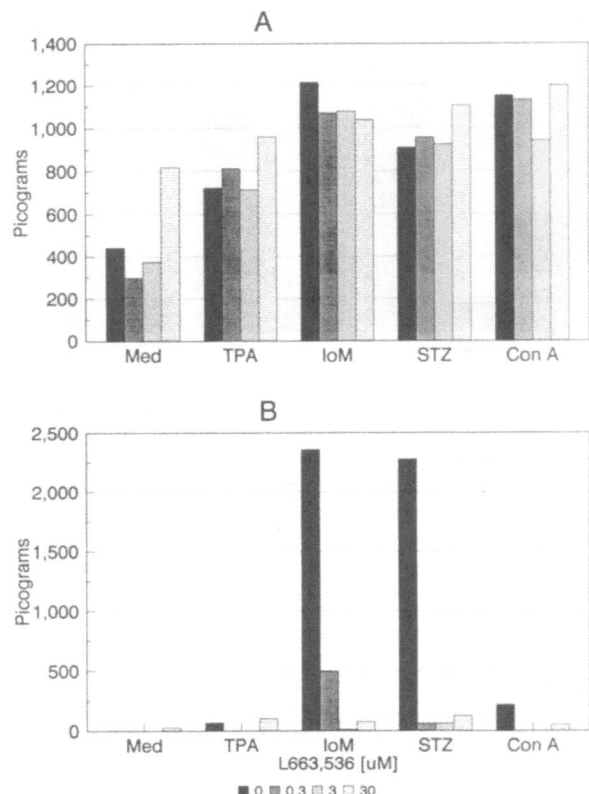

Fig. 6.   Effect of MK-886 on PGE$_2$ (A) or LtB4 (B) Release.
Monocytes were stimulated in a protocol identical to that
described in the legend to figure 4, except for omission of
the label.   Eicosanoids were determined by RIA.

        Concanavalin A treatment of human monocytes causes PLA$_2$ activation
and attendant CO product formation and release, but results in little
or no 5-LO activation.   Presumably, the levels of intracellular
calcium rise associated with this agent are insufficient, or the
duration too short, to adequately activate the enzyme.
Alternatively, some aspect of "normal" 5-LO translocation necessary
for its action, as seen from the inhibition given by MK-886, is not
initiated by perturbation of Con A receptors.

        Con A action may be inhibited by H-7 or staurosporine, indicating
that some component of PKC activation may be involved.   In the
absence of knowledge of which proteins are phosphorylated to initiate
AA release, this conclusion must remain tenuous, especially in
consideration of the known effects of these inhibitors on other
pathways.   Despite the dependence of the effect on calcium, the
action of Con A is not likely mediated via inositol trisphosphates.
This is suggested by the difference between Con A and STZ
stimulation.   Recent data from our lab (in preparation) indicated
that Con A fails to induce phosphatidylinositol bisphosphate (PIP$_2$)
hydrolysis, as has been observed in neutrophils.[17]

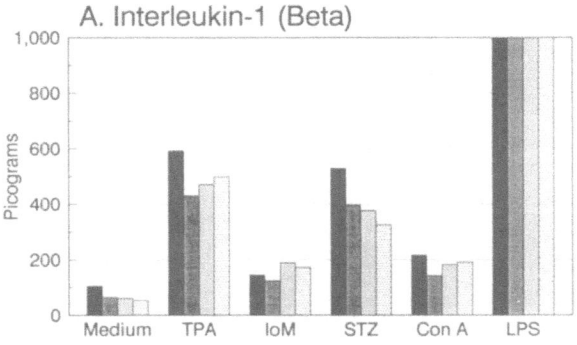

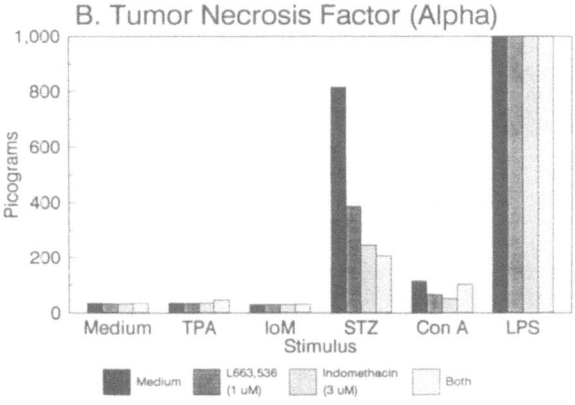

Fig. 7.   Effect of MK-886 on IL1ß (A) or TNFα (B) Release.
Monocytes were incubated overnight in the indicated
concentrations of drug in microwells. The supernatants were
assayed for immunoreactive cytokine. The drug itself did
not affect the RIA performance characteristics. Values of
IL1 or TNF were between 4000 - 4100 pg and were not changed
by the presence of drug.

        Many stimuli which promote or induce AA release likewise initiate
other biological functions of monocytes. This has made conclusions
regarding a role for PLA₂ in these tempting. Many data supporting
this have been derived through the use of inhibitors, including
those which putatively act on PLA₂ directly, as well as those which
act on pathways leading to formation of eicosanoids. Unfortunately,
many studies fail to simultaneously assess the pathways which their
drugs putatively inhibit. In the absence of a rational link between
the stimuli, their mechanism of action, and the regulation of both
PLA₂ and the function in question, these studies can not be regarded
as definitive. We have, as an example, described one system where
eicosanoid formation and superoxide or lymphokine release are not
correlatable. Similarly PLA₂ activation itself does not correlate
with either superoxide, IL1 or TNF release (Table I). Con A clearly
activates PLA₂, but is a poor stimulus for superoxide release or TNF
production. Ionomycin, arguably the "best" inducer of AA release,
fails to stimulate superoxide generation under the conditions
employed here and also is a poor initiator of lymphokine production.

TABLE I

	AA	CO	LO	O_2^-	IL1	TNF
Med	±	±	-	±	±	±
TPA	++	++	-	+++	++	++
IoM	+++	+++	+++	±	+	±
STZ	++	++	++	++	++	++
ConA	+++	+++	-	±	+	±
LPS	+	+	-	-	++++	+++

LPS is a marginal activator of AA release, has no effect on superoxide release, but is the most potent stimulus of either lymphokine release.

Labeled AA release or direct measurement of CO products remain useful tools in assessing PLA_2 activation in situ. However, an understanding of the limitations of the techniques involved, must be kept in mind. At this time we do not know precisely from which pools AA is released for one or another use. We also do not know whether activation occurs by the direct actions associated with our agents, or whether regulatory proteins are involved. It also remains to be determined whether these regulators, including lipocortins[19] and phospholipase A_2 activating proteins (PLAP)[20], are different for the various stimuli under study. Also missing is a better understanding of translational and transcriptional control of PLA_2 in different cells, including monocytes at different stages of differentiation. The advent of molecular biological innovations described at this symposium: isolation and cloning of genes for PLA_2 and its regulatory elements, new, specific, inhibitors of these enzymes, bioactive peptide fragments derived from PLA_2 and other enzymes of the arachidonic acid cascade, will soon allow definitive resolution of the role of PLA_2 in monocyte physiology and host immunity.

ACKNOWLEDGEMENT

Many thanks to D. Lake and J. Donelson for editoral assistance and to Dr. L. Harvath and Dr. K. Zoon for critical review.

REFERENCES

1. M. E. Goldyne, G. F. Burrish, P. Poubelle, and P. Borgeat, Arachidonic acid metabolism among human mononuclear leukocytes: lipoxygenase-related pathways. J. Biol. Chem. 259:8815 (1984).

2. R. J. Bonney and P. Davies, Possible autoregulatory functions of the secretory products of monnuclear phagocytes. Contemp. Top. Immunobiol. 13:199 (1984).

3. D. N. Rush and P. A. Keown, Human monocyte chemiluminescence triggered by IgG aggregates. Requirement of phospholipase

activation and modulation by Fc receptor ligands. Cell. Immunol. 87:252 (1984).

4. T. Hoffman, E. F. Lizzio, J. Suissa, and E. Bonvini, Dual stimulation of phospholipase activity in human monocytes: Role of calcium-dependent and calcium independent pathways in arachidonic acid release and eicosanoid formation. J. Immunol. 140:3912 (1988).

5. A. A. Aderem, W. Scott, and Z. Cohen, Evidence for sequential signals in the induction of the arachidonic acid cascade in macrophages. J. Exp. Med. 163:139 (1986).

6. M. J. Berridge, Inositol triphosphate and diacylglycerol as second messengers, Biochem. J. 220:345 (1984).

7. J. Moscat, M. Aracil, E. Diez, J. Balsinde, P. Garcia Barreno, and A. M. Municiom, Intracellular Ca²⁺ requirements for zymosan-stimulated phophoinositide turnover in mouse peritoneal macrophages. Biochem. Biophys. Res. Comm. 134:367 (1986).

8. A. Boyüm, Isolation of lymphocytes, granulocytes, and macrophages. Scand. J. Immunol. 5 (Suppl. 5.9),

9. H. C. Stevenson, E. Bonvini, T. Favilla, P. Miller, Y. Akiyama, T. Hoffman, R. Oldham, and D. Kanapa, Characterization of purified cryopreserved human monocyte function in assays of superoxide production, accessory cell function, chemotaxis, and in fluorescent cell sorter analysis. J. Leuk. Biol. 36:521 (1984).

10. G. Abrahamsen, C. S. Carter, M. Rubin, H. G. Goetzman, E. J. Read, E. F. Lizzio, C. Lee, M. Hanson, P. A. Pizzo and T. Hoffman. Large scale separation of peripheral blood monocytes: Influence of elutriation medium and storage conditions on inflammatory mediator release. J. Leuk. Biol. (in press)

11. E. Pick, and Y. Keisari, Superoxide anion and hydrogen peroxide production by chemically elicited peritoneal macrophages-- induction by multiple nonphagocytic stimuli. Cell. Immunol. 59:301 (1981).

12. J. P. Scully, G. B. Segal, and M. A. Lichtman, Calcium exchange and ionized cytoplasmic calcium in resting and activated human monocytes. J. Clin. Invest. 74:589 (1984).

13. M. R. Costa-Casnellie, G. B. Segal, and M. A. Lichtman, Signal transduction in human monocytes: Relationship between superoxide production and the level of kinase C in the membrane. J. Cell. Physiol. 129:336 (1986).

14. T. Tamaoki, H. Nomoto, I. Takahashi, Y. Kato, M. Morimoto, and F. Tomita, Staurosporine, a potent inhibitor of phopholipid/calcium dependent protein kinase. Biochem. Biophys. Res. Comm. 135:397 (1986).

15. A. I. Tauber, Protein kinase C and the activation of the human neutrophil NADPH-oxidase. Blood 69:711 (1987).

16. M. Rola-Plesczcynski and I. Lemaire, Leukotrienes augment interleukin-1 production by human monocytes. J. Immunol. 135-3958 (1985).

17. J. Gillard, A. W. Ford-Hutchinson, C. Chan, S. Charleson, D. Denis, A. Foster, R. Fortin, S. Leger, C. S. McFarlane, H. Morton, H. Piechuta, D. Riendeau, C. A. Rouzer, J. Rokach, R. Young. L-663-536 (MK-886)(3-[1-(4-chlorobenzyl)-3-t-butyl-thio-5-isopropylindol-2-yl]-2,2-dimethylpropanoic acid), a novel, orally active leukotriene biosynthesis inhibitor. Canad. J. Physiol. Pharmacol. 67:456 (1989).

18. H. M. Korchak, L. B. Vosshall, C. Zagon, P. Ljubich, A. M. Rich, and G. Weissman, Activation of the neutrophil by calcium-mobilizing ligands I. a chemotactivc peptide and the lectin concanavalin a stimulate superoxide anion generation but elicit different calcium movements and phophoinositide remodeling. J. Biol. Chem. 263-11090 (1988).

19. H. M. Korchak, L. B. Vosshall, C. Zagon, P. Ljubich, A. M. Rich, and G. Weissmann, Activation of the neutrophil by calcium-mobilizing ligands I., A chemotactic peptide and the lectin concavalin A stimulate superoxide anion generation but elicit different calcium movements and phophoinositide remodeling. J. Biol. Chem. 1631-11090 (1988).

20. J. S. Bomalaski, D. G. Baker, L. Brophy, N. V. Resurreccion, I. Spilberg, I. J. Muniain, and M. A. Clark. Phospholipase A2-activating protein (PLAP) stimulates human neutrophil aggregation and release of lysosomal enzymes, superoxide, and eicosanoids. J. Immunol. 142:3957 (1989).

21. R. Flower, Lipocortin, Biochem. Soc. Trans. 17:276 (1989).

INHIBITION OF PHOSPHOLIPASE A₂ BY UTEROGLOBIN AND ANTIFLAMMIN

PEPTIDES

Lucio Miele, Eleonora Cordella-Miele, Antonio Facchiano
and Anil B. Mukherjee

Section on Developmental Genetics
Human Genetics Branch, NICHD, NIH
Bethesda, Maryland 20892

INTRODUCTION

Blastokinin (1) or uteroglobin (UG; 2) is a secretory protein with a low molecular mass (15.8 kDa). This protein was originally discovered in the rabbit uterine fluid during early pregnancy (1, 2). Its synthesis and secretion in the endometrium are stimulated by progesterone (2-5). Since its discovery in 1967, UG has been one of the most thoroughly studied markers of progesterone action in the rabbit endometrium. In addition to the uterus, UG has been found in several other organs of rabbits, namely, the oviduct, the male genital tract and the tracheobronchial tree of both male and female animals (4-9). More recently, we detected UG in the circulation of the rabbit (10). The source of this protein in circulation seems to be the tracheobronchial epithelium and/or the progesterone-induced uterine endometrium (10).

UG was purified to homogeneity in 1977 (11) and its primary and quaternary structures have been described (11-14). UG is a dimer formed by two identical antiparallel chains of 70 amino acid residues each, joined by two interchain disulfide bonds between Cys-3 and 69′, and 3′ and 69. Each monomer consists of four α-helical segments and one β-turn (Lys - 26 to Glu - 29) (15, 16) but no β-structure is present (See Fig. 1).

Although UG was first discovered in the uterus of rabbits during early pregnancy, it has now been found in several other organs. These are: the oviduct, the male genital tract, the tracheobronchial tree and the digestive tract of both male and female animals (4-9). In these organs, UG is produced by epithelial cells: Endometrial nonciliated cells and bronchiolar Clara cells, for example, give intensely positive immunofluorescence for UG (17). More recently, *in situ* hybridization experiments with a UG–cDNA probe have revealed that during pregnancy nearly all the endometrial epithelial cells contain UG-mRNA (18). The signal is much stronger in glandular epithelial cells than in luminal cells. In the tracheobronchial tree UG-specific immunofluorescence has been detected in ciliated epithelial cells and in some cells of the alveolar wall, but not in goblet cells (18).

By means of RIA, Western blotting, progesterone (P) binding

Biochemistry, Molecular Biology, and Physiology of Phospholipase A₂ and Its Regulatory Factors
Edited by A. B. Mukherjee, Plenum Press, New York, 1990

and PLA$_2$ inhibition studies (see below), we find that a protein electrophoretically, immunologically, and functionally identical to UG is present in rabbit blood obtained from uteroovarian and pulmonary veins (10).

The presence of a UG-like protein in humans has been the subject of conflicting reports (19-24). However, the techniques used by most authors, such as gel electrophoresis and radial immunodiffusion, are relatively insensitive. We found that a protein, cross-reactive to monospecific anti-UG antibody with a molecular weight similar to that of UG-monomer, is indeed present in the human endometrium (25-26). Progesterone seems to be the inducing hormone in this organ, as judged from the fact

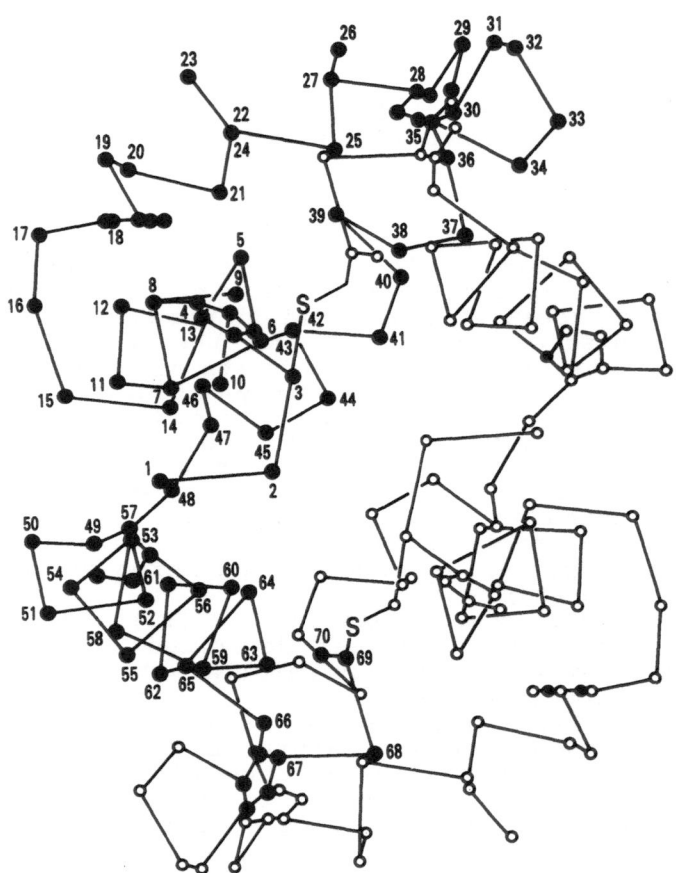

Figure 1. b-axis projection of molecular coordinates of UG dimer. Only α-carbons, disulfide bonds and prolines are represented. Note disulfide bonds between residues 3′ and 69 and 69′ and 3. Resolution 2.2 A. Reproduced with permission from ref. 15.

that this protein is most abundant in the endometrium during the mid- to late luteal period of the menstrual cycle. Additionally, we have demonstrated the presence of a similar protein in the human neonatal tracheobronchial epithelium (27) and in the adult prostate (28).

The regulation of UG synthesis by steroid hormones is

different in different organs of the rabbit. During early pregnancy, UG is detectable in the uterine fluid at day 3 of pregnancy, and its absolute amount in the uterine secretion increases until day 6 (For a review see ref. 29). When expressed as percent of total uterine fluid proteins, UG level peaks between days 4 and 5 of pregnancy, after which it starts to decrease. This pattern can be reproduced by the administration of progesterone (P) to stimulate UG synthesis. A subsequent administration of estradiol abrogates UG to disappear. Interestingly, a sharp increase in plasma estradiol after day 5 of pregnancy has been reported in rabbits (30). After day 10 of pregnancy UG is almost undetectable in uterine fluid by gel electrophoresis. However, by using more sensitive RIA techniques, we have detected trace amount of UG in uterine homogenates throughout pregnancy until two days prior to parturition (Kikukawa and Mukherjee, unpublished results). Similarly, UG mRNA levels could be detected by Northern blot in decidual cells up to 27th day of pregnancy in the rabbit (unpublished data).

Androgens can also stimulate endometrial UG synthesis, but this effect is most likely exerted through the P receptor (31). Whether androgens can stimulate UG synthesis in the male genital tract through their own receptor remains to be clarified. Recently, Chilton et al. have reported that prolactin significantly amplifies P-induced stimulation of UG synthesis in the endometrium (32). UG synthesis in the oviduct is stimulated by estradiol only (33). UG and UG mRNA are constitutive in the lung (see above), and both are indistinguishable from their uterine counterparts. However, lung UG synthesis is completely independent of ovarian steroids. Glucocorticoids can stimulate UG synthesis in the lung by about 3-fold (29).

THE FUNCTION OF UG

Although initially UG was thought to be a blastocyst growth factor (blastokinin), this hypothesis was not confirmed by further experimental scrutiny. One of the better known properties of UG has been suggested to be related to its ability to bind P and related compounds. The physiological relevance of this property has never been demonstrated, and remains uncertain. UG has been proposed to be a P carrier between the uterus and the blastocyst, a P scavenger (34), or a factor involved in the maintenance of high intraluminal P concentration in the preimplantation uterus. In fact, both a UG-dependent accumulation of P in the uterine lumen and a P-dependent uptake of UG by rabbit endometrium have been described (see ref. 29). These observations were interpreted as demonstrations of a P carrier function of UG. However, no biochemical evidence was presented that P in the uterine secretion was actually bound to UG, nor that the internalized UG was P-bound. In addition, the function of UG in the lumen of organs other than the uterus is likely to be unrelated to its steroid-binding properties, since UG does not bind androgens, estrogens or corticosteroids (29).

For the past several years our laboratory has been working to delineate the biological functions of UG. Our working hypothesis is that one possible function of UG may be to participate in the control of inflammatory and immunological responses to extraneous agents in the mucosae of organs communicating with the external environment (29, 35). Several years ago, we proposed a hypothesis that at least one of the functions of uterine UG may be to protect the embryo from maternal immune and inflammatory response during implantation

(36). In this respect, UG could be a mediator of the pregnancy-related immunosuppressive action, which had been previously suggested to be exerted by P (37). This hypothesis has been supported by the results of several experiments: in fact, we were able to demonstrate a tolerogenic effect exerted by UG on rabbit blastomeres and spermatozoa (38-40). The presence of active transglutaminase (TG) was required for a maximal tolerogenic effect. When epididymal spermatozoa or blastomeres were preincubated with UG alone or with UG plus TG and then used to challenge maternal lymphocytes, the lymphocyte response (measured by [^{3}H] thymidine incorporation by responder cells) was drastically and specifically inhibited (38). This effect was reproduced by treating blastomeres with crude pregnant uterine fluid (38, 39) and spermatozoa with crude prostatic fluid. Both these fluids contain UG and TG. Uterine fluid from nonpregnant animals (which does not contain UG in detectable amounts) was ineffective, as well as prostatic fluid treated with anti-UG, or anti-TG antibodies. Moreover, the TG inhibitor neopentylchloroethylnitrosourea (NPCNU) drastically reduced the tolerogenic effect. It should be pointed out that TG is synthesized by both prostate and the uterus. Also, uterine TG increases during pregnancy and TG-deficiency leads to habitual abortion (29). These results led us to the conclusion that TG and UG may have a physiological antigen masking role on both spermatozoa and developing embryos during fecundation and nidation. The most likely mechanism is a covalent cross linking of UG with still unidentified cell surface antigens, catalyzed by TG via the formation of γ-glutamyl-ϵ-lysine isopeptide bond(s). UG is a good substrate for TG in conventional assays (41) and, with the exception of Lys-42, all Lys and Gln residues in UG are exposed to the solvent (16) in positions readily accessible to a cross-linking reaction. Another possible mechanism for the putative sperm- and embryo-protective effect of UG, besides masking of antigenic sites, could be the inhibition of phagocyte activity and therefore, antigen processing by free or membrane-bound UG. We investigated this possibility, and found that UG can inhibit both monocyte and neutrophil chemotaxis and phagocytosis *in vitro* (42-44). Eighty percent inhibition of the internalization of labeled formyl-peptide chemoattractant was also observed at concentrations of UG that completely inhibited chemotaxis (42-44). Thus, we concluded that UG can alter the immune response against allogenic cells by masking their surface antigenic determinants and by directly impairing phagocyte functions. Subsequently, we found that UG is also a powerful inhibitor of thrombin-induced, but not of arachidonic acid-induced, platelet aggregation (45). These results suggested that UG must be inhibiting an enzymatic step proximal to arachidonic acid release from cellular phospholipids (45).

Since phospholipase A_2 (PLA$_2$) (EC 3.1.1.4) is involved in the activation of chemotaxis and in platelet activation and since other antichemotactic and antiinflammatory proteins, such as lipocortin I (42, 43, 46-48) also are inhibitors of PLA$_2$, we investigated whether or not UG also had PLA$_2$-inhibitory properties. We found that purified UG in micromolar concentrations can inhibit purified porcine pancreatic PLA$_2$ and all three PLA$_2$ activities from RAW 264.7 mouse macrophages (49). These activities have pH optima at 4.0, 7.0 and 8.5, respectively. With the porcine pancreatic enzyme we used a sonicated dispersion of pure phosphatidylcholine as the substrate. In the case of RAW 264.7 PLA$_2$ activities, the substrate was represented by membranes of the same cells, metabolically labeled with radioactive arachidonic acid. The highest inhibitory effect was seen with macrophage PLA_2 at pH 7. UG was inhibitory to this activity at concentrations as low as 0.9 μM (see Tables 1 and 2). It should be noted that all the concentrations of UG used in this system were within or below the physiological range of UG levels found in the pregnant uterus and lung respectively. It should also be mentioned that the PLA_2 assay used in these studies is not as sensitive as the one currently used in our laboratory and therefore, the concentration of UG which yielded the highest inhibition of PLA_2 is overestimated.

Table 1. Effect of UG on PLA_2 activities from RAW 264.7 mouse macrophage cell line having pH optimum 4, 7, and 8.5, respectively.

PLA_2	[a]UG	[³H]Arachidonate[b] (cpm)	% Inhibition
pH 4	-	10,706 ± 191	
	+	6,949 ± 330	34.5
pH 7	-	5,460 ± 232	
	+	1,555 ± 122	71.5
pH 8.5	-	8,010 ± 320	
	+	4,728 ± 215	41.0

[a]UG concentration 2.5 μM.
[b]N = 3; values represent the mean ± SD. Redesigned with permission from ref. 49

Since in the case of RAW 264.7 macrophages a cell lysate was used as a source of the enzyme, we cannot rule out the possibility that the more pronounced PLA_2 inhibition by UG in this system was mediated by an endogenous TG. However, pure UG alone inhibits porcine pancreatic PLA_2 in a chemically defined assay system. Thus, we proposed that UG might be involved in the modulation of PLA_2 activity *in vivo* as well. This hypothesis is further supported by preliminary experiments from our laboratory showing that P induction of UG synthesis in intact rabbit uterus is paralleled by a sharp decrease in prostaglandin (PG) levels in this organ (unpublished data). This effect is more pronounced on $PGF_{2\alpha}$ than on PGE_2.

PLA_2 catalyzes the hydrolysis of the fatty acid moiety from the *sn*-2 position of glycerophospholipids. This has been proposed to be the rate-limiting step of the arachidonate cascade, controlling the availability of free arachidonate for the cyclooxygenase and lipooxygenase pathways. In addition, by catalyzing the hydrolysis of the fatty acid from the *sn*-2 position of 1-O-alkyl, 2-acyl *sn*-glycero-3-phosphocholine, PLA_2 can generate lyso-platelet activating factor (PAF)-acether, the metabolic precursor of PAF-acether, a potent non-eicosanoid mediator of inflammation (50).

Table 2. Dose response of UG inhibition of PLA_2 activity from RAW 264.7 cells having pH optimum 7.

UG concentration (μM)	[^{3}H]Arachidonate[a] (cpm)	% Inhibition
0	1712 $\pm$ 32	0
0.06	1721 $\pm$ 23	0
0.6	1777 $\pm$ 52	0
0.9	1355 $\pm$ 83	20.9
1.25	702 $\pm$ 85	59.0
2.5	347 $\pm$ 34	79.7

[a]$n = 3$; values represent the mean $\pm$ SD. Redesigned with permission from ref. 49.

Therefore, the PLA_2 step is a key point of regulation of both eicosanoid and PAF-acether biosynthesis. The activation of PLA_2 has been reported to be involved in the control of a myriad of cellular responses, from chemotaxis and platelet activation to synaptic vesicle fusion (29). Additionally, PLA_2 has been suggested to play a role in the G-protein cycle of cellular signal transduction (51). More recently, PLA_2 has been shown to participate in the mitogenic effect of cell-matrix interaction in the *Geodia* system (52). In addition, infection of Hela cells by poliovirus has been demonstrated to cause *de novo* synthesis of an endogenous inhibitor of PLA_2 (53). Thus, understanding the mechanism of regulation of this enzyme (both activation and inhibition) is of paramount importance in several areas of cell and molecular biology.

We undertook a comprehensive study of the mechanism of the observed PLA_2 inhibitory effect of UG, and of the possible relationship of this property of UG with its physiological function(s).

UG AS A PLA_2 INHIBITOR: STRUCTURAL-FUNCTIONAL RELATIONSHIPS

A study of the structural-functional relationship of UG as a PLA_2 inhibitor was necessary in order to clarify the mechanism of inhibition and possibly to identify the region(s) of UG which are involved in inhibiting this enzyme. The initial approach we chose was to compare the amino acid sequences of mature UG and human lipocortin I (54). The reason for this choice was that, besides the PLA_2 inhibitory properties, UG and lipocortin I also share the antichemotactic and antiphagocytic effect on neutrophils and monocytes. In addition, UG is a steroid-induced protein, and a similar property has been claimed for lipocortin I (54). The latter is the only lipocortin to be found in extracellular fluids, like UG. However, UG is a "true" secretory protein, with a canonical leader peptide (see ref. 29) and most,

(54). The latter is the only lipocortin to be found in extracellular fluids, like UG. However, UG is a "true" secretory protein, with a canonical leader peptide (see ref. 29) and most, if not all lipocortins are non-secretory intracellular proteins. Lipocortin I appears to be secreted, but the mechanism of secretion is still unclear (55). Using a variety of computer programs we compared the peptide sequence of 70 amino acid-long

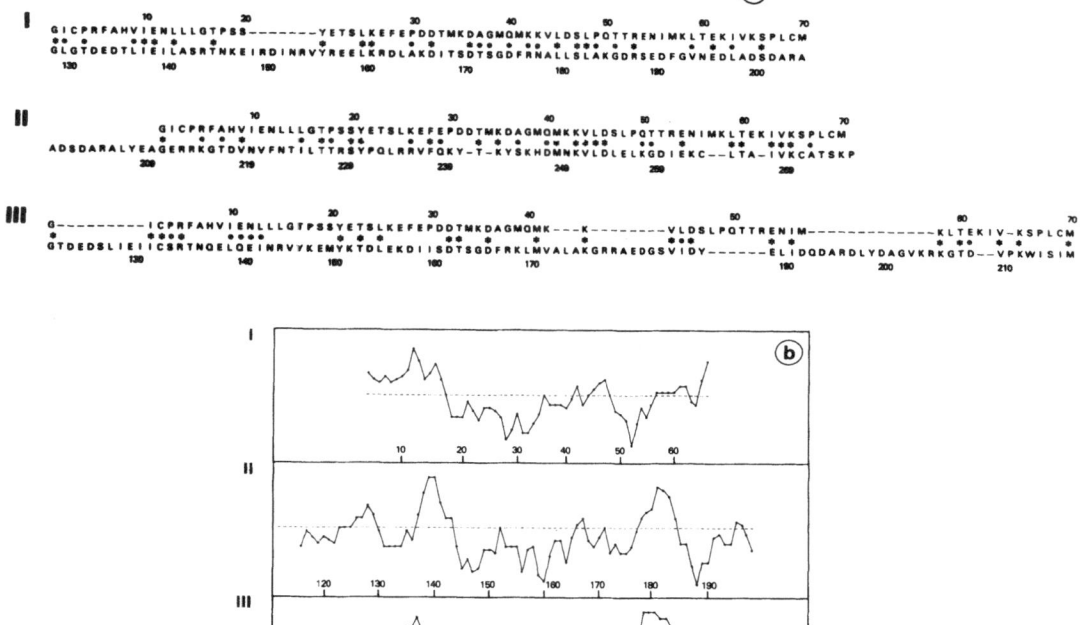

Figure 2(a). Alignment of UG with lipocortin I repeat two (1), lipocortin I repeat three (11) and lipocortin II (III), obtained using the program PRTALN (59). Identities are indicated by asterisks and conservative substitutions by dots. Conservative substitutions were added manually and are defined by pairs of residues falling into the following groups. S, T, A, G, P; N, D, E, Q; R, K, H; M, I, L, V; F, Y, W.

Figure 2(b). Hydropathy profiles of UG (I), lipocortin I repeat two (II) lipocortin I repeat three (III), lipocortin II repeat two (IV) and porcine pancreatic PLA_2, residues 24-92 (V). The profiles were obtained using a segment of seven residues. Positive peaks indicate hydrophobic regions; negative peaks indicate hydrophilic regions. Reproduced with permission from ref. 56.

mature UG monomer with those of lipocortin repeats. Figure 2 shows the alignment of mature UG monomer with human lipocortins I and II (56).

Lipocortins I and II are formed by four non-identical repeated units of about 70 amino acids each (57, 58). The program PRTALN (59) aligns UG with a region of lipocortins I and II that corresponds to the second repeat (Fig. 2a, I and III), the region of highest similarity between the two lipocortins. The total similarity (identities + conservative substitutions) is 40% in both cases. However, the highest number of identities (nineteen) is found between UG and lipocortin I repeat 3 (residues 199-275), with a total similarity of 43%. Moreover, a striking local similarity was identified between residues 40-46 of UG and residues 247-253 of lipocortin I, repeat three. In UG, this region corresponds to the C-terminal portion of α-helix three (residues 32-47) (15, 16). As it is "centered" on a relatively long region of local similarity, the alignment with repeat three is probably the most accurate. We drew four conclusions from these results. (1) There is amino-acid sequence similarity between UG and lipocortins; (2) UG is more similar to lipocortin I than to lipocortin II, although it can be aligned to the region of highest similarity between the two lipocortins; (3) compared to repeats two and three of lipocortin I, the highest density of identities and conservative substitutions is in the C-terminal half of UG (residues 33-70) and particularly in α-helix three (Fig. 3a); (4) the region of highest similarity between helix three of UG and repeat three of lipocortin I can be precisely identified as a heptapeptide spanning residues 40-46. This heptapeptide aligns with lipocortin I residues 247 - 253 (Fig. 2a).

Hydropathy profiles of UG and the corresponding regions of lipocortins are shown in Fig. 2b) (I-IV). The similarity is evident over most of the UG sequence and particularly in the region of UG between residues 37-52. Interestingly, the hydropathy profile of porcine pancreatic PLA_2 (Fig. 2b, V) shows a striking resemblance with that of the UG monomer and of the corresponding regions of lipocortins, even in the absence of significant amino acid sequence similarity. It has been independently suggested that lipocortin repeats and PLA_2 could have a similar three-dimensional organization (58). Moreover, refined crystallographic data have shown that the molecular surface of UG is strikingly similar to that of PLA_2 (16). These data may indicate that PLA_2, the UG monomer and the lipocortin repeats have a similar three-dimensional organization (57).

On the basis of our computer analyses, we synthesized oligopeptides corresponding to the C-terminal half of UG α-helix three and tested them for PLA_2-inhibitory activity. Table 3 shows the amino-acid sequences and PLA_2-inhibitory properties of synthetic peptides derived from UG and lipocortin I. Peptide one (P1), which corresponds to the nine C-terminal amino-acid residues of α-helix three, is a very potent inhibitor of PLA_2 with Peptide 2 (P2), corresponding to lipocortin I residues 246-254, is as active as P1 (Table 3). Peptide three (P3), which retains full inhibitory activity under these experimental conditions, was constructed by substituting an asparagine for a lysine residue corresponding to UG lysine 42. The removal of two amino-acid residues from the N-terminal and one residue from the C-terminal of P1 abolish the inhibitory activity (Table 3).

The "core" tetrapeptide (Lys-Val-Leu-Asp), which is common to all the active peptides, is by itself inactive as a PLA_2 inhibitor. No significant difference in PLA_2 inhibitory activity between peptides one, two and three could be detected under these experimental conditions, at peptide concentrations between 1 and 200 nM (unpublished data).

We further investigated the PLA$_2$-inhibitory properties of the peptides and their parent proteins under initial velocity conditions. Figure 3 shows that UG, P1 and P2 have comparable dose responses. However, UG had a maximal inhibitory effect at 1 nM, whereas, the peptides (P1 and P2) had a similar effect at concentrations between 4 and 8 nM. Under these conditions, both UG and the peptides had a remarkable inhibitory effect in sub-nanomolar concentrations (Fig. 3). Decreasing enzyme concentration caused a parallel decrease in the optimum inhibitory concentrations of peptides. When P1 was pre-incubated with the lipid substrate instead of the enzyme, its inhibitory effect was drastically reduced (Fig. 3). This may indicate that the inhibitory effect of P1 is exerted through an interaction with the enzyme rather than the substrate. This possibility is further supported by the observation that P1, at concentrations ranging from 1 nM to 10 μM, had no inhibitory effect on *Bacillus cereus* phospholipase C, under identical experimental conditions to those used for PLA$_2$ (data not shown). Figure 3 clearly shows that P3 under these conditions is less active than P1 and P2. This difference tends to decrease when the concentration of P3 is increased.

Table 3. Amino-acid sequences of synthetic peptides and their PLA$_2$ inhibitory activity.

Peptide	Sequence	residue number		PLA$_2$ inhibition
1	MQMKKVLDS	UG	39-47	++++
2	HDMNKVLDL	LC-I	246-254	++++
3	MQMNKVLDS	--		++++
4	KVLD	--		--
5	MKKVLD	--		--
6	MNKVLD	--		--
7	GICPRFAHVI	UG	1-10	--

PLA$_2$ was assayed according as described (56). Briefly, the reaction mixture contained 100 mM Tris HCl, pH 8.0, 100 mM NaCl, 1 mM CaCl$_2$, 1 mM sodium deoxycholate, 10 μM 1-stearoyl, 2-[1-^{14}C] arachidonyl phosphatidylcholine (58 mCi mmol^{-1}; Amersham) and 25 ng (approximately 36 nM) porcine pancreatic PLA$_2$ (Boehringer, 700 U mg^{-1}) in a total volume of 50 μl. It should be noted that this assay using 36 nM PLA$_2$ yields highly variable results. More consistent results are obtained when initial velocity conditions are used as shown in Figure 5. Reproduced with permission from ref. 56.

Recombinant lipocortin I inhibits PLA$_2$ in this system with an overall dose-response curve similar to that of P3. Also, lipocortin I seems to be less active than the lipocortin-I-derived peptide (P2). However, this difference may be artifactual, as it has been reported that purified preparations of recombinant lipocortin I are mixtures of several disulfide-bonded forms and are contaminated with bacterial proteins (60).

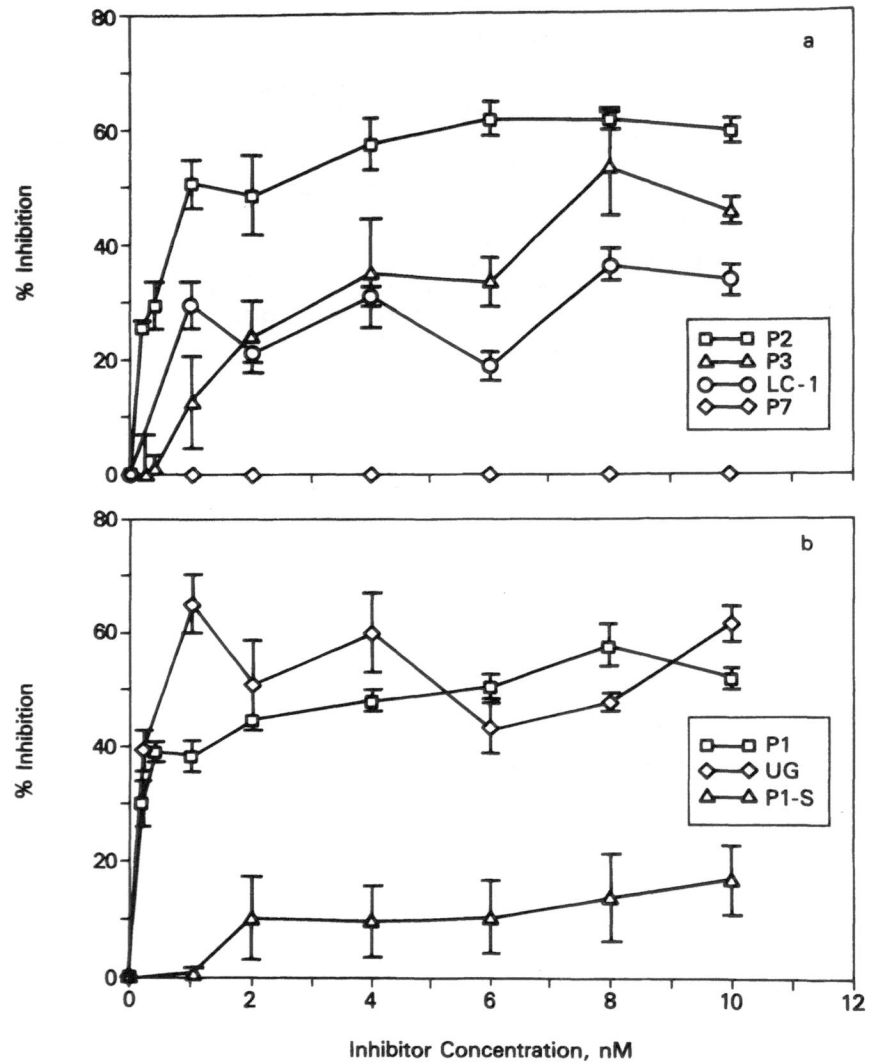

Figure 3. Inhibition of porcine pancreatic PLA$_2$ by UG, lipocortin I and synthetic oligopeptides. Each point represents the mean of at least four separate determinations ± standard deviation. a, *open squares*, peptide 2; *open triangles*, peptide 3; *open circles*, lipocortin-I; *open parallelograms*, peptide 7. b, *open squares*, peptide 1; *open parallelograms*, UG; *open triangles*, peptide 1, pre-incubated with the lipid substrate. Non-specific proteins (horse heart myoglobin and chicken egg lysozyme) were tested for PLA$_2$-inhibitory activity at concentrations between 1 nM and 10 μM and were found to be inactive. In some experiments, phospholipase C from *Bacillus cereus* (0.01 U, Boehringer, grade I) was substituted for PLA$_2$. In these experiments the reaction time was 4 min. Reproduced with permission from ref. 56.

Several conclusions can be drawn from these experiments. The first two and the last residue of P1 can be replaced by other amino acids, but not eliminated, without loss of activity. This may indicate that the length of the peptide is critical, possibly for conformational reasons; the minimum peptide length that can form two complete turns of a α-helix is eight residues. The lysine residue in P1 corresponding to lysine 42 of UG is not indispensable for the inhibitory activity. This was expected, because in UG, lysine 42 is not exposed to the solvent, as it is H-bonded to the main-chain carbonyl of glycine 15 (see ref. 15). However, hybrid peptide three (P3) seems to be less active than both parent peptides one and two, particularly at low concentrations, possibly because there is a decreased tendency to assume the active conformation in solution. Recombinant lipocortin I inhibits PLA_2 in our system although with a lower activity than UG or P1, in the absence of phosphatidylserine and in the same concentration range as the enzyme. Both the presence of phosphatidylserine and a large molar excess of lipocortin over PLA_2 have been reported to be necessary for the inhibition of PLA_2 by lipocortin in assays using autoclaved *E. coli* cells or extracted *E. coli* phospholipids as enzyme substrates (61, 62). The similarity in the PLA_2 inhibitory properties of P1 and purified UG supports the hypothesis that P1 corresponds to the active site, or at least one of the active sites, responsible for the PLA_2-inhibitory activity of UG. However, these data only gave us indirect information on the mechanism of the inhibition observed. Additional work was needed to address this problem.

A further investigation on the possible mechanism of action of UG and the antiflammin peptides as PLA_2 inhibitors had as a necessary prerequisite a systematic characterization of the assay used. Although in most cases mixed micellar substrates have been used with non-charged detergents such as Triton-X100, we felt that for the purpose of studying potential PLA_2 inhibitors with the pancreatic enzyme, the use of deoxycholate had some advantages. First, among the relatively simple substrate systems that have been used, mixed micelles of a bile acid and phosphatidylcholine are the most similar to the physiological substrate of pancreatic PLA_2. Second, it is known that the porcine pancreatic enzyme has a very low affinity for mixed micelles containing Triton (63), as well as for phospholipid bilayers and vesicles, and that it has a clear preference for negatively charged interfaces. With most substrates except those containing a negatively charged detergent the binding of PLA_2 to the interface is relatively slow and it is often the limiting step. Under conditions which result in a slow binding of PLA_2 to the substrate, it is easier to observe artifactual inhibition by substances that simply bind to phospholipids or to lipid-water interfaces. On the other hand, conditions resulting in instantaneous interfacial activation of PLA_2, in the presence of a molar excess of a detergent, should minimize non-specific effects due to lipid-binding, particularly if the putative inhibitors are used at very low molar concentration, in the same range as that of PLA_2. Thus, we thoroughly characterized and the conditions of the PLA_2 inhibition assay using deoxycholate-phosphatidylcholine mixed micelles. Optimum values were found for the following variables: pH, temperature, Ca^{++} concentration, enzyme concentration, deoxycholate concentration and time of reaction. In addition, the timing and technique of substrate preparation and enzyme dilution were carefully standardized. The results of these experiments are the subject of a manuscript which is currently in preparation (Facchiano et al.,

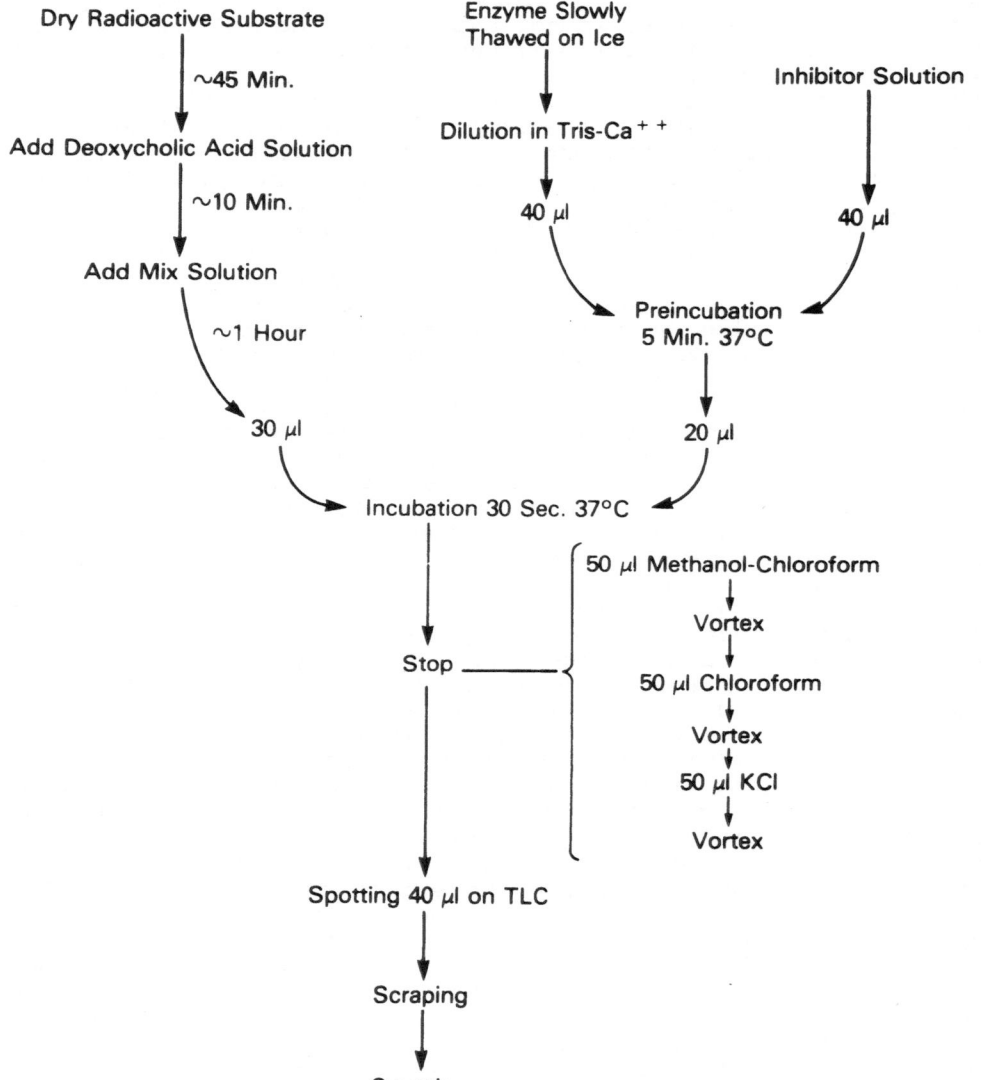

Figure 4. Schematic representation of the procedure currently used in our laboratory for the assay of PLA_2.

in preparation). Several minor details turned out to be very important for the reproducibility of the assay, such as the use of HPLC grade water for the solutions. An additional problem that can be encountered in testing the activity of PLA_2-inhibitory proteins and peptides is that the commercially available pancreatic PLA_2 enzyme preparations are contaminated with proteases in variable amounts. Thus, it is very difficult to obtain reproducible results if the protease activity in these enzyme preparations is not inhibited or eliminated. We routinely dialyze commercial PLA_2 preparations against several changes of 10 mM Tris-HCl, pH 8.0 and include 0.25 mM PMSF in the first change. The procedure which is currently used in our laboratory for the assay of porcine pancreatic PLA_2 is schematically summarized in Fig. 4.

Under these conditions, we investigated the dose-response curves of UG and the peptides as PLA_2 inhibitors. The results obtained were very similar to those originally reported (56), although the average percent inhibition obtained were slightly lower. The mechanism of inhibition was initially investigated by Michaelis-Menten kinetic studies. We obtained hyperbola-shaped velocity vs. substrate concentration curves by varying the concentration of phosphatidylcholine between 10 and 90 μM and keeping the concentration of deoxycholate constant at 1 mM. Under these conditions, the molar ratio of detergent to phosphatidylcholine is always >10. Thus, we assumed that the "quality of the interface" did not dramatically change over this range of phosphatidylcholine concentrations. In addition, the total molarity of detergent + substrate, which is roughly proportional to the total area of the interface (64) changes from 1.029 to 1.090 mM, while the molar fraction of phospholipid/phospholipid + deoxycholate increases from 2.8×10^{-3} to 8.3×10^{-2}. Thus, for practical purposes we assumed the total area of the interface to be constant and the molar concentration of phosphatidylcholine to be approximately proportional to the interfacial concentration of phosphatidylcholine in moles x surface area unit. Under these assumptions we obtained straight double reciprocal plots. The data were analyzed by means of ENZYME (65), a program based on an iterated weighted fit algorithm which is specifically designed to evaluate different models of enzyme inhibition. Based on the results of several experiments, we found that UG and antiflammins 1 and 2 inhibit PLA_2 in a way which is apparently independent of the substrate concentration. When seven different inhibition models were evaluated with ENZYME, the program estimated that the most likely mechanism of inhibition was pure noncompetitive inhibition. This observation might support the possibility that UG and the peptides exert their inhibitory effect in this system by interacting with the enzyme rather than with the substrate. This hypothesis was further supported by several other observations (Facchiano et al., manuscript in preparation): i) UG and both peptides affect the fluorescence spectrum of PLA_2 in the absence of substrate; ii) the increase in PLA_2 fluorescence observed upon addition of mixed micelles of deoxycholate and phosphatidylcholine is reduced in the presence of UG and the peptides; iii) UG can be crosslinked to PLA_2 in solution by glutaraldehyde iv) the formation of glutaraldehyde-crosslinked complexes of UG and PLA_2 in solution can be prevented by peptides 1 and 2 and v) ^{125}I-UG does not bind bacterial membranes in the presence or absence of Ca^{++}. The results of these experiments will be reported in detail elsewhere (Facchiano et al., manuscript in preparation). On the basis of these data, we formulated a working hypothesis which is schematically described in Fig. 5. In this model, UG and the peptides interfere with the process of interfacial activation of PLA$_2$, possibly by interfering with the PLA$_2$ dimerization step which is associated with the process (66-70). However, several questions remain to be answered and this model has to be presently considered tentative. It is unclear, for example, whether the interaction between PLA$_2$ and UG affects the binding of the enzyme to the interface or the self-association of lipid-bound PLA$_2$. More detailed information might be obtained if the conditions are found for a co-crystallization of PLA$_2$ with UG or with the peptides. Work is currently in progress on this problem. Our data do not exclude the possibility that UG or the peptides might also interact with lipids under different conditions, such as in eukaryotic biological membranes.

Although the oligopeptides described above are potent inhibitors of PLA$_2$ activity, the major drawback seems to be their instability in solution and their short shelf life. We do not understand completely the reasons for this instability. However, there are several theoretical (or as yet unproven) possibilities: (1) aggregation and subsequent inactivation of the peptides in a concentrated aqueous solution, (2) destruction in solution (or by hydration of lyophilized peptides) of internal salt bridges which stabilize the structure of these peptides and (3) sulfoxide formation of methionine residues which may be essential for PLA$_2$ inhibitory activity. The latter possibility is supported by data from Camussi et al. (71, see also the chapter by Camussi et al. in this volume), which show that these peptides, particularly P1,

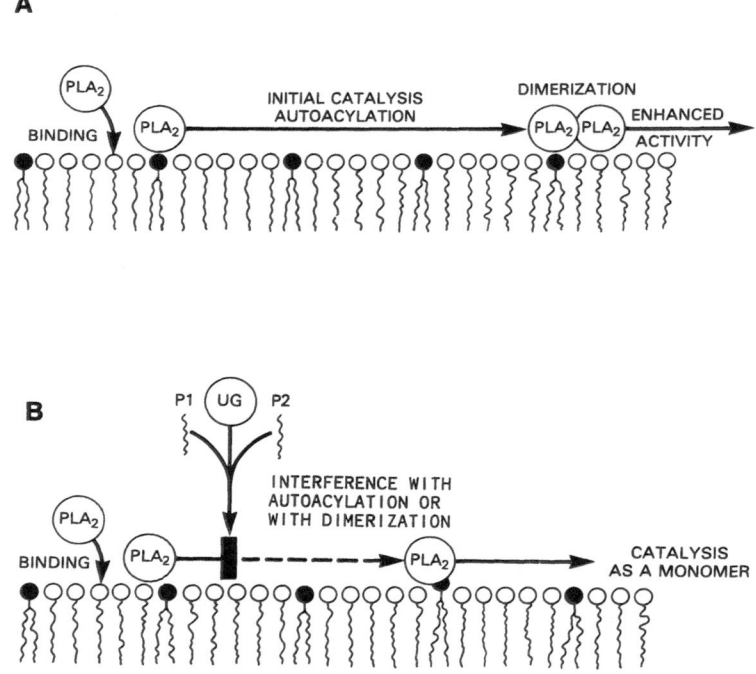

Figure 5. Schematic representation of our current working hypothesis on the mechanism of action of UG and derived peptides as PLA$_2$ inhibitors. (A) interfacial activation of PLA$_2$; (B) putative mechanism of action of UG and derived peptides.

can be easily inactivated by oxidation and are protected from inactivation by dithiothreitol and mercaptoethanol. Because of their instability the biological activity of the peptides (i.e., antiinflammatory action) is somewhat variable and unless the experiments with these peptides are very carefully controlled, inconsistent results may be obtained.

BIOLOGICAL PROPERTIES OF UG-DERIVED PEPTIDES

Inhibitors of PLA_2 have been considered for a long time as potentially ideal antiinflammatory agents, since they would ensure a dual inhibition of both the cyclooxygenase and lipooxygenase pathways of the arachidonate cascade. Thus, we tested the PLA_2-inhibitory synthetic peptides for antiinflammatory activity *in vivo* and found that both UG-derived P1 and lipocortin I-derived P2 have very potent antiinflammatory effect on the carrageenan-induced rat footpad oedema (56). Table 3 shows the antiinflammatory effects of P1 and P2 compared to those of known antiinflammatory agents. Both P1 and P2, when injected locally, immediately after carrageenan administration, inhibited the formation of inflammatory edema over a wide range of doses. The dose-response curves of P1 and P2 do not differ significantly, except perhaps in the range between 0.2 and 2 mg kg^{-1}. At 2 mg kg^{-1} both peptides caused a virtually complete suppression of the inflammatory response. Both parent proteins, UG and lipocortin I, when used at 100 μg kg^{-1} had anti-inflammatory activity comparable to that of a 10-fold higher dose of indomethacin. The effect of both P1 and P2, at the highest dose used, is drastically reduced by a concomitant local administration of 10 μg of arachidonic acid (Table 3). This dosage of arachidonic acid had no inflammatory effect when administered alone and did not significantly increase the inflammatory effect of carrageenan when administered with it. These observations support the hypothesis that an important *in vivo* mechanism of these peptides may be the inhibition of arachidonate release from the cell membrane by PLA_2. However, our data do not rule out a minor contribution from other unknown mechanism(s) at the highest dosages of P1 and P2. Preliminary data indicate that P1 and P2 also have potent anti-inflammatory effects when administered by intraperitoneal and intravenous routes. Because of their potent antiinflammatory effect, we have denominated these peptides "antiflammins" (56).

In Figure 6 the morphological and histological appearance of rat foot pads before and after injection with carrageenan and reduction of edema by P1 is shown. It is quite evident that P1 is very effective in reducing carrageenan-induced rat footpad edema. The histology of the specimens shows that migration of inflammatory cells into the site where carrageenan was injected is dramatically reduced as compared to saline injected controls. UG is known to have antichemotactic effect on neutrophils and monocytes (42-44). Thus, we speculated that UG derived P1 might have similar effects on these cells (56). Recently, compelling evidence has been presented by Dr. Baglioni and his coworkers (see ref. 71 and the chapter by Camussi et al. in this volume) demonstrating that P1 and P2 inhibit neutrophil PLA_2 and the generation of PAF from neutrophils, as well as neutrophil chemotaxis and intradermal Arthus reaction in rats. Collaborative efforts between our laboratory and other groups at the NIH are currently ongoing to characterize the antiinflammatory activity of peptides 1 and 2 in various disease models. Preliminary results indicate that both peptides have

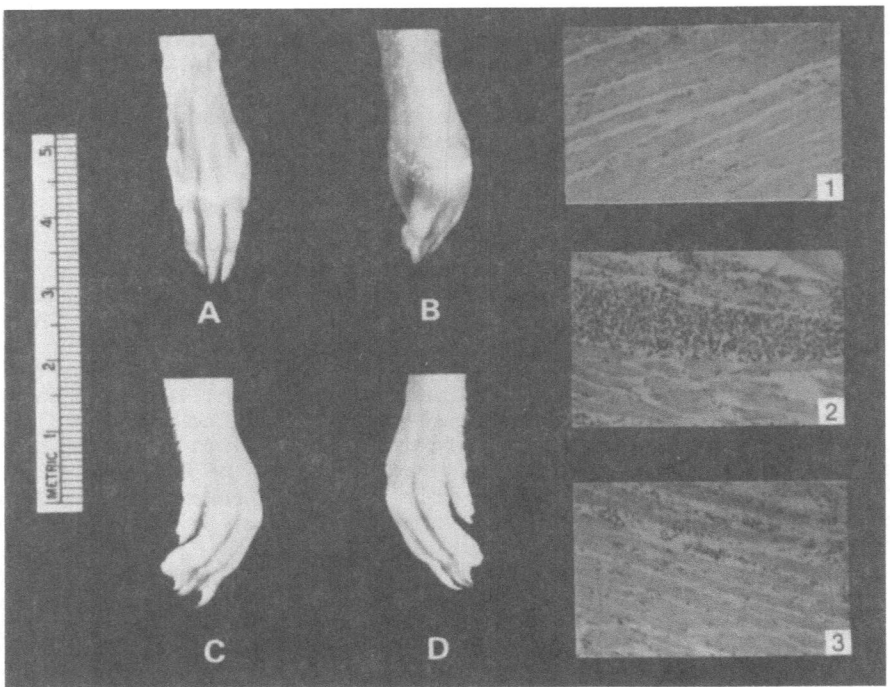

Figure 6. Morphology and histology of rat foot pads before and after peptide 1 treatment following carrageenan administration. A = Control; B = Carrageenan injected (normal saline treatment); C & D = Carrageenan treatment followed by intravenous peptide 1 treatment (100 μg/kg). 1, 2 & 3 are histology of footpad A, B and C respectively. Note significantly lower number of inflammatory cells in 3 as compared to 2.

therapeutic effects on lipopolysaccharide-induced uveitis in rats (72). In addition, Ialenti et al. (73) have recently shown that antiflammin peptides are potent inhibitors of carrageenan-induced, but not of dextran-induced inflammation. Since dextran-induced inflammation is known not be mediated by PLA_2 activation, the authors suggest that this may be a further indirect evidence that the mechanism of action of antiflammins *in vivo* involves inhibition of PLA_2(s).

Another biological property of UG is its ability to inhibit thrombin-induced platelet aggregation (45). Thus, it was of interest to test whether P1 and P2 also have any effects on platelet aggregation. Vostal et al. (74) have recently shown that both peptides 1 and 2 inhibit thrombin-induced aggregation of platelets. P1 also inhibits serotonin secretion induced by thrombin and ADP. Also the "core" tetrapeptide KVLD (peptide 4), which is common to both peptides 1 and 2 but does not inhibit PLA_2, inhibits thrombin- and ADP-induced platelet aggregation but not serotonin secretion. The effect of peptides 1, 2 and 4 on thrombin-induced platelet aggregation is due to an inhibitory effect exerted by all three peptides on the esterolytic activity of thrombin.

The effect of peptide 1 on ADP-induced aggregation and secretion is virtually abolished in the presence of 10^{-4} M arachidonate and it is presumably due to inhibition of PLA_2. The effect of peptide 4 on ADP-induced aggregation is due to inhibition of fibrinogen binding to platelets, possibly by a mechanism similar to that of RGDS peptide. P1, although it contains a KVLD sequence, does not inhibit fibrinogen binding to platelets. When peptides 1 and 4 were used together, they had an additive inhibitory effect on ADP-induced aggregation and serotonin secretion. These observations lend further support to the hypothesis that an active site of UG which is responsible for some of the biological activities of the protein corresponds to the nonapeptide identified by us in α-helix 3 of UG.

Table 4. anti-inflammatory effect of peptides 1 and 2 on carrageenan-induced rat paw oedema

Treatment	Number of animals	% Inhibition (mean ± standard deviation)
P1, 2 mg kg^{-1}	7	82.6 ± 5 (P<0.01)
P1, 200 μg kg^{-1}	9	57.5 ± 7 (P<0.01)
P1, 20 μg kg^{-1}	9	36.0 ± 8 (P<0.01)
P1, 2 μg kg^{-1}	9	31.5 ± 9 (P<0.01)
P1, 0.2 μg kg^{-1}	9	24.0 ± 5 (P<0.01)
P1, 0.02 μg kg^{-1}	9	n.s. (P>0.05)
P2, 2 mg kg^{-1}	8	96.2 ± 2 (P<0.01)
P2, 200 μg kg^{-1}	8	30.5 ± 7 (P<0.01)
P2, 20 μg kg^{-1}	8	36.3 ± 7 (P<0.01)
P2, 2 μg kg^{-1}	8	23.3 ± 3 (P<0.01)
P2, 0.2 μg kg^{-1}	8	20.0 ± 4 (P<0.01)
P2, 0.02 μg kg^{-1}	8	12.2 ± 3 (P<0.05)
DEX,1 μg kg^{-1}	5	79.9 ± 6 (P<0.01)
IND, 1 mg kg^{-1}	5	26.1 ± 4 (P<0.01)
P1, 2 mg kg^{-1}+AA	4	21.1 ± 5 (P<0.01)
P2, 2 mg kg^{-1}+AA	4	20.2 ± 1 (P<0.01)
UG, 100 μg kg^{-1}	4	35.5 ± 7 (P<0.01)
LC-1, 100 μg kg^{-1}	4	24.5 ± 1 (P<0.01)
P7, 2 mg kg^{-1}	4	n.s. (P>0.05)
P7, 10 μg kg^{-1}	4	n.s. (P>0.05)
BSA, 2 mg kg^{-1}	4	n.s. (P>0.05)
BSA, 100 μg kg^{-1}	4	n.s. (P>0.05)
LSZ, 2 mg kg^{-1}	4	n.s. (P>0.05)
LSZ, 100 μg kg^{-1}	4	n.s. (p>0.05)

P1, peptide 1; P2, DEX, dexamethasone; IND, indomethacin; AA, arachidonic acid; UG, uteroglobin; LC-1, lipocortin I; P7, peptide 7; BSA, bovine serum albumin; LSZ chicken egg lysozyme; n.s. not significant. Animals were injected in the sub-planter space with 1.0 mg lambda-carrageenan (Sigma, type IV) in 0.1 ml sterile saline. Inhibitors, control substances or vehicle alone were injected about 30s after carrageenan, in a volume of 0.1 ml, to avoid non-specific interactions *in vitro* between carrageenan and inhibitors. Controls which received saline alone and saline plus vehicle were included. Peptides were dissolved in sterile 10mM Tris, pH 8, because this yielded more consistent and reproducible dose responses compared to peptides dissolved in saline. Dorsoplantar paw thickness was measured with a vernier caliper immediately before the carrageenan injection and 4 h after treatment. When AA was used, it was administered 10 μg per paw. At this dosage, AA did not cause appreciable paw swelling when administered alone, nor did it increase the swelling caused by carrageenan when administered with it (N, four animals per group). Inhibitory effects were assessed by comparing the dorsoplantar paw thickness of inhibitor-treated groups to that of vehicle-treated groups. The results were analyzed by a one-tailed Student's t test for groups of unpaired observations and by one-way ANOVA (56) and differences were considered statistically significant when P values were < 0.05. Reproduced with permission from ref. 56.

CONCLUSIONS

The evolutionary origin of PLA$_2$, inhibitory proteins is unknown. It has been suggested that lipocortins and calelectrins are derived from a single repeat protein by gene duplication (75). However, we cannot deduce from our data whether or not the similarities between UG and lipocortin I arose from divergent evolution. It is entirely possible that the structural similarity between the UG monomer and the lipocortin/ calelectrin "repeat" unit may be the result of convergent evolution. One fact that may support this hypothesis is the absence in UG of the "endonexin fold", a motif which occurs in all Ca^{++}-phospholipid binding proteins of the lipocortin family (76). It is noteworthy that lipocortin I repeat 3, which is the most similar to UG, has a defective endonexin motif (58). The absence of the endonexin consensus, in fact, excludes UG by definition from the lipocortin/calelectrin/ endonexin family.

One important point that needs to be mentioned is that, despite the structural similarity, there are important differences between PLA$_2$ inhibition by UG and lipocortin I. Notably, UG does not seem to require lipid binding for its inhibitory activity in the mixed-micelle assay system we used and does not bind *E. coli* membranes. On the contrary, lipocortins I and II have been shown to bind acidic lipids in *E. coli* membranes and in vesicles in the presence of Ca^{++} (62, 63). This phenomenon has been suggested to be the basis for the PLA$_2$ inhibitory activity of these proteins *in vitro* (61, 62). The situation is more complex with phosphatidylcholine sonicated vesicles, in which case surfactant effects or competitive effects have been suggested to cause the PLA$_2$ inhibition observed with lipocortin I and lipocortin II (calpactin I) (77). It is entirely possible that the inhibition we observe with UG and antiflammin peptides in our mixed micellar assay is different from, and possibly unrelated to, these lipid-binding phenomena. Besides the different physical state of the substrate and the presence of a large excess of detergent, the PLA$_2$ reaction is virtually instantaneous in our system (about 0.7% of the substrate is hydrolyzed in 30 sec.) while reaction times of several minutes to one hour are used with non-micellar substrates (61, 62, 77). In general, it is still unclear whether any of these *in vitro* observations can be safely extrapolated to biologically relevant conditions. These conditions include Ca^{++} concentrations in the nano- to micromolar range, cell membranes as the substrate, presence of cytoskeletal structures etc. One interesting point, for example, is whether the weak interaction detected in solution between some proteins of the lipocortin family and porcine pancreatic PLA$_2$ (K$_D$ $\approx$ 10^{-5} M for lipocortin I) (78) has any biological significance. This phenomenon has been suggested to be mechanistically irrelevant *in vitro* (78). It might be pointed out that such interactions could be important under conditions in which both interacting proteins are membrane-bound. In this case, the concentration of the two proteins in moles per surface area unit would be far higher than the molar concentrations. In addition, membrane binding itself could influence the affinity of the protein-protein interaction through effect(s) on protein conformation. For example, while porcine pancreatic PLA$_2$ shows virtually no tendency to self-associate in solution, dimerization of lipid-bound enzyme, after autocatalyzed self-acylation of Lys 56, is suggested to be the basis for interfacial activation of this PLA$_2$ (66-70).

Clearly, more work will be needed to elucidate the

physiological role of the various putative PLA$_2$ inhibitory proteins described so far. Future investigations should take into account some potential problems. First, it is unclear to what extent the observation that a protein inhibits porcine pancreatic or snake venom PLA$_2$ acting on autoclaved bacterial cells can be interpreted as an indication that the function of this protein *in vivo* is to inhibit PLA$_2$. This conclusion is to be considered tentative, particularly if assays based on autoclaved *E. coli* cells are used under unphysiological experimental conditions (for example, low incubation temperature). We feel that only if inhibitory properties on cellular PLA$_2$ activity(ies) are demonstrated, and if there is evidence of biological activity related to PLA$_2$ inhibition in cellular systems *in vitro* or in animal models, then the hypothesis of a physiological PLA$_2$ inhibitory role of a substance should be proposed. In this respect, it should be noted. that UG inhibits mouse 264.7 macrophage PLA$_2$ activities with cell membranes as the substrate (see above). Lipocortin I also inhibits alveolar macrophage PLA$_2$ (79) and prevents the release of eicosanoids from isolated guinea pig lungs (60). Both proteins are antiinflammatory *in vivo* (56, 80). The *in vivo* effects of these two proteins, as well as the effects of UG on intact neutrophils and platelets, can be reproduced with the antiflammin peptides. To our knowledge, evidence for similar biological activity has not been obtained for other putative PLA$_2$ inhibitors of the lipocortin family. This consideration leads us to think of another potential problem: labeling a family of proteins with a common name, based on sequence similarity among the members of the family, tends to generate the unwarranted assumption that all the proteins in a family have the same biological properties and physiological function. One important difference that is often overlooked between the two most characterized lipocortins (I and II) is that lipocortin II is actually a subunit of a heterotetrameric protein (calpactin) (81) which interacts with actin and phospholipids, and which can aggregate exocytotic vesicles at Ca^{++} concentrations almost two orders of magnitude lower than lipocortin I (81). Isolated lipocortin II, in the absence of the 11 kDa subunit which is part of calpactin, does not bind actin. On the contrary, lipocortin I is a monomeric protein, which is secreted in the extracellular fluid by an as yet unknown mechanism. Original reports of the isolation of lipocortin I described it as a glycoprotein containing approximately 10 % carbohydrates (82). It is unknown whether differences in post-translational modifications can influence the cellular localization or the function of natural lipocortin I. At least 6 different "lipocortins" have been described so far (83), and a number of related Ca^{++}-phospholipid binding proteins known as endonexins, synexin, calelectrins etc. are suggested to be part of the same evolutionary family (75, 76). This family in turn is part of a larger group of 23 proteins originally described as chromobindins (84), which also includes phosphatidylinositol-specific phospholipase C, protein-kinase C, calmodulin-like proteins, etc. The "chromobindins" can all bind to chromaffin granules in the presence of Ca^{++} and are candidates for the role of "fusion proteins" inducing the fusion of exocytic granules with the cell membrane. While it seems reasonable to think that these proteins are membrane-directed, it is unlikely that they all have the same function. It is possible that the structural similarities between proteins in this group are due essentially to the fact that all of them interact with cell membranes in the presence of Ca^{++}. Some of these proteins have been shown to inhibit pancreatic PLA$_2$ activity in assays based on autoclaved bacterial cells (83). However, as we mentioned above, it is unclear whether the information obtained with such procedures has any physiological relevance. An intriguing observation is that vesicles aggregated by calpactin can actually fuse if *cis*-unsaturated fatty acids, including arachidonic acid, are added to the system (81). Since such fatty acids can be products of a PLA$_2$-catalyzed reaction, it is not unreasonable to speculate that a membrane PLA$_2$ may participate in the regulation of vesicle fusion, and that a vesicle bound "fusion" protein may well be also an inhibitor of PLA$_2$. Such an inhibitory activity may prevent an activated PLA$_2$ from causing premature lysis of the secretory granule membrane, or from damaging the integrity of the cell membrane. Thus, it is also possible that a fusion-regulatory activity and a PLA$_2$ inhibitory activity may not be mutually exclusive and may coexist in the same protein(s). Furthermore, actin microfilaments, to which some of the lipocortins bind, have been also shown to participate in the regulation of PLA$_2$ activity in human platelets (85).

In conclusion, our data indicate that UG and lipocortin I have a region of high local sequence similarity. Synthetic peptides generated on the basis of this region were found to inhibit PLA$_2$ *in vitro* and to have a variety of biological activities that mimic those of the entire proteins from which they are derived (56, 71-74). Whether these observations reflect the physiological role(s) of these proteins remains to be clarified. Our data do not allow to conclude that there is evolutionary relationship between UG and lipocortins, and there is some indication that the structural similarity between them might be the result of convergent evolution. Also, our observations do not support any conclusions as to the possible biological role(s) of the other proteins of the lipocortin/calelectrin/endonexin family.

The antiflammin peptides can provide a useful model to study the role of PLA$_2$ in inflammation, and to develop more stable antiinflammatory agents of therapeutic importance. In addition, they can provide useful information on the structure-function relationship of the proteins from which they are derived, particularly in the case of UG, which is biochemically and biologically a better characterized protein.

A fundamental question in the biology of inflammation is how the local inflammatory response to injurious agents coming in contact with wet epithelia is modulated. The mucosae of organs such as the tracheobronchial, gastrointestinal and genitourinary tracts come into contact with myriads of foreign and potentially harmful agents, of microbial and non-microbial origin. These agents can damage the integrity of tissues by a direct effect, or if they are immunogenic, by eliciting an immune response which in turn can cause inflammation, smooth muscle spasm etc. Male gametes and embryos carrying paternal histocompatibility antigens can be considered a special case of extraneous immunogenic agent with which the female genital tract comes in contact. The constant presence of defensive mechanisms is necessary to scavenge inert foreign agents and to prevent microbial invasion of the mucosae. On the other hand, inflammatory responses in mucosal organs must be strictly regulated under normal circumstances, in order to preserve vital physiological functions, such as respiratory exchanges, absorption of nutrients, fecundation and implantation. The ill consequences of uncontrolled inflammation are evident in a large number of human diseases. For the past ten years we have been suggesting that small molecular weight proteins such as UG in the rabbit may be responsible for the modulation of inflammatory responses in the mucosal organs in mammals. In mammalian species other than the rabbit, several investigators have demonstrated that proteins similar to UG also serve as modulators of immune and inflammatory phenomena (86-88). These observations, together with our data, seem to indicate that UG and UG-like proteins may participate in maintaining the integrity of the wet epithelia of various organ systems, including those of the reproductive tracts of mammals, thereby protecting the function of these organs against damage resulting from uncontrolled inflammatory response(s).

REFERENCES

1. Krishnan, R.S., and Daniel, Jr. J.C. 1967 Science 158:490.
2. Beier, H.M. 1969 Biochem Biophys. Acta 160:289.
3. Arthur, A.T., and Daniel, Jr. J.C. 1972. Fertil. Steril. 23:115.
4. Kay, E., and Feigelson, M. 1974. Biochem. Biophys. Acta. 271:436.
5. Petzoldt, U., Dames, W., Grottschweski, G.H.M, and Neuhoff, V. 1972. Cytobiology 5: 272.
6. Beier, H.M., Bohn, H., and Muller, W. 1975 Cell Tissue Res. 165:1.
7. Kirchner, C. 1976. Cell Tissue Res. 170:415.
8. Kirchner, C., and Schroer, H.G. 1976. J. Reprod. Fertil. 47: 325.
9. Noske, I.G., Feigelson, M. 1976 Biol. Reprod. 15: 704.
10. Kikukawa, T., and Mukherjee, A.B. 1989. J. Cell Mol. Endocrinol. 62: 177.
11. Nieto, A., Ponstingl. H., and Beato, M. 1977. Arch. Biochem. Biophys. 180:82.
12. Ponstingl, H., Nieto, A, and Beato, M. 1978. Biochemistry 17: 3908.
13. Popp, R.A., Foresman, K.R., Wise, L.D., and Daniel, Jr., J.C. 1978. Proc. Natl. Acad. Sci. U.S.A. 75:5516.
14. Atger, M., Mercier, J.C., Haze, G., Fridlansky, F., and Milgrom, E. 1979. Biochem. J. 177:985.
15. Mornon, J.P., Fridlansky, F., Bally, R., and Milgrom, E. 1980. J. Mol. Biol. 127:415.
16. Morize, I., Surcouf, E., Vaney, M.C., Epelboin, Y., Buehner, M., Fridlansky, R., Milgrom E., and Mornon, J.P., 1987. J. Mol. Biol. 194:725.
17. Aumuller, G., Seitz, J., Heyns, W., and Kirchner, C. 1985. Histochemistry 83:413.
18. Warembourg, M., Tranchant, O., Atger, M., and Milgrom, E. 1986. Endocrinology 119: 1632.
19. Mayol, R.F., and Longenecker, D.E. 1974. Endocrinology 95: 1534.
20. Shirai, E., Izuka, R., and Notake, Y. 1972. Fertil. Steril. 23:522.
21. Daniel, Jr. J.C. 1971. Adv. Biosci. 6:191.
22. Daniel, Jr. J.C. 1972. Fertil. Steril. 23:522.
23. Wolf, D.P., and Mastroianni, Jr. L. 1975. Fertil. Steril. 26:240.
24. Voss, H.G., and Beato, M. 1977. Fertil. Steril. 28: 972.
25. Cowan, B.C., North, D.H., Whitworth, N.S., Fujita, R., Schumacher, U.K., and Mukherjee, A.B. 1986. Fertil. Steril. 45:820.

26. Kikukawa, T., Cowan, B.C., Tejada, R.I., and Mukherjee, A.B. 1988. J. Clin. Endocrinol. Metab. 67:315.
27. Dhannireddy, R., Fujita, R., and Mukherjee, A.B. 1988 Biochem Biophys. Res. Commun. 152:1447.
28. Manyak, M., Kikukawa, T., and Mukherjee, A.B. 1988, J. Urol. 140:176.
29. Miele, L., Cordella-Miele, E., and Mukherjee, A.B. 1987. Endocrine Rev. 8:474.
30. Challis, J.R.G., Davies, I.J., and Ryan, K.J. 1973. Endocrinol. 93:971.
31. Janne, O., Isomaa, V.V., Torkkeli, T.K., Isotalo, H.E., and Kopu, H.T. 1983. In: Wayne Bardin, C., Milgrom, E., and Mauvais-Jarvis, P. (eds) Progesterone and Progestins. Raven Press New York, p 33.
32. Chilton, B. S., Mani, S. K. and Bullock, D. W. 1988. Mol. Endocrinol. 2:1169.
33. Goswami, A., and Feigelson, M. 1974. Endocrinology 95: 669.
34. Bochskanl, R., and Kirchner, C. 1981. Wilh. Roux Arch. Dev. Biol. 190:127.
35. Mukherjee, A. B., Cordella-Miele, E., Kikukawa, T., and Miele, L. 1988. In: Zappia, V., Galletti, P., Porta R., and Wold, F. (eds) Advances in Post-translational Modifications of Proteins and Ageing. Plenum PUblishing, New York, pp. 135-152.
36. Mukherjee, A.B., Laki, K., and Agrawal, A.K. 1980. Med. Hypoth. 6:1043.
37. Siiteri, P.K., Febris, F., Clemens, L,E., Chang, R.J., Gondos, B., and Stites, D. 1977. Ann. N.Y. Acad. Sci. 286:384.
38. Mukherjee, D.C., Ulane, R.E., Manjunath, R., and Mukherjee, A.B. 1983. Science 219:989.
39. Mukherjee, A.B., Ulane, R.E., and Agrawal, A.K. 1982. Amer. J. Reprod. Immunol. 2:135.
40. Mukherjee, A.B., Cunningham, D., Agrawal, A.K., and Manjunath, R. 1982. Ann. N.Y. Acad. Sci. 392:401.
41. Manjunath, R., Chung, S.I., and Mukherjee, A.B. 1984. Biochem. Biophys. Res. Commun. 121:400.
42. Schiffman, E., Geetha, V., Pencev, D., Warabi, H., Mato, J., Hirata, F., Brownstein, M., Manjunath, R., Mukherjee, A.B., Liotta, L., and Terranova, V.P. 1983. In: Keller, H., T i l l E.D. (eds) Agents and Actions Supplements. Birkhauser Verlag, Basel, Vol. 12:106.
43. Schiffman, E., Geetha, V., Pencev. D., Mato, J., Garcia-Castro, I., Chiang, P.K., Manjunath R. and Mukherjee, A.B. 1984. In: Kay, A.B., Frank Austen K., Lichtenstein, L.M. (eds). Asthma: Physiology, Immunopharmacology and treatment. Third International Symposium. Academic Press, London. p. 173.
44. Vasanthakumar, A., Manjunath, R., Mukherjee, A.B., Warabi, H., and Schiffman, E. 1988. Biochem. Pharmacol. 37:389.
45. Manjunath, R., Levin, S.W., Kumaroo, K.K., Butler, J.DeB., Donlon, J., McDonald, H., Fujita, R. Schumacher, U.K., and Mukherjee, A.B. 1987. Biochem. Pharmacol. 36:741.
46. Flower, R.J., and Blackwell, G.J. 1979. Nature 275:456.
47. Hirata, F., Schiffman, E., Venkatasubramanian, K., Solomon, D. and Axelrod, J.A. 1980. Proc. Natl. Acad. Sci. USA 77: 2535.
48. DiRosa, M., Flower, R.J., Hirata, F., Parente, L., and Russo-Marie, F. 1984. Prostaglandins 28:441.
49. Levin, S.W., Butler, J. DeB, Schumacher, U.K., Wightman, P.D., and Mukherjee, A.B. 1986. Life Sci. 38:1813.

50. Benveniste, J. and Pretolani, M. 1985. In: Russo-Marie, F., Mencia-Huerta J.M., and Chignard, M. (eds). Advances in Inflammation. Research. Raven Press, New York, Vol. 10:7.
51. Bourne, H. R. 1989. Nature 337:504.
52. Gramzow, M., Schroeder, H., Fritsche, U., Kurelec, B., Robitzki, A., Zimmermann, H., Friese, K., Kreuter, M. H., and Muller, W. E. G. 1989. Cell 59:939.
53. Guinea, R., Lopez-Rivas, A., and Carrasco, L. 1989. J. Biol. Chem. 264:21923.
54. Wallner, B. P., Mattaliano, R.J., Hession, C., Cate, R. L., Tizard, R., Sinclair, L. K., Foeller, C., Chow, E. P., Browning, J. L., Ramachandran, K. L., Pepinsky, R. B. 1986. Nature 320:77.
55. Crompton, M. R., Moss, S. E. and Crumpton, M. J. 1988. Cell 55:1.
56. Miele, L., Cordella-Miele, E., Facchiano, A., and Mukherjee, A.B. 1988. Nature 335:726.
57. Weber, K., and Johnson, N. 1986 FEBS Lett. 203:95.
58. Geisow, M.J. 1986 Febs Lett. 203:99.
59. Wilbur, W.J., and Lipman, D.J. 1983. Proc. Natl. Acad. Sci. USA. 80:726.
60. Cirino, G., Flower, R.J., Browning, J.L., Sinclair, L.K., and Pepinsky, R.B. 1987. Nature, 328:270.
61. Davidson, F.F., Dennis E.A., Powell, M., and Glenney, J.R. Jr. 1987. J. Biol. Chem. 262:1698.
62. Haigler, H.T., Schlaepfer, D.D., and Burgess, W.H. 1987 J. Biol. Chem. 262:6921.
63. Slotboom, A. J., Verger, R., Verheij, H. M., Baartsman, P. H. M., Van Deenen, L. L. M., and De Haas, G. H. 1976. Chem. Phys. Lipids 17:128.
64. Deems, R. A., Eaton, B., and Dennis, E. A. 1975. J. Biol. Chem. 250:9013.
65. Lutz, R. A., Bull, C. and Rodbard, D. 1986. Enzyme 36:197.
66. Romero, G., Thompson, K. and Biltonen, R. L. 1987. J. Biol. Chem. 262:13476.
67. Gheriani-Gruszka, N., Almog, S., Biltonen, R. L. and Lichtenberg, D. 1988. J. Biol. Chem. 263:11808.
68. Bell, J. D. and Biltonen, R. L. 1989. J. Biol. Chem. 264:12194.
69. Cho, W., Tomasselli, A. G., Heinrikson, R. L. and Kezdy, F. J. 1988. J. Biol. Chem. 263:11237.
70. Tomasselli, A. G., Hui, J., Fisher, J., Zurcher-Neely, H., Reardon, I. M., Oriaku, E., Kezdy, F. J. and Heinrikson, R. L. 1989. J. Biol. Chem. 264:10041.
71. Camussi, G., Tetta, C., Bussolino, F., and Baglioni, C. 1990. J. Exp. Med. 171:913.
72. Chan, C., Ni, M., Miele, L., Mukherjee, A. B., Nussenblatt, R. B. 1990. 5th International Symposium on the Immunology and Immunopathology of the Eye. Tokyo, March 13-15, 1990.
73. Ialenti, A., Doyle, P. M., Hardy, G. N., Simpkin, D. S. E., and DiRosa, M. 1990. Agents and Actions 29:48.
74. Vostal, J., Mukherjee, A. B., Miele, L., and Shulman, N. R. 1989. Biochem. Biophys. Res. Commun.165:27.
75. Sudhoff, T.C., Slaughter, C.A., Leznicki, I., Barion, P., and Reynolds, G.A. 1988. Proc. Natl. Acad. Sci. USA 85:664.
76. Kretsinger, R. H., and Creutz, C. E. 1986. Nature 320:573.
77. Davidson, F. F., Lister, M. D., and Dennis, E. A. 1990. J. Biol Chem. 265:5602.
78. Ahn, N. G., Teller, D. C., Bienkowski, M. J., McMullen, B. A., Lipkin, E. W., and De Haen, C. 1988. J. Biol. Chem. 263:18657.

79. Errasfa, M., Bachelet, M., and Russo-Marie, F. 1988. Biochem. Biophys. Res. Commun. 153:1267.
80. Cirino, G., Peers, S. H., Flower, R. J., Browning, J. L., and Pepinsky, R. B. 1989. Proc. Natl. Acad. Sci. USA 86:3428.
81. Drust., D. S., and Creutz, C. E. 1988. Nature 331:88.
82. Hirata, F. 1983. Adv. Prostaglandin, Thromboxane, Leukotriene Res. 11:73.
83. Pepinsky, R. B., Tizard, R., Mattaliano, R., Sinclair, L. K., Miller, G. L., Browning, J. L., Chow, E. P., Burne, C., Huang, K., Pratt, D., Watcher, L., Hession, C., Frey, A. Z., and Wallner, B. P., 1988. J. Biol. Chem. 263:10799.
84. Creutz, C. E., Zaks, W. J., Hamman, H.C., Crane, S., Martin, W. H., Gould, K. L., Oddie, K. M., and Parsons, S. J. 1987. J. Biol. Chem. 262:1860.
85. Nakano, T., Hanasaki, K. and Arita, H. 1989. J. Biol. Chem. 264:5400.
86. Metafora, S., Facchiano, F., Facchiano, A., Esposito, C., Peluso, G., and Porta, R. 1987. J. Prot. Chem. 6:353.
87. Porta, R., Esposito, C., Persico, P., Peluso, G. and Metafora, S. 1988. In: Zappia, V. Galletti, P., Porta, R., and Wold, F., (eds). Advances in Posttranslational modification of Proteins and Aging. Plenum Press, New York. Vol. I.
88. Paonessa, G., Metafora, S., Tajana, G., Abrescia, P., De Santis, A., Gentile, V. and Porta, R. 1984. Science 226:852.

ANTIFLAMMINS INHIBIT SYNTHESIS OF PLATELET-ACTIVATING FACTOR AND

INTRADERMAL INFLAMMATORY REACTIONS

Giovanni Camussi[1,2], Ciro Tetta[2], and Corrado Baglioni[3]

Cattedra di Nefrologia Sperimentale, Dipartimento di Biochimica
e Biofisica, 1ª Facoltà di Medicina, Università di Napoli[1]; Cattedra di Nefrologia, Università di Torino[2]; Department of Biological Sciences, State University of New York at Albany, N.Y.[3]

INTRODUCTION

Lipocortins belong to a family of related proteins that are thought to mediate the anti-inflammatory activity of corticosteroids (Flower et al., 1984). Lipocortins inhibit phospholipase A_2 (PLA_2) activity in vitro with a mechanism still unclear (Davidson et al., 1987; Haigler et al., 1987). Furthermore, recombinant lipocortin I inhibits eicosanoid synthesis in vivo in perfused lungs (Cirino at al., 1987). Another steroid-induced protein with PLA_2 inhibitory activity is uteroglobin, a rabbit secretory protein (Levin et al., 1986; Morize et al., 1987). Cloning and sequencing of lipocortins cDNAs has provided the amino acid sequence of these proteins (Pepinsky et al., 1986; Wallner et al., 1986). Lipocortin I and II comprise four non-identical repeats of 70 amino acids each; two identical subunits of 70 amino acids form uteroglobin. Miele et al. (1988) noticed a striking sequence similarity between amino acids 40-46 of uteroglobin and 247-253 of lipocortin I, repeat 3, and discovered that synthetic peptides corresponding to such sequences show potent inhibitory activity on isolated pancreatic PLA_2 and an anti-inflammatory effect on carrageenan-induced rat foot pad edema in vivo. Miele et al. (1988) proposed to call these peptides "antiflammins".

In the present investigation, we examined the effect of antiflammins on PLA_2 activity in intact cells. We monitored the synthesis of a phospholipid, platelet-activating factor (PAF), that is a mediator of inflammation and endotoxic shock synthesized by monocytes/macrophages, polymorphonuclear neutrophils (PMN), basophils, platelets and endothelial cells after appropriate stimuli (Camussi, 1986). PAF is synthesized in response to inflammatory stimuli via PLA_2 cleavage of 1-$\underline{O}$-alkyl-2-$\underline{sn}$-acyl-glycero-3-phospho-choline and acetylation of the lyso-PAF generated (Hanahan, 1986). The cytokine tumor necrosis factor (TNF), phagocytosis and proteinases induce macrophages, PMN and vascular endothelial cells to synthesize and release PAF (Camussi et al., 1987a). In other experiments, we measured the PLA_2-dependent release of arachidonic acid induced by TNF or phagocytosis. We report here that antiflammins inhibit PAF synthesis and release of arachidonic acid. Furthermore, the antiflammins inhibit inflammatory reactions induced in rat skin by in situ formation of immune complexes in an Arthus reaction or by intradermal injection of TNF.

The experiments shown in Fig. 1 established that rat peritoneal macrophages release PAF in response to TNF. The macrophages were incubated for 1 h with 20 ng/ml of TNF and PAF was assayed after purification from cells and medium, as described by Camussi et al. (1987a). The macrophages produced PAF in amounts comparable to those obtained by stimulation with the calcium ionophore A23187 or by phagocytosis of complement-activated zymosan C. The level of cell-associated PAF was approximately equal to that in the medium, suggesting that TNF was a good inducer of both PAF synthesis and release.

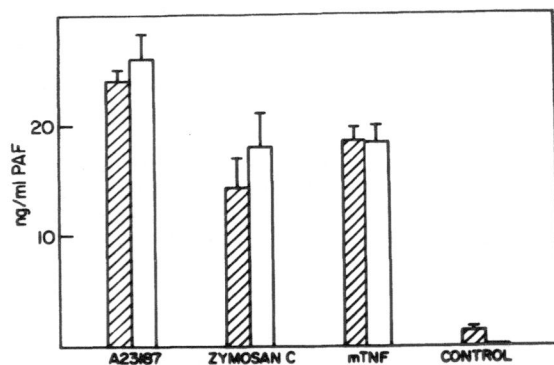

Fig. 1. Synthesis and release of PAF by rat peritoneal macrophages
treated for 1 h with 1 μg/ml of A23187, 0.2 mg/ml of zymosan
C, 20 ng/ml of mTNF or untreated. 10^6 cells were used for each
experiment. Cell-associated PAF, shaded columns; supernatant
PAF, blank columns. PAF concentration is referred to 1 ml of
cell culture medium and to a corresponding cell aliquot. Verti-
cal bars indicate the SD of the mean of three experiments.
Reproduced from the Journal of Experimental Medicine (1987) 166:
1390-1404, by copyright permission of the Rockefeller University
Press.

The following experiments measured the time course of PAF synthesis and release into the supernatant by macrophages treated with 20 ng/ml of TNF (Fig. 2). After 30 min PAF was mainly cell-associated, whereas after 1 h it was present in equal amounts in cells and supernatant. Cell-associated PAF decreased afterwards, whereas that in the supernatant increased sharply after 1 h but declined slightly afterwards (Fig. 2). This indicated that maximal synthesis of PAF occurred within 1 h of TNF addition and that PAF was gradually released into the supernatant. PAF was also metabolized during this incubation, since its total amount (cell-associated + supernatant) decreased. These results show that the amount of PAF synthesized by TNF-treated rat peritoneal macrophages is comparable to that produced in response to other stimuli. The response to TNF is rapid, since already at 30 min relatively large amounts of PAF are detected in these cells. The subsequent decrease of cell-associated PAF in macrophages continuously treated with TNF indicates that its synthesis is a transient response because of specific regulatory mechanisms.

PAF synthesis is inhibited by treating macrophages with two inhibitors of serine proteinases: the competitive inhibitor N-acetyl-DL-phenyl-alanine-β-naphtyl ester and L-1-tosylamide-2-phenylethyl chloromethyl ketone, an alkylating agent that is an irreversible inhibitor (Camussi et al., 1987a). Conversely, treatment with proteinases such as elastase and cathepsin G stimulates macrophages and neutrophils to synthesize and release PAF (Camussi et al., 1988). The inhibition of PAF synthesis by antiproteinases and its stimulation by proteinases suggests that a proteolytic activity is required to induce PAF synthesis. PLA_2 activity is thought to be regulated in vivo by membrane-associated lipocortins and synthesis of PAF may require proteolytic cleavage of lipocortins to activate PLA_2.

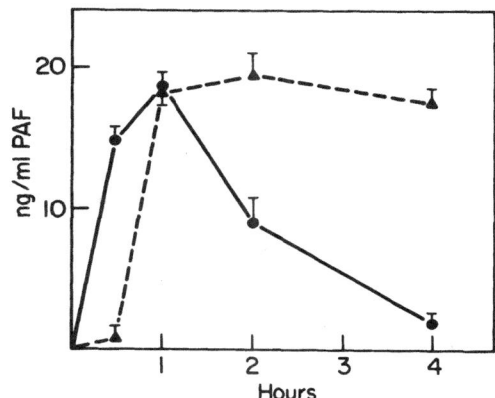

Fig. 2. Time course of PAF synthesis and release by rat peritoneal macrophages stimulated with 20 ng/ml of TNF. Cell-associated PAF (●) and PAF released in the supernatant (▲) by 10^6 cells were used for each time point. Reproduced from the Journal of Experimental Medicine (1987) 166:1390-1404, by copyright permission of the Rockefeller University Press.

ANTIFLAMMINS INHIBIT PAF SYNTHESIS

Peptides MQMKKVLDS (antiflammin-1) and MQMNKVLDS (antiflammin-3) were a gift of Dr. Anil B. Mukherjee of the National Institutes of Health (Bethesda, MD). Peptide HDMNKVLDL (antiflammin-2) was purchased from Peptide Biotechnologies (Washington, DC). These peptides were stored under nitrogen in sealed glass vials and dissolved in saline containing 10 mM β-mercaptoethanol (ME) or 1 mM dithiothreitol (DTT) to prepare 0.1 mM stock solutions. Dilutions were prepared immediately before use. The effect of antiflammins on PAF synthesis was examined in rat peritoneal macrophages stimulated either by phagocytosis of opsonized yeast spores prepared as described by Camussi et al. (1988) or by human recombinant TNF (Fig. 3). Antiflammin-2 (AF-2) inhibited PAF synthesis and release at much lower concentrations than antiflammin-1 (AF-1). A similar inhibition was observed for the synthesis of PAF that remained cell-associated (data not shown). In these experiments, the IC_{50} for AF-2 was ~25 nM, whereas that for AF-1 was >200 nM.

Antiflammin-2 inhibited PAF synthesis also in human PMN stimulated by TNF, phagocytosis of opsonized yeast spores (BYS-C3b) or by treatment with elastase (Table 1). This finding suggests that AF-2 inhibits PAF synthesis by a common mechanism, i.e. by inhibiting enzymatic activities required for the conversion of membrane phospholipids to PAF. The IC_{50} for AF-2 in these experiments was ~50 nM. However, 200 nM AF-1 did not inhibit PAF synthesis in a similar experiment with PMN (data not shown). It should be pointed out that AF-2 was not inactivated by elastase, since its inhibitory activity did not decrease after 30 min incubation with 1 μg/ml of this proteinase.

163

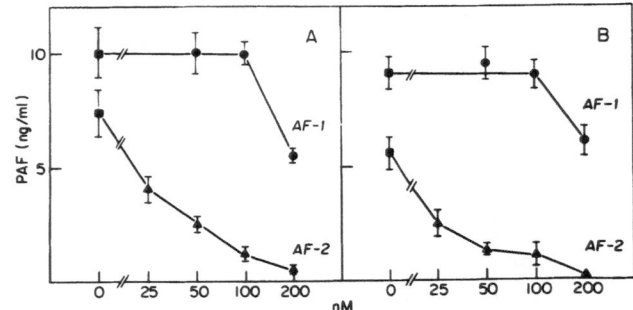

Fig. 3. Antiflammins inhibit PAF synthesis induced by phagocytosis of
yeast spores (A) or by 20 ng/ml TNF (B) in rat peritoneal macrophages preincubated for 30 min at 22°C with the concentration of
AF-1 or AF-2 shown in the abscissa and then stimulated at 37°C.
Four separate experiments are shown with the amount of PAF synthesized by control stimulated macrophages indicated for each
experiment (■). The mean ± SD of PAF released into the medium
in triplicate assays is reported. Control incubations received
the same amount of ME as incubations with antiflammins.

Table 1. Antiflammin-2 inhibits PAF synthesis induced by
different stimuli in human neutrophils.

INDUCER	ANTIFLAMMIN-2 (nM)	PAF (ng/ml) Released	Cell-bound
None	–	0.4±0.2	0.4±0.3
TNF	–	6.2±1.3	4.5±0.9
TNF	100	1.6±0.7	1.1±0.9
TNF	50	3.5±0.9	2.8±1.2
BYS-C3b	–	10.4±2.3	7.8±2.5
BYS-C3b	100	4.5±0.8	2.1±1.1
BYS-C3b	50	6.3±1.1	5.9±1.3
Elastase	–	7.4±1.1	4.7±1.2
Elastase	100	2.5±0.5	1.9±1.3

AF-2 was added for 30 min to PMN kept at 22°C; control
PMN were preincubated without additions. PMN were stimulated
with 10 ng/ml of TNF, 1 μg/ml of elastase for 10 min and
BYS-C3b for 20 min at 37°C. PAF released from 2×10^6 PMN and
cell-associated was measured as described in Fig 1. The mean
± SD of three experiments is shown in all the tables.

INHIBITION OF PHOSPHOLIPASE A_2 BY ANTIFLAMMINS

The finding that AF-1 was much less inhibitory than AF-2 for PAF
synthesis was surprising since Miele et al. (1988) have reported that
these peptides inhibit isolated porcine pancreatic PLA_2 at similar
concentrations. However, when we measured the effect of antiflammins
on PLA_2 activity in PMN homogenates, AF-1 was again much less inhibitory than AF-2 (Table 2). The IC_{50} for AF-2 was ~100 nM, whereas this
concentration of AF-1 inhibited PLA_2 activity only 16%.

The PLA$_2$ inhibitory activity of antiflammins was also examined in intact PMN by measuring the release of labeled arachidonic acid (Fig. 4). After preincubation with [^{14}C]arachidonic acid, PMN released a basal amount of label. PMN stimulated by TNF or phagocytosis released 2.3 and 5.4-fold more label, respectively, than control cells. This release was presumably due to activation of PLA$_2$ and it was inhibited by much lower concentrations of AF-2 than AF-1 (Fig. 4). The IC$_{50}$ for AF-2 was ~100 nM, whereas that for AF-1 was ~1 μM.

Table 2. Antiflammins inhibit phospholipase A$_2$ activity in human neutrophils homogenate.

ADDITIONS	nM	[^{14}C]OLEIC ACID	INHIBITION (%)
None	—	1.59±1.2	—
AF-1	100	1.34±0.3	16
AF-1	1000	0.31±0.1	81
AF-2	10	1.71±0.6	0
AF-2	100	0.75±0.2	53
AF-2	200	0.41±0.1	74

The PLA$_2$ activity was measured in sonicated human PMN according to Blackwell et al. (1982). The reactions contained 0.1 ml of PMN sonicate, 500 nM α-palmitoyl-β-1-[^{14}C]oleoyl-L-phosphatidyl choline and 0.9 ml of 0.5 M Tris buffer, pH 8, with 50 mM CaCl$_2$. After 1 h at 37°C, 2 ml of ethanol and of chloroform were added. The samples were extracted and the [^{14}C]oleic acid was separated from unhydrolyzed substrate by TLC on silica gel using chloroform: methanol: acetic acid (70:10:1) as solvent. The [^{14}C]oleic acid hydrolyzed per μg of protein is indicated as a % of the substrate input.

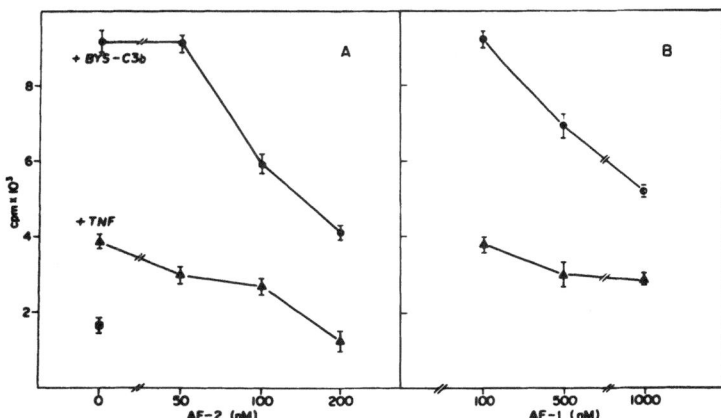

Fig. 4. AF-2 (A) and AF-1 (B) inhibit the release of arachidonic acid induced by TNF or phagocytosis. PLA$_2$ activity in intact cells was estimated according to Hirata et al. (1979). 10^7 PMN were incubated in 5 ml of modified Gey's solution with 1.25 μCi of [^{14}C]arachidonic acid at 37°C for 1 h, washed and resuspended in this solution. The release of arachidonic acid was measured in 0.5 ml incubations after 20 min at 37°C by centrifuging the cells and counting the supernatant. The label released after 20 min at 37°C is indicated for untreated cells (■) and for TNF (▲) or BYS-C3b-treated cells (●). The experiments were carried out in triplicate and the mean ± SD is shown.

The above experiments show a good correlation between the inhibition of PAF synthesis and that of PLA$_2$ activity by AF-2. Further evidence that this inhibition is responsible for the activity of AF-2 was obtained by adding porcine pancreatic PLA$_2$ to incubations with human PMN stimulated by TNF or phagocytosis. In previous experiments, addition of PLA$_2$ to PMN was shown to restore the synthesis of PAF that had been inhibited by staurosporine (Camussi et al., 1989). Pancreatic PLA$_2$ was effective in reversing the inhibition of PAF synthesis in PMN treated with AF-2 (Table 3). Therefore, by adding a relatively large amount of pancreatic PLA$_2$, we could reverse the inhibition of PLA$_2$ by AF-2.

Table 3. Addition of phospholipase A$_2$ reverses the inhibitory activity of antiflammin-2 in human neutrophils.

INDUCER	ADDITIONS		PAF (ng/ml)	
	AF-2	PLA$_2$	Released	Cell-bound
None	–	–	0.4±0.2	0.3±0.2
TNF	–	–	5.1±1.3	4.5±1.8
TNF	–	+	5.8+2.1	5.1±1.2
TNF	+	–	1.5±1.3	1.8±1.2
TNF	+	+	4.5±1.9	4.1±1.1
BYS-C3b	–	–	10.2±2.5	9.7±1.7
BYS-C3b	–	+	9.6±3.2	10.4±2.3
BYS-C3b	+	–	5.1±1.1	3.2±1.2
BYS-C3b	+	+	9.4+0.3	7.2±1.3

Where indicated, 100 nM antiflammin-2 (AF-2) was added for 20 min to PMN kept at 22°C. PMN were induced to produce PAF by 10 ng/ml of TNF for 10 min or by phagocytosis of opsonized yeast spores (BYS-C3b) for 20 min at 37°C. Where indicated, 1 µg/ml of porcine pancreatic PLA$_2$ was added together with the inducer.

INACTIVATION OF ANTIFLAMMINS

There may be several explanations for the discrepancy between the results obtained with isolated PLA$_2$ by Miele et al. (1988) and the present results obtained in incubations with intact cells or cell homogenates. We routinely dissolved these peptides either in 10 mM ME or in 1 mM DTT to protect methionine residues from oxidation. These solutions were stored at 4°C and used over few weeks without apparent loss of activity. However, aliquots of these solutions that had been frozen no longer inhibited PAF synthesis by PMN (Table 4). Therefore, in spite of the presence of reducing agents, the antiflammins were apparently inactivated by freezing and thawing. In an attempt to reactivate the antiflammins, we treated the frozen solutions with reducing agents for different times at various temperatures. We discovered in this way that the inhibitory activity of AF-2 was in large part recovered by heating at 45°C for 15 min (Table 4).

As a consequence of the inclusion of ME or DTT in antiflammins solutions, these reducing agents were present in the incubations with cell homogenates or intact cells. This was another major difference between our assay conditions and those previously reported by Miele et al.(1988). In incubations with human PMN, AF-2 in 2 µM ME and AF-3 in 0.5 µM DTT were less inhibitory for PAF synthesis than when much higher concentrations of these reducing agents were present (Table 3). This finding suggested that relatively high concentrations of ME or DTT had to be added to the incubations to obtain inhibition of PLA$_2$ activity and PAF synthesis by antiflammins. A possible reason for this requirement is the release by macrophages and PMN of oxidizing agents that inactivate the antiflammins. Such sensitivity to oxidation was shown by treating anti-flammin solutions with H$_2$O$_2$ that was then hydrolyzed with catalase before addition to incubations with PMN (Table 5). The inhibitory activity of AF-1 was completely abolished by this treatment, whereas that of AF-2 was greatly reduced. This finding suggests that AF-1 may be more sensitive to inactivation by oxidizing agents than AF-2.

Table 4. Effect of reducing agents on antiflammins inhibition of PAF synthesis.

ANTFLAMMIN	STORAGE	REDUCING AGENT (μM)		PAF (ng/ml)
None	–	ME	20	10.1±1.2
AF-2	frozen	ME	20	9.1±1.2
AF-2	frozen → 45°C	ME	20	1.4±1.2
AF-2	frozen → 45°C	ME	2	8.5±1.1
AF-3	4°C → 45°C	ME	50	0.3±0.2
AF-3	4°C → 45°C	DTT	0.5	3.5±1.2
AF-3	4°C → 45°C	DTT	50	0.8±0.2

The antiflammins were prepared as 0.1 mM stock solutions in 10 mM ME or 1 mM DTT, except for one sample of AF-2 that was in 1 mM ME and one sample of AF-3 that was in 0.1 mM DTT. The concentration of AF-2 and AF-3 in the assays with human PMN was 200 and 500 nM, respectively. The reducing agents were added to the incubations with the antiflammin solutions and their final concentration is shown. The PMN were stimulated by phagocytosis immediately after the addition of antiflammins.

Table 5. Effect of oxidation on the activity of antiflammins.

ADDITIONS	PAF (ng/ml)	
	RELEASED	CELL-ASSOCIATED
None	0.4±0.2	0.3±0.1
BYS-C3b	9.8±3.1	7.1±1.5
BYS-C3b + 1 μM AF-1	1.2±1.4	1.9±1.5
BYS-C3b + 1 μM H$_2$O$_2$-treated AF-1	9.1±2.1	9.0±1.3
BYS-C3b + 200 nM AF-2	1.6±1.4	1.5±1.3
BYS-C3b + 200 nM H$_2$O$_2$-treated AF-2	7.5±2.1	5.7±2.1

Human PMN were stimulated to synthesize PAF by phagocytosis of yeast spores (BYS-C3b) as described in Fig. 3. Where indicated, antiflammins were added together with yeast spores and the PMN were incubated for 20 min before measuring released and cell-associated PAF. The antiflammins were also treated with 10 mM H$_2$O$_2$ for 10 min and then with 500 units/ml of catalase for 10 min at 37°C before addition to PMN.

In the above experiments, we pretreated macrophages or PMN with antiflammin solutions prior to the stimulation of PAF synthesis. However, such preincubation was unnecessary to inhibit PAF synthesis. AF-2 was almost as inhibitory when added at the same time that PMN were stimulated by TNF or phagocytosis as when added after 5 to 30 min preincubation with PMN (Fig. 5). A different result was obtained with AF-1 that was inhibitory only when added together with the agents stimulating PAF synthesis, but was almost inactive after 5 min preincubation (Fig. 5). This finding suggests that PMN inactivate AF-1 much faster than AF-2. The only differences in amino acid sequence between AF-1 and AF-2 are the MQ→HD substitutions in position 1-2, and the S→L substitution in position 9. It is possible that the N-terminal methionine of AF-1 is oxidized by PMN secretory products. However, this methionine is not essential for the biological activity of antiflammins, since it is substituted by histidine in AF-2. Miele et al. (1988) reported that the first two residues of AF-1 can be replaced but not deleted without loss of activity and suggested that the length of antiflammins is critical, possibly for conformational reasons. Presumably, oxidation of the N-terminal methionine disrupts the conformation of AF-1 and may account for the lower inhibitory activity of this peptide, but we cannot exclude that AF-1 is less inhibitory than AF-2 for other reasons. The oxidation of the Met residue in position 3 may result in a complete loss of activity for all antiflammins, as suggested by the inactivation of these peptides after the incubation with H_2O_2 (Table 4). This Met residue may be less sensitive to oxidation than the N-terminal Met since reducing agents protect only AF-2 in incubations with PMN.

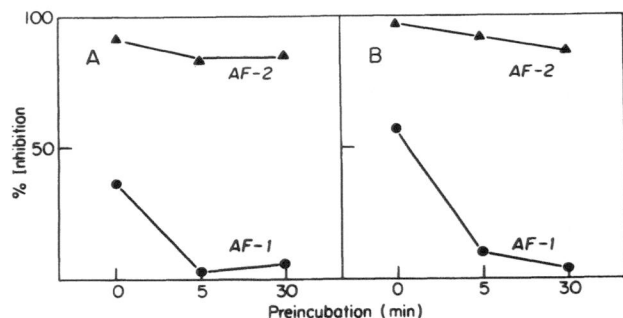

Fig. 5. The effect of preincubation with human neutrophils on the inhibitory activity of antiflammins. Human PMN were preincubated with 500 nM AF-1 (●) or 200 nM AF-2 (▲) for the time shown in the abscissa and then stimulated with 10 ng/ml TNF (A) or by phagocytosis of yeast spores (B). The % inhibition of PAF synthesis is shown relative to control cells not treated with antiflammins.

The finding that AF-1 was inactivated during the preincubation with PMN provided an explanation for the discrepancy between the inhibition by this peptide of purified PLA_2 reported by Miele et al. (1988) and its low inhibitory activity in assays with intact cells. Furthermore, this experiment shows that AF-2 inhibits PAF synthesis without an appreciable lag. Therefore, antiflammins are inhibitors that interact rapidly with their target, but their activity is entirely dependent on the presence of reducing agents, such as ME or DTT, in incubations with cells involved in inflammatory responses, such as macrophages or PMN. This observation suggests that antiflammins are readily inactivated by oxidation, although we cannot exclude that antiflammins and reducing agents inhibits PLA_2 synergistically. This seems unlikely in view of the finding that ME had no effect on PAF synthesis when added up to 1 mM concentration (Fig. 6).

In this experiment, we examined the inhibitory activity of AF-3 in the presence of ME and the effect of this reducing agent alone. Furthermore, AF-3 was not preincubated with PMN, but it was added together with the stimulus for PAF synthesis. Under these conditions, AF-3 was quite inhibitory with an IC_{50} ~40 nM. It should be pointed out that AF-3 differs from AF-1 only in a K→N substitution in position 4 and contains Met at the N-terminus.

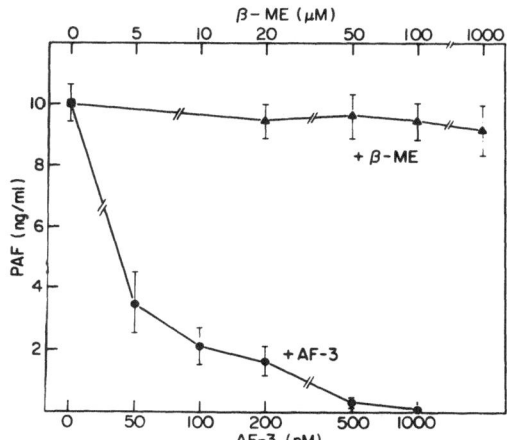

Fig. 6. Inhibition of PAF synthesis by AF-3 and lack of effect of mercaptoethanol (β-ME). Synthesis of PAF was stimulated by phagocytosis in PMN as described in Table 1. Incubations contained both AF-3 and ME (●) or only ME (▲).

MECHANISM OF ACTION AND BIOLOGICAL ACTIVITY OF ANTIFLAMMINS

The following experiment was designed to establish whether the continuous presence of antiflammins was required to inhibit PAF synthesis. Human PMN and rat macrophages were preincubated for 30 min with 100 nM AF-2, washed and then stimulated by phagocytosis of yeast spores. Control cells were not treated with AF-2 but were handled in the same way, since the preincubation and washes were found to reduce PAF synthesis in response to different stimuli. As shown in Table 6, PMN and macrophages pretreated with AF-2 synthesized PAF in amounts comparable to control cells when this peptide was removed by washing. This result shows that the inhibition by AF-2 is reversible and that the continuous presence of antiflammins is required to obtain an inhibition of a PLA_2-dependent activity.

To evaluate the inhibitory activity of antiflammins in a complex inflammatory response, we induced an Arthus reaction in Lewis rats by an intravenous injection of BSA and Evans blue followed after 30 min by the intradermal injection of anti-BSA antibody (Camussi et al., 1987b). The rats were sacrificed 3 h after the last injection and the area of Evans blue extravasation was first measured (Table 7). Circular skin areas were then excised and processed for light microscopy examination. PMN present around vessels at the center of the intradermal injection were counted to quantitate leukocyte infiltration. The Arthus reaction was characterized by severe inflammatory lesions in dermis around vessels, by edema and focal interstitial hemorrhage associated with increased vascular permeability, as judged by the extravasation of Evans blue (Table 7). The intradermal injection of 100 ng of AF-2 together with the anti-BSA antibody suppressed this increase in vascular permeability and the leukocyte infiltration. The intravenous injection of 1 μg of AF-2 (5 min before that of anti-BSA antibody) inhibited much less the effects of the Arthus reaction, but the increase in vascular permeability was significantly lowered.

Table 6. Loss of the inhibitory effect on PAF synthesis by washing neutrophils or macrophages after a pretreatment with AF-2.

CELLS	ANTIFLAMMIN-2 (nM)	PAF (ng/ml) Released	Cell-bound
Neutrophils	--	8.2±1.3	7.1±1.9
	100	7.8±1.2	6.9±3.1
Macrophages	-	5.2±1.3	3.1±1.9
	100	4.2±0.5	3.9±1.8

The cells were preincubated with or without AF-2 for 30 min at $22^{O}C$, washed twice with medium and stimulated with BYS-C3b for 20 min. PAF released into the culture medium or cell-associated was measured as described in Table 1.

In subsequent experiments, we examined the effect of AF-2 on the increased vascular permeability and leukocyte infiltration induced by TNF injected intradermally 30 min after Evans blue (Table 7). The injection of TNF increased vascular permeability and leukocyte infiltration much less than in the Arthus reaction and leukocytes were mainly accumulated in the lumen of vessels as intravascular aggregates adherent to the endothelium. The intravenous injection of 100 ng of AF-2 inhibited both effects of the intradermal injection of TNF. These results show that AF-2 inhibits the increase in vascular permeability and leukocytes infiltration induced by an Arthus reaction or by intradermal injection of TNF.

Table 7. Effect of AF-2 on the increase in vascular permeability and leukocyte infiltration induced in rats by a reverse passive Arthus reaction (BSA + anti-BSA) and by TNF.

TREATMENTS	-	I.D.	I.V.	-	I.D.	I.V.
	(diameter of blueing, mm)			(PMN/0.032 mm^2)		
BSA+anti-BSA	17±2	2±1	9±2	46±12	3±2	33±11
TNF	7±2	0	1±1	12±4	3±1	2±1
Control	3±1	N.D.	N.D.	3±1	N.D.	N.D.

Treatments, additions and examination of skin samples are described in the text. The BSA was injected intravenously. The anti-BSA antibody, 0.2 μg of TNF, and 1 μg of AF-2 was injected intradermally (I.D.) or intravenously (I.V.).

CONCLUSIONS

The antiflammins are promising anti-inflammatory agents. The observations reported here suggest that nM concentrations of these peptides may produce a striking pharmacological effect. In our experiments, an intradermal injection of 100 ng AF-2 suppresses the inflammatory response in an Arthus reaction (Table 7). This effect is observed with lower antiflammin doses than those used to inhibit the carrageenan-induced rat paw edema by Miele et al. (1988). Injection of ~500 μg AF-2 in the rat subplantar space inhibits 96% of the swelling caused by carragenaan, but lower amounts are much less inhibitory. This inflammatory reaction is quite different from an Arthus reaction and the effect of antiflammins cannot be meaningfully compared. Furthermore, in our experiments the antiflammins were injected together with 10 μM ME that may enhance their stability. The consensus sequence derived from the comparison of the antiflammins described so far is xxMxKVLDx, where x is a position occupied by two unrelated amino acid residues, such as His and Met in position 1, or Asn and Lys in position 4. The Met residue in position 3 is apparently conserved and may have an important role in the inhibitory activity of antiflammins. At the same time, this Met residue may be responsible for the instability of these peptides. One goal of future research is to obtain peptides with specific substitutions of amino acid residues that render the antiflammins unstable. For example, peptides with substitutions of the methionine in position 3 will allow to establish whether this residue is essential for the inhibitory activity of antiflammins. In conclusion, AF-2 or another peptide of similar activity, but with the methionine replaced by another amino acid residue to improve its stability, are good candidates as anti-inflammatory agents in the therapy of acute and chronic inflammatory diseases.

REFERENCES

Blackwell, G.J., R. Carnuccio, M. Di Rosa, R.J. Flower, C.S.J. Langham, L. Parente, P. Persico, N.C. Russel-Smith, and D. Stone. 1982. Glucocorticoids induce the formation and release of anti-inflammatory and anti-phospholipase proteins into the peritoneal cavity of the rat. Br. J. Pharmac. 76:185.
Camussi, G. 1986. Potential role of platelet-activating factor in renal pathophysiology. Kidney Int. 29:469.
Camussi, G., F. Bussolino, G. Salvidio, and C. Baglioni. 1987a. Tumor necrosis factor/cachectin stimulates rat peritoneal macrophages and human endothelial cells to synthesize and release platelet activating factor. J. Exp. Med. 166:1390.
Camussi, G., I. Pawlowsky, R. Saunders, J. Brentjens, and G. Andres. 1987b. Receptor antagonist of platelet activating factor inhibits inflammatory injury induced by in situ formation of immune complexes in renal glomeruli and in skin. J. Lab. Clin. Med. 110:196.
Camussi, G., C. Tetta, F. Bussolino, and C. Baglioni. 1988. Synthesis and release of platelet-activating factor is inhibited by plasma $\alpha_1$proteinase inhibitor or $\alpha_1$antichymotrypsin and stimulated by proteinases. J. Exp. Med. 168:1293.
Camussi, G., C. Tetta, F. Bussolino, and C. Baglioni. 1989. Tumor necrosis factor stimulates human neutrophils to release leukotriene B4 and platelet-activating factor. Induction of phospholipase A$_2$ and acetyl-CoA:1-alkyl-sn-glycreo-phosphocholine O$_2$-acetyltransferase activity and inhibition by antiproteinases. Europ. J. Biochem. 182:661.

Cirino, G., R.J. Flower, J.L. Browning, L.K. Sinclair, and R.B. Pepin-
    sky. 1987. Recombinant human lipocortin 1 inhibits thromboxane
    release from guinea-pig isolated perfused lung. Nature 328.
Davidson, F.F., A.E. Dennis, M. Powell, and J.R. Glenney. 1987. Inhi-
    bition of phospholipase A_2 by "lipocortins" and calpactins. An
    effect of binding to substrate phospholipids. J. Biol. Chem.
    262:1698.
Flower, R.J., J.N. Wood, and L. Parente. 1984. Macrocortin and the
    mechanism of action of the glucocorticoids. Adv. Inflammation
    Res. 7:61.
Haigler, H.T., D.D. Schlaepfer, and W.H. Burgess. 1987. Characteriza-
    tion of lipocortin I and an immunologically unrelated 33-kDa
    protein as epidermal growth factor/kinase substrates and phospho-
    lipase A_2 inhibitors. J. Biol. Chem. 262:6921.
Hanahan, D.J. 1986. Platelet activating factor: a biologically active
    phosphoglyceride. Ann. Rev. Biochem. 55:483.
Hirata, F., B.A. Corcoran, K. Venkatasubramanian, E. Schiffmann, and
    J. Axelrod. 1979. Chemoattractants stimulate degradation of
    methylated phospholipids and release of arachidonic acid in rab-
    bit leukocytes. Proc. Natl. Acad. Sci. USA 76:2640.
Huang, K.-S., B.P. Wallner, R.J. Mattaliano, R. Tizard, C. Burne, A.
    Frey, C. Hesslon, and P. McGray. 1986. Two human 35 kd inhibitors
    of phospholipase A_2 are related to substrates of pp60v-src and of
    the epidermal growth factor receptor/kinase Cell 46:191.
Levin, S.W., J.D. Butler, U.K. Schumacher, P.D. Wightman, and A.B.
    Mukherjee. 1986. Uteroglobin inhibits phospholipase A_2 activity.
    Life Sci. 38:1813.
Miele, L., E. Cordella-Miele, A. Facchiano, and A.B. Mukherjee. 1988.
    Novel anti-inflammatory peptides from the region of highest simi-
    larity between uteroglobin and lipocortin I. Nature 335:726.
Morize, I., E. Surcuf, M.C. Vaney, Y. Epelboin, M. Buehner, F. Frid-
    lansky, E. Milgrom, and J.P. Mornon. 1987. Refinement of the C222
    crystal form of oxidized uteroglobin at 1.34 A resolution. J.
    Mol. Biol. 194:725.
Pepinsky, R.B., L.K. Sinclair, J.L. Browning, R.J. Mattaliano, J.E.
    Smart, E.P. Chow, T. Falbel, and A. Ribolini. 1986. Purification
    and partial sequence analysis of a 37-kDa protein that inhibits
    phospholipase A_2 activity from rat peritoneal exudates. J. Biol.
    Chem. 261:4239.
Wallner, B.P., R.J. Mattaliano, C. Hession, R.L. Cate, R. Tizard, L.K.
    Sinclair, C. Foeller, E.P. Chow, J.L. Browning, K.L. Ramachan-
    dran, and R.B. Pepinsky. 1986. Cloning and expression of human
    lipocortin, a phospholipase A_2 inhibitor with potential anti-
    inflammatory activity. Nature 320:77.

ISOLATION AND CHARACTERIZATION OF cDNA CLONES

FROM HUMAN PLACENTA CODING FOR PHOSPHOLIPASE A$_2$

Robert Crowl[1], Cheryl Stoner[1], Tim Stoller[1],
Yu-Ching Pan[2] and Robert Conroy[1]

Departments of Molecular Genetics[1], and Protein
Biochemistry[2], Roche Research Center
Hoffmann-La Roche Inc.
Nutley, NJ 07110

INTRODUCTION

From the point of view of protein structure and catalytic mechanism, phospholipases A$_2$ (PLA$_2$'s) are among the best characterized enzymes. Most of what is known about the structure and enzymology of PLA$_2$ is based on studies using the enzymes purified from snake venom and mammalian pancreas (1,2). Although it has been known for some time that PLA$_2$ activity can be found in virtually every tissue and cell type (3,4), these cellular PLA$_2$'s are present in only trace amounts, and, until recently, efforts to purify and characterize them have been impeded. Thus, despite an extensive knowledge of the structure and enzymology of PLA$_2$, based on the paradigm of the pancreatic and snake venom enzymes, much remains to be learned regarding different cellular forms of PLA$_2$, their physiological roles, and how PLA$_2$ activity is regulated in cells.

Much of the recent interest in PLA$_2$ can be ascribed to its presumed role in the arachadonic acid (AA) "cascade", i.e., the release of AA from membrane phospholipids and the subsequent generation of secondary metabolites (e.g., prostaglandins (PG's), leukotrienes, and platelet activating factor) which can elicit a diverse set of cellular responses, including inflammation (4,5). It has been suggested that a membrane-bound form of PLA$_2$ would be the likely candidate for being involved in AA release (4-6). Membrane-bound forms of PLA$_2$ which have been purified include those from rat spleen (7), rat liver mitochondria (8), rat platelet (9,10), sheep erythrocytes (11), and the mouse macrophage-like cell line P388D$_1$ (6). A comparison of the amino-terminal 25 residues of the rat platelet, mitochondrial, and spleen PLA$_2$'s shows complete identity with the exception of the first residue, which is Ser in platelet, Asp in mitochondria, and unknown in spleen (7). The PLA$_2$'s from sheep erythrocytes and the mouse macrophage-like cell line P388D$_1$ differ from the other enzymes in that they exhibit characteristics of integral membrane proteins and have molecular weights of 18-18.5 Kd, instead of 14-15 Kd (6,11).

Among the biological responses to prostaglandin synthesis and release is the initiation of human parturition. In addition to the finding that free AA and PG levels are elevated in amniotic fluid during active labor, an increase in phospholipase (A_2 and C) activity is observed in decidual tissue as gestation advances (12). There is some evidence that AA release may be regulated by proteins which exhibit PLA_2 inhibitory activity, such as uteroglobin (13) and gravidin (14), present in uterine and placental tissues. The PLA_2 thought to play a role in parturition, however, has not been identified. Recently, the purification of a PLA_2 from the crude membrane fraction of placenta was reported (15). In this report, we describe the molecular cloning and sequence analysis of cDNA clones encoding PLA_2 from human placenta.

MATERIALS AND METHODS

Materials. A human placenta cDNA library made in phage lambda gt11 was obtained from Clontech. Restriction endonucleases, T4 DNA ligase, T4 polynucleotide kinase were purchased from New England Biolabs or Boehringer-Mannheim. Gamma ^{32}P-ATP was obtained from ICN. Other radioactive nucleotides, ^{3}H-oleic acid, and ^{35}S-cysteine were from Amersham. The plasmid pGEM-7Zf(+) and the Riboprobe system for generating SP6 transcripts were obtained from Promega. Reagents for DNA sequence analysis were from U.S. Biochemical Corp. Poly A^+ mRNA preparations used for Northern analysis were obtained from Clontech.

Isolation of cDNA Clones. The placenta library was screened following the method of Wood, et al. (16). Phage (5×10^5) were plated on E. coli strain LE392 (17) at a density of 5×10^4 PFU per 150 mm petri plate. Plaques from each plate were transferred in duplicate to nitrocellulose filters (S&S, BA85). The filters were denatured, neutralized, baked at 80° under vacuum, and pre-hybridized in 6X SSC (1X SSC is 0.15 M NaCl, 0.015 M sodium citate) , 5X Denhardt's solution (1X is Ficoll/ polyvinylpyrrilodone/bovine serum albumin, each at 0.2 mg/ml), and 100 ug/ml yeast tRNA at 37° for 8 hrs. Oligonucleotides were labeled at the 5' end with ^{32}P by phosphorylation with T4 polynucleotide kinase and gamma ^{32}P-ATP. The filters were then hybridized at 37° for 16 hrs in the same solution containing 50 pmoles of a ^{32}P-labeled mixed probe (64-fold degenerate) at 5×10^7 cpm per filter. The filters were washed twice (for 20 min each) at room temperature in 6X SSC and twice (for 30 min each) at 50° in 3M TMAC (tetramethylammonium chloride), 50 mM Tris-HCl (pH8), 2 mM EDTA. After air-drying, the filters were subjected to autoradiography with an intensifying screen for 3 days at -70°. Plaques corresponding to hybridization signals on duplicate filters were taken through further rounds of screening until isolated single plaques produced positive signals with probes corresponding to different regions of PLA_2.

Subcloning and DNA Sequence Analysis. DNA was prepared from recombinant phage according to standard procedures (18). The cDNA inserts were excised from the phage vector by digestion with Eco RI and isolated by preparative agarose gel electrophoresis. The fragments were subcloned into either pUC19 (19) or pGEM-7Zf(+) at the Eco RI site. DNA sequence analysis was performed by the dideoxy-chain termination method (20), using ^{35}S-labeled deoxyadenosine 5'-[alpha-thio]triphosphate and the universal and reverse M13 primers. Additional primers were synthesized based on the sequences determined within the cDNA insert to obtain additional sequence information.

PLA_2 Assay. ^{3}H-oleate labeled membranes from E. coli K12 were prepared essentially as described (21). To measure PLA_2 activity, the labeled substrate was added to a 50 ul reaction mixture containing enzyme in 50 mM Tris-HCl (pH 8), 2 mM $CaCl_2$, and 150 mM NaCl. The enzyme reaction was incubated at 37° for 30 min, and the reaction was terminated by the addition of 25 ul of 1 N HCl and 15 ul of 100 mg/ml BSA. Following centrifugation, 15 ul of the supernatant was removed for scintillation counting to measure the level of fatty acid liberated.

Northern Blot Analysis. Samples (2 ug) of polyA$^+$ mRNA were subjected to gel electrophoresis in 1.4% agarose containing 6.6% formaldehyde. The resolved RNA was transferred to a membrane filter (Hybond-C, Amersham), prehybridized in a solution containing 50% formaldehyde, 5X SSC, 0.3% SDS, and 0.25 mg/ml denatured salmon sperm DNA at 42° for 1 hr, and hybridized in the same solution with the EcoRI fragment from clone 9-1 that had been labeled with ^{32}P-CTP by random-priming. Hybridization was carried out at 45° for 3 days. The filter was washed with increasing degrees of stringency, beginning with 2 X SSC, 0.1%SDS at room temperature for 10 min and ending with a 30 min wash in 0.1X SSC, 0.1% SDS at 65°. The filter was subjected to autoradiography with an intensifying screen at -70°.

Genomic Blot Analysis. Membrane filters containing electrophoretically resolved genomic DNA that had been digested with various restriction enzymes were obtained from Clontech. The filters were pre-hybridized in 6 X SSC, 5X Denhardt's solution, and 50 ug/ml denatured salmon sperm DNA at 65° for 2-3 hrs. Hybridization was carried out in the same solution containing ^{32}P-labeled EcoRI fragment from clone 9-1 for 16 hrs at 65°. The filters were washed twice for 15 min each at room temperature in 2X SSC and then twice for 30 min each at 65° in 2X SSC prior to autoradiography.

COS Cell Transfection and Immunoprecipitation Analysis. COS-7 cells ($2x10^5$ cells in 35mm dishes) were transfected with 2 ug of the expression vector pBC12BI/hpPLA$_2$ 9-1 following a DEAE-Dextran protocol previously described (22). Sixty hrs post-transfection, cells were washed with PBS and incubated for 90 min in 2 ml of medium without serum and cysteine. The medium was removed, and the cells were incubated in 1 ml of serum-free medium containing 400 uCi ^{35}S-cysteine for 3 hrs. After labeling, the medium was removed and the cells were washed in PBS and lysed in 1 ml of RIPA buffer (22). 10 ug of a cocktail of 4 different monoclonal antibodies made against purified placental PLA$_2$ was added to the lysate and incubated at 4° overnight. The medium was adjusted to 1X RIPA buffer (without SDS) and incubated with the antibodies at 4° overnight. Antibody-antigen complexes were recovered by precipitation with Protein G beads (Genex) and subjected to SDS polyacrylamide gel electrophoresis followed by fluorography.

RESULTS AND DISCUSSION

Screening the Human cDNA Library

Based on protein sequence information obtained from a PLA$_2$ purified from the crude membrane fraction of human placenta (15; Wada, et al., unpublished), oligonucleotide probes were designed and synthesized (Fig. 1). A mixed-pool of oligonucleotides were made (23 nt, 128-fold degenerate) corresponding to the sequence of a peptide derived from a tryptic digestion of succinylated PLA$_2$ protein. In addition, a 50 nt "best guess" probe corresponding to the amino terminus was designed taking into consideration codon utilization and dinucleotide frequencies in human genes (23). The mixed-sequence probes were labeled at the 5' end with ^{32}P by phosphorylation with polynucleotide kinase and used to screen a human placental cDNA library by hybridization. Among $5x10^5$ plaques

A)

```
     AsnLysThrThrTyrAsnLysLysTyrGlnTyrTyrSerAsnLysHis

          3' ATATTATTTTTCATAGTCATAAT  5'
             G  G  C  T  G  T  G
```

B)

```
AsnLeuValAsnPheHisArgMetIleLysLeuThrThrGlyLysGluAlaAlaLeuSerTyrGlyPheTyrGly
AACCTGGTGAACTTCCACAGGATGATCAAGCTGACCACAGGCAAGGAGGC
5'                                                              3'
```

FIG. 1. Design of DNA probes for cloning the cDNA for hpPLA$_2$. A) A mixed
probe corresponding to a region within the sequence of a 16
residue tryptic peptide of succinylated hpPLA$_2$. B) A "best
guess" 50mer coding for the first 17 of the amino-terminal 25
residues determined for hpPLA$_2$.

screened, 12 hybridization signals were observed on duplicate filters.
Five of these also showed specific hybridization with the amino-terminal
probe (50mer). The cDNA insert in each of the clones was shown to be
800-850 bp. Two of these clones, designated 9-1 and 9-2, were
characterized further.

Nucleotide Sequence and Deduced Amino Acid Sequence of PLA$_2$

The cDNA inserts were excised from the phage vector by EcoRI digestion
and sub-cloned into plasmids for DNA sequence determination using the
dideoxy-chain termination method (20). The nucleotide sequence and
deduced amino acid sequence of clone 9-1 is shown in Fig. 2. The cDNA
contains an open reading frame of 144 amino acids. The translated
sequence codes for amino acid residues identical with those determined for
the amino-terminus (residues 1-25), as well as for 2 tryptic peptides
(63-67 and 101-116), of the PLA$_2$ protein purified from human placenta.
The Asn residue at the amino-terminus of the purified enzyme (residue +1)
is preceded by 20 mostly hydrophobic residues encoded by the cDNA that are
likely to represent a signal peptide. There is a potential glycosylation
site near the C-terminus (residues 101-103), however protein sequence data
for the tryptic fragment containing this region showed no evidence of
glycosylation.

An alignment of the human placental (hp) PLA$_2$ sequence (excluding
the signal sequence) with other PLA$_2$'s shows 35% identity with human
pancreatic PLA$_2$ (24,25), 35% with a cobra venom (Naja) PLA$_2$ (26), 46%
with the Crotalus atrox enzyme (27), and 69% with rat platelet PLA$_2$
(10) (see Fig. 3). The hpPLA$_2$ contains 14 cysteine residues in an
arrangement that is characteristic of Group II snake venom PLA$_2$'s
(28). When the hpPLA$_2$ sequence is modeled onto the crystal structure
of the Crotalus PLA$_2$ (data not shown), it is evident that most of the
residues in the interior of the enzyme, including the catalytic network
(His48, Asp49, Tyr 51, Tyr 66, and Asp91) are conserved, and that most of
the differences between the two occur on the enzyme surface. The number
of basic residues that appear on the surface of the placental enzyme is
striking, in particular, within the alpha helix at the amino terminus and
within the loop from 52 to 57. It is interesting to note that

176

GAATTCGGGATACAACTCTGGAGTCCTCTGAGAGAGCCACCAAGGAGGAGCAGGGGAGCGACGGCCGGGG
CAGAAGTTGAGACCACCCAGCAGAGGAGCTAGGCCAGTCCATCTGCATTTGTCACCCAAGAACTCTTACC


```
ATG AAG ACC CTC CTA CTG TTG GCA GTG ATC ATG ATC TTT GGC CTA CTG CAG GCC
 M   K   T   L   L   L   L   A   V   I   M   I   F   G   L   L   Q   A
-20                                         -10
```

```
CAT GGG AAT TTG GTG AAT TTC CAC AGA ATG ATC AAG TTG ACG ACA GGA AAG GAA
 H   G   N   L   V   N   F   H   R   M   I   K   L   T   T   G   K   E
    -1   1                                10
```

```
GCC GCA CTC AGT TAT GGC TTC TAC GGC TGC CAC TGT GGC GTG GGT GGC AGA GGA
 A   A   L   S   Y   G   F   Y   G   C   H   C   G   V   G   G   R   G
            20                              30
```

```
TCC CCC AAG GAT GCA ACG GAT CGC TGC TGT GTC ACT CAT GAC TGT TGC TAC AAA
 S   P   K   D   A   T   D   R   C   C   V   T   H   D   C   C   Y   K
            40                              50
```

```
CGT CTG GAG AAA CGT GGA TGT GGC ACC AAA TTT CTG AGC TAC AAG TTT AGC AAC
 R   L   E   K   R   G   C   G   T   K   F   L   S   Y   K   F   S   N
                        60                              70
```

```
TCG GGG AGC AGA ATC ACC TGT GCA AAA CAG GAC TCC TGC AGA AGT CAA CTG TGT
 S   G   S   R   I   T   C   A   K   Q   D   S   C   R   S   Q   L   C
                                80
```

```
GAG TGT GAT AAG GCT GCT GCC ACC TGT TTT GCT AGA AAC AAG ACG ACC TAC AAT
 E   C   D   K   A   A   A   T   C   F   A   R   N   K   T   T   Y   N
    90                              100
```

```
AAA AAG TAC CAG TAC TAT TCC AAT AAA CAC TGC AGA GGG AGC ACC CCT CGT TGC
 K   K   Y   Q   Y   Y   S   N   K   H   C   R   G   S   T   P   R   C
        110                             120                     124
```

TGA GTCCCCTCTTCCCTGGAAACCTTCCACCCAGTGCTGAATTTCCCTCTCTCATACCCTCCC
TCCCTACCCTAACCAAGTTCCTTGGCCATGCAGAAAGCATCCCTCACCCATCCTAGAGGCCAGGCAGGAG
CCCTTCTATACCCTCCCAGAATGAGACATCCAGCAGATTTCCAGCCTTCTACTGCTCTCCTCCACCTCAA
CTCCGTGCTTAACCAAAGAAGCTGTACTCCGGGGGGTCTCTTCTGAATAAAGCAATTAGCAAATCAAAAA
AAACCCGAATTC (857)

FIG. 2. Nucleotide sequence of cDNA 9-1 and the deduced amino acid
sequence of human placenta PLA$_2$. The amino acid residues
determined by protein sequence analysis are underlined.

```
                1
hu panc   avwQFrkMIKCvIPGsdpfleYnnYGCYCGlGGSGTPVDeLDkCCQTHDNCYdqAkKldsCkflld
Naja      NLyQFknMIqCtvPsRS wwdfadYGCYCGrGGSGTPVDdLDRCCQVHDNCYNeAEKiSGC
C.atrox   SLVQFetlIm kIaGRSglLwYsaYGCYCGwGGhGlPqDATDRCCfVHDCCY      gKatdC
hpPLA2    NLVnFhrMIK ltTGKeAaLSYGFYGCHCGVGGRGSPKDATDRCCVTHDCCYkRLEK rGC
rat       SLleFgqMIl fkTGKrAdvSYGFYGCHCGVGGRGSPKDATDeCCVTHeCCYNRLEK SGC
               *    *            *** ** ** *  * *  * **  *  **     *    *

hu panc   NPYThTYSYSCSGSaITCSsKNkeCe AfICnCDRnAAICFSk APYNkahkNLdtKysCQ
Naja      wPYFkTYSYeCSqGtlTCkGGNn CaaAavCdCDRlAAICFAg APYNDNdYNinlKarCQE
C.atrox   NPKTvSYtYSeenGeIiC GGdDpCg tQICECDKAAAICFrdNiPSYDNKYwLfPpKdCREepepC
hpPLA2    GTKFLSYKFSnSGSrITC aKQDSCR sQLCECDKAAAtCFARNKttYnkKYQyYsNKhCRgsTPrC
rat       GTKFLTYKFSYrGGqIsCStnQDSCR kQLCqCDKAAAeCFSRNKkSYslKYQfYPNKfCk??TPsC
               *          *       *    * **   ** **              *     124
```

FIG. 3. Amino acid sequence comparison of selected PLA$_2$'s. Residues that are identical among the 5 sequences are indicated (*). Amino acids that are in common with any of the other proteins are denoted by capitalization. See text for the sources of the sequences other than hpPLA$_2$.

autocatalytic acylation of lysines occurs at positions 7 and 10 within the amino terminal alpha helix of a basic PLA$_2$ from <u>Agkistrodon piscivorus piscivorus</u> (29) and at lysine 56 in porcine pancreatic PLA$_2$ (30). Acylation of these normally monomeric enzymes results in dimerization and activation of the PLA$_2$'s. The oligomeric form of hpPLA$_2$ has not yet been determined, nor is it known whether acylation of this enzyme occurs.

Expression in Mammalian Cells

To demonstrate that the cDNA encodes an enzymatically active PLA$_2$, a restriction fragment from cDNA 9-1 containing the entire coding region, including the putative signal peptide, was inserted into pBC12BI (22), and the resulting expression plasmid was transfected into COS cells. Samples of the medium were removed after 72 hrs and tested for PLA$_2$ enzymatic activity. The results indicate that, compared to mock-transfected controls, PLA$_2$ activity is produced and secreted by COS cells transfected with the expression plasmid (Fig. 4). To quantitate the amounts produced, serial dilutions of purified PLA$_2$ were prepared in medium and assayed for activity to generate a standard curve. Based on this determination, which assumes that the expressed PLA$_2$ has the same specific activity as the purified enzyme, the level of hpPLA$_2$ secreted is 1.8 to 3.6 ng/ml. To follow PLA$_2$ protein, transfected cells were metabolically labeled with ^{35}S-Cysteine, and the radio-labeled proteins were subjected to immunoprecipitation analysis using monoclonal antibodies made against the purified enzyme (Fig. 5). The results from SDS-gel electrophoresis of the immunoprecipitates show that a band co-migrating with the purified enzyme is detected in the culture medium; however, most of the labeled protein is found in the cell lysate. It is possible that the PLA$_2$ is effectively secreted from the cell but that it adheres to the cell membrane.

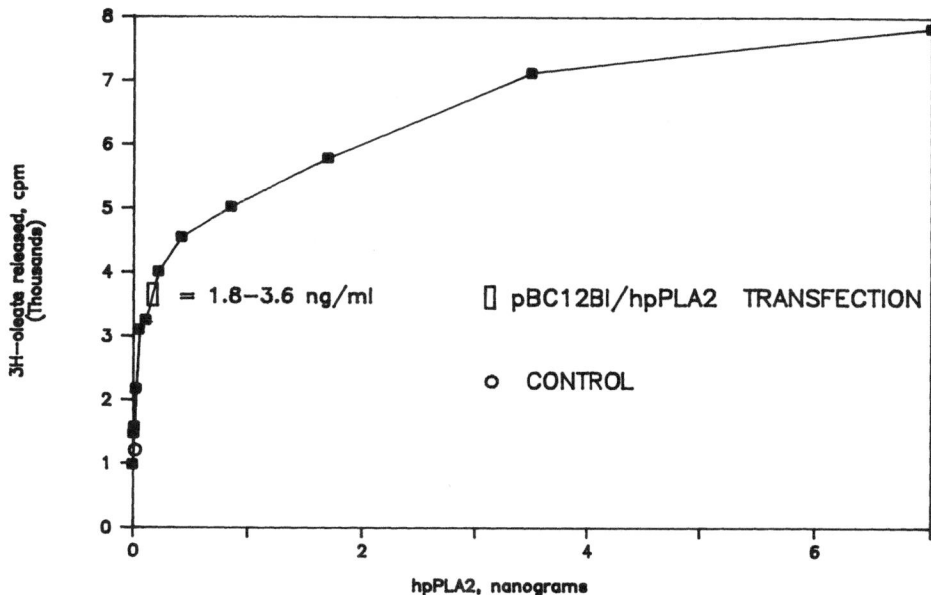

FIG. 4. PLA₂ assay of COS cell media. Samples (30 ul) of the medium from COS cell cultures at 72 hrs post-transfection were assayed for PLA₂ activity as described in METHODS. A standard curve (solid black squares) was generated by adding purified PLA₂ to medium. Each point represents the average of two determinations.

hpPLA₂ is Identical to Human Synovial Fluid PLA₂

While this work was underway, two reports appeared which described the isolation of genomic and cDNA clones coding for an extracellular, non-pancreatic PLA₂ purified from human rheumatoid synovial fluid (31,32). The hpPLA₂ cDNA described here encodes the identical amino acid sequence as that reported. The only difference noted in the DNA sequence is the presence of 3 additional nucleotides (ATA) at the 5' end of cDNA 9-1 compared to that reported by Seilhamer et al. (31). Thus, the PLA₂ that is found in a soluble, extracellular form in inflammatory exudate, also occurs as a membrane-associated enzyme in placenta.

Hayakawa, et al. (9) compared the amino-terminal sequence of the membrane-bound form of PLA₂ from rat platelets with that of a PLA₂ secreted from platelets upon stimulation with thrombin and found that residues 2-17 are identical (residue 1 could not be determined). Human platelet PLA₂ has also been characterized, and it is identical to the enzyme purified from synovial fluid (32). It is not yet clear whether the PLA₂ in synovial fluid is derived from platelets or released by inflammatory cells such as neutrophils or activated macrophages. The cellular source of the placental PLA₂ also remains unknown. We cannot rule out the possibilty that the extracellular form of PLA₂ from serum adheres to cells within placental tissues. However, the RNA blot data (discussed below) indicates that the PLA₂ gene is actively transcribed in placental tissue.

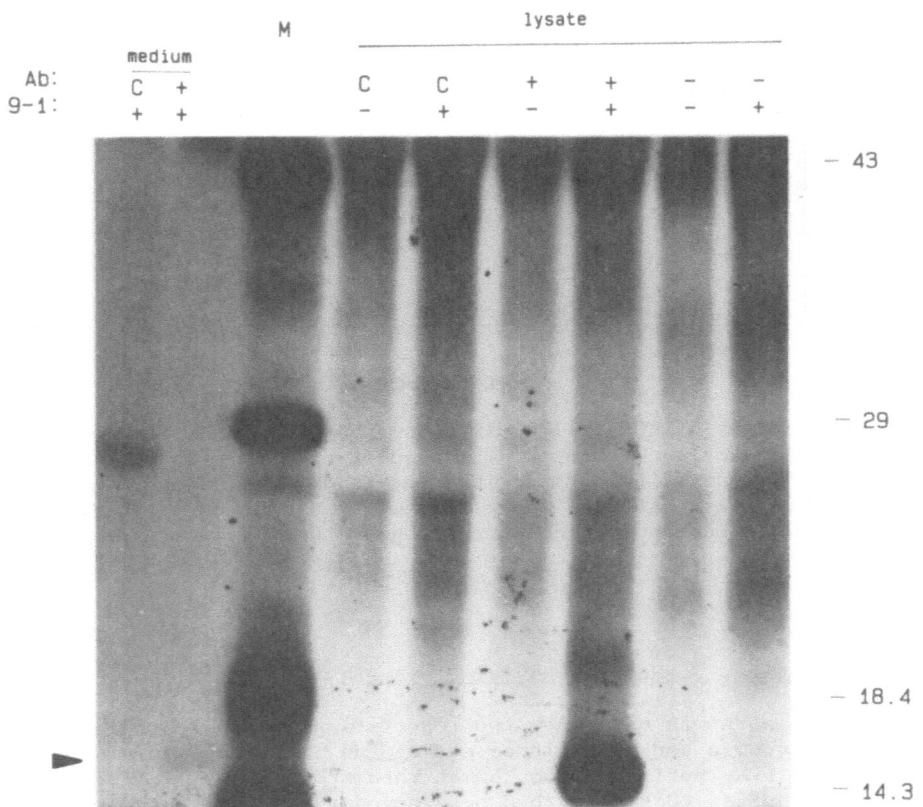

FIG. 5. Immunoprecipitation analysis of PLA_2 expressed in COS cells. The figure shows the autoradiorah of the immunoprecipitated proteins resolved by SDS-PAGE. The molecular weights of ^{14}C-labeled protein standards (M) are indicated in kilodaltons. Samples from the medium and cell lysate from COS cells transfected with pBC12BI/hpPLA$_2$ 9-1 (+) or mock-transfected cells (-) are indicated. Monoclonal antibodies (Ab) used for immunoprecipitation are (+), a cocktail of 4 different monoclonal Ab's made against purified placental PLA$_2$; (C), a control Ab against an unrelated protein; (-), no Ab. The arrow marks the postion of the PLA$_2$ protein band.

DNA and RNA Blot Analysis

As discussed above, the protein sequence comparison in Fig. 3 shows that hpPLA$_2$ is closely related to rat platelet PLA$_2$. To determine the extent of homology of hpPLA$_2$ at the DNA level with rat and other mammalian non-pancreatic PLA$_2$'s, ^{32}P-labeled cDNA 9-1 was used to probe a genomic blot containing EcoRI-digested DNA from human, monkey, rat, and mouse (Fig. 6A). Bands that specifically hybridize with the probe are evident in the lanes corresponding to both human and monkey. Very weak hybridization was detected for rat and mouse under the conditions used. A similar analysis was done for human genomic DNA that had been digested with various restriction endonucleases (Fig. 6B). The pattern observed is what would be expected from the sequence of the genomic clone of non-pancreatic PLA$_2$ (33,34). These results are consistent with there being only one gene corresponding to this PLA$_2$ cDNA.

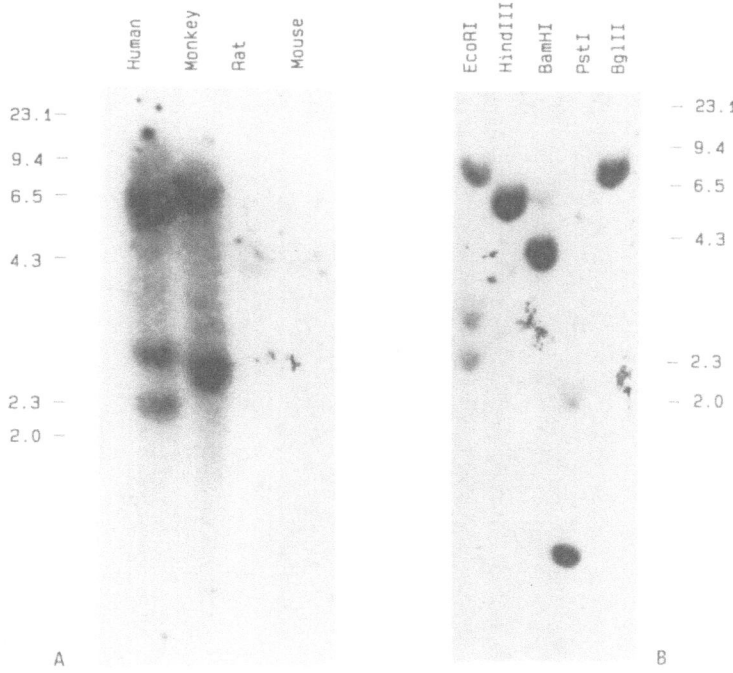

FIG. 6.   Genomic DNA analysis.   A) EcoRI-digested genomic DNA's as
indicated were probed as described in METHODS.   B) Human genomic
DNA digested with the indicated restriction endonucleases.   The
molecular weight standards, indicated in kilobase pairs, are
lambda DNA HindIII fragments.

To survey the tissue distribution of PLA$_2$ mRNA, polyA$^+$ RNA's from various monkey tissues were subjected to Northern blot analysis (Fig. 7). The experiments were done with monkey RNA's instead of human because of the availability of mRNA from a greater variety of tissues, and the data from the genomic blot indicated a high degree of homology. Serial dilutions of RNA from in vitro transcription of the cDNA clone 9-1 using SP6 polymerase were included as standards on the blot to determine approximate levels of mRNA in each sample. Compared to these standards the level of PLA$_2$ mRNA in human placenta is approximately 0.001% of total poly A$^+$ RNA. The results show that PLA$_2$ mRNA is present in liver and spleen at levels comparable to human placenta, present in much lower amounts in lung and heart, and not detectable in kidney or brain. It interesting to note that the levels of PLA$_2$ mRNA in the different monkey tissues parallel the levels of phospholipase activity determined several years ago in the respective tissues of rat (35). These results suggest that the product of a single gene encoding a non-pancreatic PLA$_2$ is present in a variety of tissues. The questions of what role(s) this PLA$_2$ plays in different cell types and how the transcription of the PLA$_2$ gene is regulated can now be addressed using the nucleic acid probes and monoclonal antibodies now available.

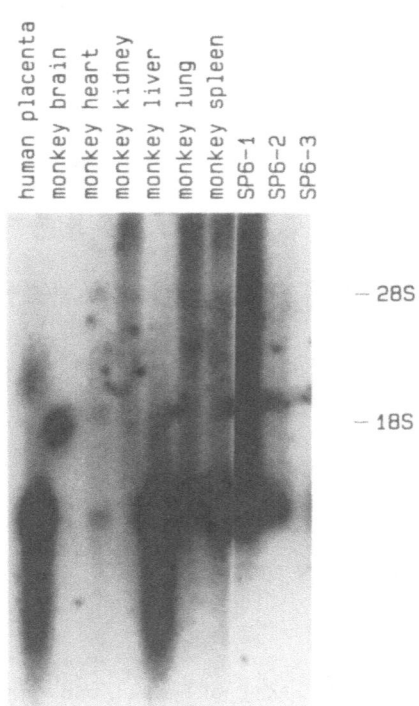

FIG. 7. RNA blot analysis. Samples of poly A$^+$ RNA as indicated and
described in the text were probed with ^{32}P-labeled cDNA 9-1.
The positions of 28S and 18S ribosomal RNA's are indicated.
Ten-fold serial dilutions of SP6 transcription product of clone
9-1 were included on the blot (SP6-1 to 3).

Are Isoforms of PLA$_2$ Generated by Alternative Splicing?

To date, only two genetically distinct PLA$_2$'s have been identified in both rat and human by cDNA cloning, the pancreatic (25,34) and non-pancreatic forms (this study and references 31,32,35). Can the diversity of physiologcal functions attributed to PLA$_2$ be due to the enzymes encoded by these two genes, and if so how does one account for the different molecular weight forms of PLA$_2$? One mechanism for generating different functional forms of a protein from a single gene is alternative splicing of precursor RNA. In recent years numerous examples of differential splicing events giving rise to protein isoforms have been reported (36). One of the five cDNA clones isolated in this study, designated 9-2, provides limited evidence that this mechanism may be operational in the case of the PLA$_2$ gene. Clone 9-2 contains the identical coding sequence for residues 1-124 of placental PLA$_2$; however, 9-2 differs from 9-1 by the presence of additional coding sequence within the signal peptide of the 9-1 sequence (Crowl, in preparation). This sequence is derived from part of the intron that separates the signal sequence coding region (32), utilizing an alternative splice acceptor site. The additional sequence maintains the translational reading frame with encoded amino acids that result in an alternative signal peptide. Assuming that this signal peptide is processed, the amino-terminus of the "mature" PLA$_2$ is extended by 35 residues. The predicted molecular weight of the resulting translation product would be 18,557, a size reminiscent of the membrane-bound forms from sheep erythrocytes and the mouse macrophage-like cell line P388D$_1$. Whether the protein product encoded by cDNA 9-2 is actually expressed in placenta, or any tissues, is currently under investigation.

ACKNOWLEDGEMENTS

The authors wish to thank J. Hulmes and M. C. Miedel for help with protein sequence analysis, W. McComas for oligonucleotide synthesis, K. Wada and W. Levin for providing purified hpPLA$_2$, L. Reik for supplying the monoclonal antibodies against hpPLA$_2$, D. Gash for help with the initial transfection experiments, V. Madison for the molecular modeling work, and P. Lomedico for advice and encouragement.

REFERENCES

1. Dennis, E.A. (1983) In: The Enzymes, Third Edition, Vol. 16, Boyer, P. (ed.). Academic Press, New York, pp.307-353. .
2. Verheij, H.M., Slotbloom, A.J. and de Haas, G.H. (1981) Rev. Physiol. Biochem. Pharmacol. 91, 91-203.
3. Waite, M. (1985) J. Lipid Res. 26, 1379-1388.
4. Van den Bosch, H. (1980) Biochim. Biophys. Acta 604, 191-246.
5. Dennis, E.A. (1987) Drug Dev. Res. 10, 205-220.
6. Ulevitch, R.J., Watanabe, Y., Sano, M., Lister, M.D., Deems, R.A. and Dennis, E.A. (1988) J. Biol. Chem. 263, 3079-3085.
7. Ono, T., Tojo, H., Kuramitsu, S., Kagamiyama, H. and Okamoto, M. (1988) J. Biol. Chem. 263, 5732-5738.
8. Aarsman, A.J., de Jong, J.G.N., Arnoldussen, E., Neys, F.W., van Wassenaar, P.D., and Van den Bosch, H. (1989) J. Biol. Chem. 2264, 10008-10014.
9. Hayakawa, M., Kudo, I., Tomita, M. and Inoue, K. (1988) J. Biochem. 103, 263-266.
10. Hayakawa, M., Kudo, I., Tomita, M., Nojima, S., and Inoue, K. (1988) J. Biochem. 104, 767-772.

11. Kramer, R.M., Wuthrich, C., Bollier, C., Allegrini, P.R. and Zahler, P. (1978) Biochim. Biophys. Acta 507, 381-394.
12. Okazaki, T., Sagawa, N., Bleasdale, J.E., Okita, J.R., MacDonald, P.C. and Johnston, J.M. (1981) Biol. of Reproduction 25, 103-109.
13. Levin, S.W., Butler, J.D., Schumacher, U.K., Wightman, P.D. and Mukherjee, A.B. (1986) Life Sciences 38, 1813-1819.
14. Wilson, T. Liggins, G.C. and Joe, L. (1989) Am. J. Obstet. Gynecol. 160, 602-606.
15. Lai, C.-Y., Wada, K. (1988) Biochem. & Biophys. Res. Commun. 157, 488-493.
16. Wood, W.I., Gitschier, J., Laskey, L.A. and Lawn, R.M. (1985) Proc. Natl. Acad. Sci. 82, 1585-1588.
17. Silhavy, T.J., Berman, M.L. and Enquist, L.W. (1984) In: Experiments with Gene Fusions. Cold Spring Harbor Press, Cold Spring Harbor, N.Y.
18. Maniatis, T. Fritsch, E.F. and Sambrook, J. (1982) In: Molecular Cloning: a Laboratory Manual. Cold Spring Harbor Press, Cold Spring Harbor, N.Y.
19. Yanisch-Perron, C., Vieira J., and Messing J. (1985) Gene 33, 103-119.
20. Sanger, F., Nicklen, S. and Coulson, A.R. (1977) Proc. Natl. Acad. Sci. 74, 5463-5469.
21. Patriarca, P., Beckerdite, S. and Elsbach, P. (1972) Biochim. Biophys. Acta 260, 593-600.
22. Cullen, B.R. (1987) Methods in Enzymology 152, 684-704.
23. Lathe, R. (1985) J. Mol. Biol. 183, 1-12.
24. Verheij, H.M., Westerman, J., Sternby, B., and De Haas, G. (1983) Biochim. Biophys. Acta 747, 93-99.
25. Seilhamer, J.J., Randall, T.L., Yamanaka, M., and Johnson, L.K. (1986) DNA 5, 519-527.
26. Tsai, I.H., Wu, S.H. and Lo T.B. (1981) Toxicon 19, 141-152.
27. Randolph, A. and Heinrikson, R.L. (1982) J. Biol. Chem. 257, 2155-2161.
28. Heinrikson, R.L., Kreuger, E.T. and Keim, P.S. (1977) J. Biol. Chem. 252, 4913-4921.
29. Cho, W., Tomasselli, A.G., Heinrikson, R.L., and Kezdy, F.J. (1988) J. Biol. Chem. 263, 11237-11241.
30. Tomasselli, A.G., Hui, J., Fisher, J., Zürcher-Neely, H., Reardon, I.M., Oriaku, E., Kézdy, F.J. and Heinrikson, R.L. (1989) J. Biol. Chem. 264, 10041-10047.
31. Seilhamer, J.J., Pruzanski, W., Vadas, P., Plant. S., Miller, J.A., Kloss, J., and Johnson, L.K. (1989) J. Biol. Chem. 264, 5335-5338.
32. Kramer, R.M., Hession, C., Johanson, B., Hayes, G., McGray, P., Chow, E.P., Tizard, R., and Pepinsky, R.B. (1989) J. Biol. Chem. 264, 5768-5775.
33. Gallai-Hatchard, J.J. and Thompson, R.H.S. (1965) Biochim. Biophys. Acta 98, 128-135.
34. Ohara, O., Tamaki, M., Nakamura, E., Tsuruta, Y., Fujii, Y., Shin, M., Teraoka, H. and Okamoto, M. (1986) J. Biochem. 99, 733-739.
35. Ishizaki, J., Ohara, O., Nakamura, E., Tamaki, M., Ono, T., Kanda, A., Yoshida, N., Teraoka, H., Tojo, H. and Okamoto, M. (1989) Biochem. & Biophys. Res. Commun. 162, 1030-1036.
36. Andreadis, A., Gallego, M.E. and Nadal-Ginard, B. (1987) Ann. Rev. Cell Biol. 3, 207-242.

G PROTEIN REGULATION OF PHOSPHOLIPASE A$_2$:

PARTIAL RECONSTITUTION OF THE SYSTEM IN CELLS

Ronald M. Burch

Nova Pharmaceutical Corporation
6200 Freeport Centre
Baltimore, Maryland 21224

INTRODUCTION

Many receptor agonists stimulate eicosanoid synthesis very rapidly, in many instances the entire process being complete within 5 - 10 minutes (e.g., Burch, 1989a). The rate-limiting step is thought to be release of free arachidonate from phospholipids (Irvine, 1982). The mechanism for arachidonate release is still uncertain. Many early studies concluded that phospholipase A$_2$ (PLA$_2$) is the enzyme responsible, releasing arachidonate from the sn-2 position of phospholipids. This mechanism is supported by the finding that in mammalian cells, the sn-2 position of phosphatidylcholine, phosphatidylethanolamine, and phosphatidylinositol is enriched in arachidonate (e.g., Prescott and Majerus, 1981). Other investigators have suggested that phosphatidylinositol-specific phospholipase C (PI-PLC) is the enzyme that initiates arachidonate release (Majerus et al., 1986). In this pathway, PI-PLC releases diacylglycerol, from which arachidonate is subsequently released by a diacylglycerol lipase (Bell et al., 1979). However, it has become clear that in many cells arachidonate release can be readily separated from phosphatidylinositol metabolism (e.g., Burch et al., 1986; Burch and Axelrod, 1987; Slivka and Insel, 1987, 1988; Welsh et al., 1988).

Several mechanisms have been suggested to explain rapid activation of PLA$_2$ (Irvine, 1982; Axelrod et al., 1988; Burch, 1989a, 1989b). Commonly it is suggested that receptor occupation causes a rise in intracellular free calcium. The calcium may originate from outside the cell, entering through specific channels, or it may arise by release from intracellular stores, mediated by inositol phosphates. This type of mechanism has been proposed because studies have found that many purified PLA$_2$'s require calcium for activity. However, calcium-dependent PLA$_2$'s often are reported to require calcium concentrations in the mM range, far higher than result from receptor activation. Many PLA's require alkaline pH, and it has been proposed that activation of Na$^+$/H$^+$ exchange may cause alkalinization of the cytoplasm sufficient to initiate PLA$_2$ activity. This mechanism may provide arachidonate in platelets activated with α_2 adrenergic agonists (Sweatt et al., 1986; Banga et al., 1986) but does not appear to be widespread. Recently, it was suggested that receptor activation leads to rapid synthesis of a PLA$_2$ activator protein (Crooke et al., 1989). The problem of PLA$_2$ activation has been further compounded by findings of multiple PLA$_2$'s in any given cell. These enzymes prefer various substrates including phospholipids specifically containing or not containing ether-linked fatty acids (e.g., Ballou et al., 1986; Loeb and Gross, 1986), different pH values, from 5 to 9 (e.g., Wightman et al., 1981) and different calcium concentrations, from 100 nM to 10 mM (Wightman et al., 1981; Loeb and Gross, 1986).

Many of the receptors that cause arachidonate release are coupled to GTP-binding regulatory proteins (G proteins). G proteins couple receptors to a variety of effector proteins including adenylate cyclase, ion channels, and phospholipase C (Casey and Gilman, 1988). Since arachidonate release in response to agonists often occurs as rapidly as PI-PLC activation and can be readily separated from inositol phosphate formation in many cell types, it seemed possible, even likely, that PLA$_2$ activated by these receptors should also be coupled through G proteins.

Biochemistry, Molecular Biology, and Physiology of Phospholipase A$_2$ and Its Regulatory Factors
Edited by A. B. Mukherjee, Plenum Press, New York, 1990

185

G PROTEINS

G proteins serve to couple numerous receptors to their effectors. The most-studied G proteins belong to a genetic family of closely related members. These proteins are heterotrimeric, composed of α, β, and γ subunits (Casey and Gilman, 1988). Each α subunit has a binding site for guanine nucleotides and exhibits GTPase activity. Many α subunits also have binding sites for NAD-dependent ADP-ribosylation by cholera toxin and/or pertussis toxin. Cholera toxin activates G_s while pertussis toxin inactivates G_i, G_o and G_t (Casey and Gilman, 1988). There are numerous other G proteins in this family that are only beginning to be characterized.

The $\beta\gamma$ subunits are tightly associated and can be substituted among the various G_α subunits in assays *in vitro*. The $\beta\gamma$ complex usually terminates the activity of the α subunit when the three become associated. However, evidence has accumulated suggesting that the $\beta\gamma$ subunits have activity of their own (Neer and Clapham, 1988, and see below).

The mechanism of receptor-effector coupling mediated by G proteins has been characterized in some detail. Briefly, in the inactive state, G proteins exist as the $\alpha\beta\gamma$ complexes with GDP bound to the α subunit (Casey and Gilman, 1988). Upon binding of an agonist to its receptor, the agonist-receptor complex interacts with the G_α to cause dissociation of GDP. GTP then binds to the guanine nucleotide site, decreasing affinity of the agonist for the receptor, resulting in dissociation of the agonist and receptor from the G-GTP complex. The G-GTP complex binds to the effector, and the activated G_α-GTP probably dissociates from the $\beta\gamma$ subunits (Gilman, 1987). Deactivation of G_α-GTP is caused by the endogenous GTPase of the α subunit. This results in G_α-GDP reassociating with $\beta\gamma$.

G PROTEIN REGULATION OF PLA₂

Evidence suggesting that G proteins may couple receptors to ultimate arachidonate release and metabolism was first suggested by the observations that pertussis toxin blocked arachidonate release from neutrophils in response to the calcium ionophore A23187 (Bokach and Gilman, 1984) and chemotactic peptides (Okajima and Ui, 1984). Several additional reports soon followed (reviewed in Burch, 1989a,b).

In the first study to directly assess the effects of GTP analogs on PLA_2 activity, it was demonstrated that the α_1 adrenergic receptor in the FRTL5 thyroid cell is coupled to a pertussis toxin-sensitive G protein to stimulate PLA_2 and a pertussis toxin-insensitive G protein to stimulate PI-PLC (Burch et al., 1986). In membranes isolated from FRTL5 cells, GTPγS, a poorly hydrolyzed GTP analog, was found to stimulate metabolism of exogenous phosphatidylcholine by PLA_2.

Using bovine rod outer segments, Jelsema (1987) found that rhodopsin is coupled to PLA_2 by G_t. Stimulation of rhodopsin by light is well-established to be coupled to cGMP-phosphodiesterase by G_t (Stryer, 1987). Jelsema found that light shone onto dark-adapted rod outer segments caused activation of PLA_2. G protein mediation was suggested by the observation that GTPγS mimicked light in activating PLA_2. That G_t is the specific G protein involved was suggested by the observation that both pertussis toxin and cholera toxin blocked light-induced activation of PLA_2 as well as light-induced activation of cGMP-phosphodiesterase. Further studies in the rod outer segment preparation have demonstrated that it is the $\beta\gamma$ subunit, rather than the α subunit of G_t that activates PLA_2 (Jelsema and Axelrod, 1987). $\beta\gamma$ subunits also activate PLA_2 in other systems (Kim et al., 1989).

RECONSTITUTION OF G PROTEIN-PLA₂ INTERACTIONS

The ultimate proof that a G protein interacts directly with PLA_2 to increase its activity requires reconstitution of a purified G protein with a purified PLA_2. While certain G proteins are easily purified in good yield, no purification of a PLA_2 known to be hormonally-regulated has been reported. Thus, we have chosen to attempt to define conditions under which cells may be permeabilized to allow well-defined exogenous substrates to be used to assess PLA_2 activity.

3T3 FIBROBLASTS

We have demonstrated that 3T3 fibroblasts express B_2 bradykinin receptors (Burch et al., 1988a). Activation of bradykinin receptors on these cells leads to independent activation of PLA_2, to release arachidonate, and PI-PLC (Burch and Axelrod, 1987). That the receptors are linked to G proteins is suggested by the observations that bradykinin stimulates GTPase activity (Burch et al., 1988a) and increases

GTPγ[^{35}S] binding (Burch and Mahan, 1989) to membranes from the cells.

In previous studies we have used several methods to permeabilize 3T3 cells. Hypo-osmotic shock was used to transiently permeabilize the cells and was successful in allowing us to incorporate GTPγS (Burch and Axelrod, 1987). Later, we used digitonin to permeabilize the cells (Burch, 1989a). This method allowed demonstration of metabolism of exogenous phospholipids by the endogenous PLA$_2$. However, using this method we have not consistently been able to permeabilize cells without lysis, perhaps because of varying purity of digitonin preparations commercially available. Another agent that has been successfully used to permeabilize cells is streptolysin O (Buckingham and Duncan, 1983; Howell and Gomperts, 1987; Ali et al., 1989). This bacterial product interacts with membrane cholesterol to produce stable "pores" through which normally cell-impermeant compounds can be introduced.

Methods. SV-T$_2$ fibroblasts (3T3 cells) were cultured as previously described (Burch et al., 1988a). Cells were removed from the culture dishes by addition of 2 mM EDTA to Ca^{2+},Mg^{2+}-free Hank's balanced salt solution. Suspensions were centrifuged at 400 x g for 5 minutes, then resuspended at 5 x 10^6 cells/ml in KCl (140 mM), Pipes (10 mM), pH 7.4, MgCl$_2$ (5 mM), ATP (5 mM), EGTA (1 mM), an ATP-generating system composed of phosphocreatine (7 mM) and creatine phosphokinase (50 U/ml), and free calcium as described in the figure legends (calculated using EQCAL, Biosoft, and sometimes confirmed with a calcium electrode). Streptolysin O (2000 U/ml, Sigma) was added for 10 minutes at 37°C (resulting in > 90% of cells taking up trypan blue). Then, 1-stearoyl-2-[^{3}H]arachidonyl-phosphatidylcholine (New England Nuclear, 110 Ci/mmol) (100,000 dpm, 100 μM) or phosphatidyl[^{3}H]inositol-4,5-bisphosphate (New England Nuclear, 3 Ci/mmol) (100,000 dpm, 100 μM), hereafter referred to as PIP$_2$, was added along with guanine nucleotide and/or bradykinin and the suspensions were incubated at 37°C for 15 minutes. Reactions were terminated as follows. For PLA$_2$ activity on phosphatidylcholine, 2 ml of isopropanol:heptane:acetic acid (40:10:1) was added, the solution vortexed, then 1 ml of heptane was added. The solution was vortexed, and finally centrifuged to separate phases. The upper phase was collected, 100 mg of silica gel (200 mesh, Sigma) was added, the solution was vortexed, then centrifuged. An aliquot of the supernatant was collected and radioactivity was determined by liquid scintillation spectrometry. When PIP$_2$ was used as substrate, 100 μl of 100% trichloracetic acid were added, the solutions incubated in an ice

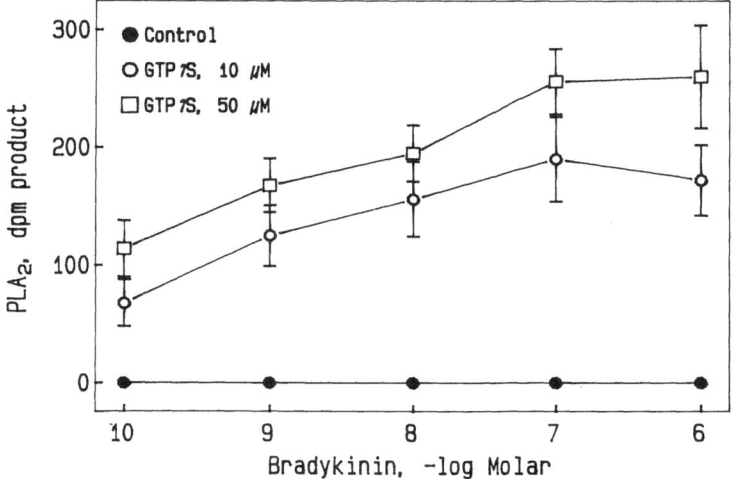

Fig. 1. GTPγS stimulates PLA$_2$ in permeabilized 3T3 fibroblasts, and augments bradykinin-stimulated PLA$_2$. Cells were resuspended in media containing 100 nM free calcium. Radiolabeled phosphatidylcholine was added simultaneously with bradykinin and/or GTPγS. Incubations were for 15 minutes at 37°C. Data are mean ± SEM for 5 separate experiments, performed in triplicate.

bath for 30 minutes, then extracted 4 times with 4 ml ether to remove the trichloracetic acid. Next, 4 ml of 5 mM sodium tetraborate-0.5 mM EDTA (pH 9) was added, and the solution was added to a column containing 0.5 ml AG 1-X8 resin (formate form, 100-200 mesh). The column was washed with 25 ml of water, followed by collection of 3 times 0.5 ml 5 mM sodium tetraborate/60 mM sodium formate to collect glycerophosphoinositol (GPI). The column was next washed with 25 ml of 0.1 mM formic acid/0.4 M ammonium formate to elute IP_1 and IP_2, then IP_3 was recovered by collection of 3 times 0.5 ml 0.1 M formic acid/1 M ammonium formate (Berridge, 1983).

Results. When 3T3 cells were permeabilized with streptolysin O, then PLA_2 activity assayed, bradykinin did not stimulate PLA_2 when free calcium was maintained at 100 nM (Figure 1). Addition of GTPγS to the media stimulated PLA_2 in a dose-dependent manner, and, in the presence of GTPγS, bradykinin stimulated PLA_2 (Figure 1).

PLA_2 activity in the permeabilized cells depended on free intracellular calcium concentration, being greater at higher calcium concentrations (Figure 2). At low free calcium concentrations (100 nM to 1 μM), bradykinin did not stimulate PLA_2. At higher calcium concentrations, 10 μM to 10 mM, PLA_2 activity in the presence of bradykinin was actually decreased compared to control (Figure 2). GTPγS stimulated PLA_2 when free calcium was lower, 100 nM to 10 μM, but when free calcium was held at 10 mM, stimulation by GTPγS was not apparent. GTPγS augmented bradykinin-induced PLA_2 activity, which was more pronounced at lower free calcium concentrations (Figure 2).

In permeabilized 3T3 cells, PI-PLC was also found to be calcium-dependent, and GTPγS, and bradykinin plus GTPγS stimulated PI-PLC (data not shown).

Thus, in permeabilized 3T3 fibroblasts, PLA_2 activity was dependent on free calcium concentration. GTPγS stimulated PLA_2 activity, while GDPβS (not shown) did not. GTPγS-stimulated PLA_2 activity was apparent only at free calcium concentrations less than 10 μM. The reason for this is unclear. Perhaps both GTPγS and calcium can stimulate the enzyme maximally, such that augmentation by both is only apparent at submaximal concentrations of either stimulator. Alternatively, several PLA_2's may be present in the cells. One may be calcium-dependent while another is G protein-dependent. Because in these cells PLA_2 is not pertussis toxin-sensitive (Burch and Axelrod, 1987), we chose to not pursue them further in the intact cell reconstitution experiments.

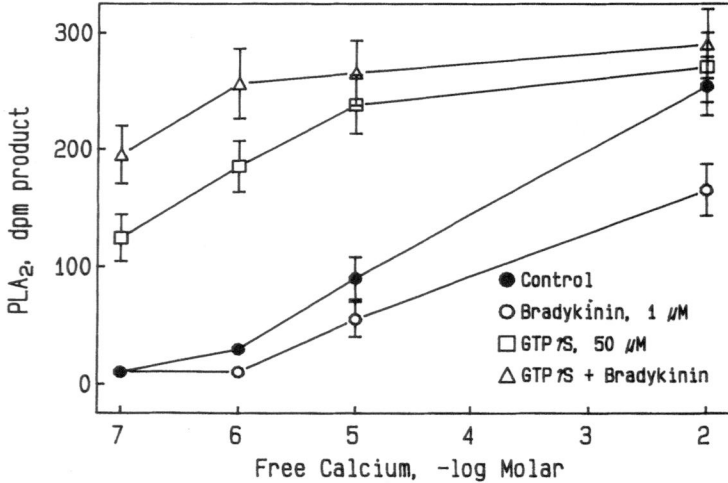

Fig. 2. GTPγS lowers the calcium requirement for PLA_2 activation. Incubations with radiolabeled phosphatidylcholine were for 15 minutes at 37°C. Bradykinin, where present, was 1 μM, GTPγS was 50 μM. Data are mean ± SEM for 3 separate experiments, performed in triplicate.

Table 1. PLA$_2$ Activity in RAW264.7 Macrophages in Response to GTPγS, PMA, LPS, or IgG$_{2b}$

Agent	Free Calcium	PLA$_2$, dpm Product
		[^{3}H]arachidonate
Control	100 nM	53 ± 16
	1 μM	64 ± 18
	10 μM	128 ± 16
	10 mM	640 ± 88
GTPγS, 50 μM	100 nM	48 ± 21
	1 μM	72 ± 15
	10 μM	142 ± 35
	10 mM	665 ± 76
Control	1 μM	75 ± 12
	10 mM	596 ± 55
PMA, 100 nM	1 μM	112 ± 18
	10 mM	944 ± 85
LPS, 10 μg/ml	1 μM	187 ± 32
	10 mM	1220 ± 185
plus GTPγS	1 μM	246 ± 28
IgG$_{2b}$, 100 μg/ml	1 μM	95 ± 16
	10 mM	865 ± 74
		[^{3}H]GPI, dpm
Control	1 μM	175 ± 18
	10 mM	168 ± 12
GTPγS	1 μM	218 ± 26
	10 mM	229 ± 21
PMA	1 μM	186 ± 20
	10 mM	428 ± 65
LPS	1 μM	208 ± 15
	10 mM	844 ± 96
plus GTPγS	1 μM	235 ± 28
IgG$_{2b}$	1 μM	166 ± 24
	10 mM	325 ± 40

GTPγS was added for 15 minutes. In experiments examining PMA, LPS, and IgG$_{2b}$, agents were added to intact cells for 18 hours prior to measuring PLA$_2$ in permeabilized cells for 15 minutes. IgG$_{2b}$ was heat aggregated by incubating at 63°C for 15 minutes at 10 mg/ml. Data are mean ± SEM from 3 separate experiments.

We next examined the potential coupling of G protein to PLA$_2$ in RAW264.7 macrophages. We have previously found that both cholera toxin and pertussis toxin increase PLA$_2$ activity and eicosanoid synthesis in these cells. The preferred substrate in intact cells appeared to be phosphatidylcholine. PI-PLC was not activated (Burch et al., 1988b).

Methods. RAW264.7 macrophages were cultured as previously described (Burch et al., 1988b). Cells were removed from culture dishes as described above (with the addition of gentle scraping) and resuspended at 5×10^6 cells/ml. Permeabilization was carried out as described above. Phospholipase activities were assayed as described above.

Results. When 1-stearoyl-2-[^{3}H]arachidonoyl-phosphatidylcholine was used as substrate, GTPγS did not activate PLA$_2$ when free calcium was maintained at 100 nM, 1 μM, 10 μM, or 10 mM (Table 1). However, 10 mM calcium did increase PLA$_2$ activity compared to the lower concentrations. If cells were pretreated with phorbol-myristate-acetate (PMA), lipopolysaccharide (LPS), or heat-aggregated IgG$_{2b}$ for 16 hours, then cells prepared, PLA$_2$ activity was marginally increased compared to nontreated cells at a free calcium concentration of 1 μM (Table 1), but was greatly increased compared to control cells in the presence of 10 mM free calcium. When PIP$_2$ was used as substrate, GPI, a product of PLA$_2$ followed by lysopholipase activity (Burch et al., 1988), was not increased by GTPγS, but was increased by 18 hour treatment with PMA, LPS, or IgG$_{2b}$ (Table 1).

In the experiments examining GPI formation when PIP$_2$ was used as substrate, we also examined IP$_3$ formation. In the presence of 100 nM, 1 μM, or 10 μM free calcium, GTPγS greatly augmented IP$_3$ formation (Table 2). High calcium concentration, 10 mM, also stimulated IP$_3$ formation.

In summary, we have previously found in RAW264.7 macrophages that both cholera toxin and pertussis toxin stimulate PLA$_2$ (Burch et al., 1988). The stimulation required several hours to become apparent. This was considered consistent with the time required for the active protomer to reach the cytosol to ADP-ribosylate G protein α subunits (Ui, 1984), then activate PLA$_2$. The present results make that hypothesis less certain. PMA, LPS, and IgG$_{2b}$ also stimulate eicosanoid synthesis in these cells (Burch, 1987). These agents, too, require hours for stimulation to become maximum. Pretreatment of RAW264.7 macrophages with PMA, LPS, or IgG$_{2b}$ increased PLA$_2$, especially when assayed at 10 mM calcium. In the pretreated cells, GTPγS still did not augment PLA$_2$. Thus, the extended time course required to increase eicosanoid synthesis in RAW264.7 macrophages may be related to induction of synthesis of additional PLA$_2$ molecules, or, perhaps, PLA$_2$ regulatory molecules.

Table 2.   GTPγS Stimulated Formation of [^{3}H]IP$_3$ in RAW264.7 Macrophages

Agent	Free Calcium	[^{3}H]IP$_3$, dpm
Control	100 nM	55 $\pm$ 20
	1 μM	68 $\pm$ 18
	10 μM	78 $\pm$ 22
	10 mM	2842 $\pm$ 425
GTPγS, 50 μM	100 nM	236 $\pm$ 48
	1 μM	412 $\pm$ 78
	10 μM	735 $\pm$ 146
	10 mM	3012 $\pm$ 345

Data are mean $\pm$ SEM for 3 experiments, performed in triplicate.

Compound 48/80 and IgE have been shown to stimulate release of arachidonate from rat peritoneal mast cells (Nakamura and Ui, 1983). Unlike RAW264.7 cells, arachidonate release in response to stimulation is prompt, being nearly complete within 10 minutes. The IgE receptor and compound 48/80 are coupled to a G protein in these cells as evidenced by pertussis toxin inhibition of inositol phosphate and arachidonate release (Nakamura and Ui, 1983; 1984; 1985). In addition, GTP analogs have been used to directly stimulate PI-PLC in mast cells (Nakamura and Ui, 1984; Cockcroft and Gomperts, 1985). Streptolysin O has been used to permeabilize mast cells in studies examining signal transduction (Howell and Gomperts, 1987). Thus, we reasoned that these cells might be more likely than RAW264.7 macrophages to experience PLA$_2$ activation by direct coupling to a G protein.

Methods. Rat peritoneal mast cells were isolated on albumin gradients as described (Nakamura and Ui, 1983). Cells were suspended to 5×10^6/ml. Permeabilization and stimulation were performed as described above for fibroblasts.

Results. GTPγS stimulated PLA$_2$ activity using phosphatidylcholine as substrate (Figure 3). Stimulation was evident at free calcium concentrations of 100 nM to 10 μM, but when calcium concentration was increased to 10 mM, stimulation by GTPγS was less apparent.

Pretreatment of cells with pertussis toxin (List), 100 ng/ml, for 2 hours, completely blocked GTPγS-stimulated PLA$_2$ activity (Table 3). G$_i$ and G$_o$ were purified from bovine brain as described (Katada et al., 1986). Addition of either G protein to pertussis toxin-pretreated cells in a ratio of 10,000 or 100,000 molecules per cell, partially reconstituted GTPγS-stimulated PLA$_2$ activation (Table 3).

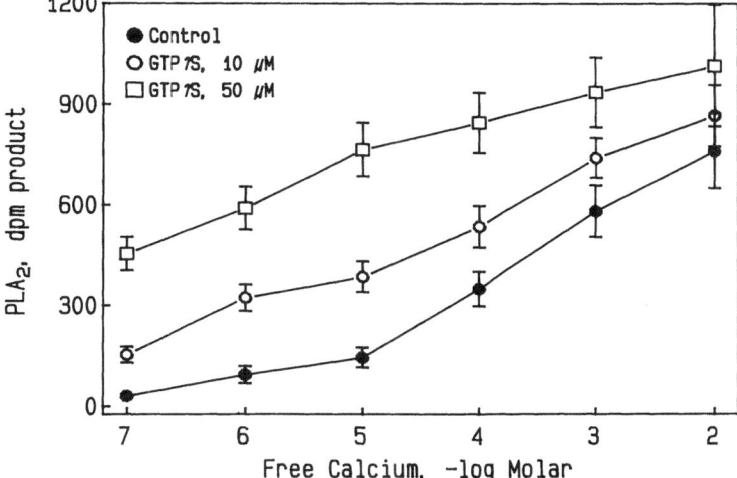

Fig. 3. GTPγS stimulates PLA$_2$ in rat mast cells. Cells were isolated on albumin gradients. Cells were permeabilized, incubated with radiolabeled phosphatidylcholine at 37°C for 10 minutes. Data are mean ± SEM for 3 separate experiments, performed in duplicate.

Table 3. Pertussis Toxin-Sensitive G Proteins in Activation of PLA$_2$ in Rat Peritoneal Mast Cells

Agent	PLA$_2$, dpm Product
Control	92 ± 42
GTPγS, 50 μM	852 ± 115
Pertussis toxin, 100 ng/ml	126 ± 38
Pertussis toxin plus GTPγS	118 ± 45
Pertussis toxin plus G$_o$, 100,000/cell	86 ± 36
Pertussis toxin plus G$_i$, 100,000/cell	102 ± 44
GTPγS plus Pertussis toxin	
plus G$_o$, 10,000/cell	365 ± 68
plus G$_o$, 100,000/cell	526 ± 108
GTPγS plus Pertussis toxin	
plus G$_i$, 10,000/cell	286 ± 78
plus G$_i$, 100,000/cell	672 ± 93

Free calcium was maintained at 1 μM in these experiments. Results are mean ± SEM of 3 experiments performed in triplicate.

p-Bromophenacyl bromide (pBPB) covalently inactivates PLA$_2$, as well as other enzymes (Hofmann et al., 1982). In several experiments mast cells were preincubated with pBPB for 1 hour prior to permeabilization. In these cells, PLA$_2$ activity was not apparent at any calcium concentration from 100 nM to 10 mM (data from 1 μM free calcium is shown in Table 4). When porcine pancreatic PLA$_2$ (Sigma), previously dialyzed against the assay buffer, was added to the pBPB-inactivated cells, little activity was apparent at 1 μM free calcium (Table 4). However, in the presence of GTPγS, pancreatic PLA$_2$ activity was apparent. GTPγS did not increase pancreatic PLA$_2$ activity in cells that had also been pretreated with pertussis toxin (Table 4).

In additional experiments pancreatic PLA$_2$ activity was assessed in the cellular incubation media, but in the absence of cells. In these experiments addition of GTPγS, plus or minus G$_o$ or G$_i$, at concentration ratios of 1 to 1000 G protein to PLA$_2$, was without effect on PLA$_2$ activity (data not shown).

In summary, arachidonate release in mast cells has been previously shown to be pertussis toxin-sensitive (Nakamura and Ui, 1984; 1985). In addition, GTPγS increased PLA$_2$ activity, which was blocked by pretreatment of the cells with pertussis toxin. When G$_i$ or G$_o$, G proteins that are inactivated by pertussis toxin, were added to pertussis toxin-treated cells, GTPγS-stimulated PLA$_2$ activity was reconstituted. pBPB inactivated cellular PLA$_2$. Addition of pancreatic PLA$_2$, which was not stimulated by GTPγS, G$_i$, or G$_o$ *in vitro*, resulted in reconstitution of GTPγS-stimulated PLA$_2$ activity.

MECHANISM OF G PROTEIN COUPLING TO PLA$_2$

The mechanism by which G proteins activate PLA$_2$ is unknown. The mechanism for G protein activation of PI-PLC appears to be direct coupling of the 2 proteins, resulting in a reduction in the enzyme's requirement for free calcium (Smith et al., 1986). A similar mechanism for G protein-coupled PLA$_2$ was suggested by a study in which GTPγS lowered the calcium requirement for chemotactic peptide stimulated arachidonate release from saponin-permeabilized neutrophils (Nakashima et al., 1988). The present study

Table 4. pBPB Inactivates Mast Cell PLA$_2$, but Pancreatic PLA$_2$ Can Reconstitute GTPγS-Stimulated PLA$_2$ Activity in pBPB-Treated Cells

Treatment	PLA$_2$, dpm Product
Control	125 ± 15
GTPγS, 50 μM	682 ± 48
pBPB pretreatment	46 ± 40
pBPB plus GTPγS	85 ± 46
pBPB plus pancreatic PLA$_2$, 1 nM	210 ± 36
pBPB plus pancreatic PLA$_2$ plus GTPγS	550 ± 110
pBPB plus pertussis toxin, 100 ng/ml	38 ± 25
pBPB plus pertussis toxin, 100 ng/ml plus GTPγS	44 ± 35
pBPB plus pertussis toxin, 100 ng/ml plus GTPγS plus pancreatic PLA$_2$	296 ± 54

All experiments were performed in the presence of 1 μM free calcium. GTPγS was 50 μM. pBPB pretreatment was for 1 hour. Pertussis toxin pretreatment was for 2 hours, with pBPB being added after 1 hour where indicated. Activity of pancreatic PLA$_2$ in the presence of 10 mM free calcium was 7480 ± 225 dpm. Data are mean ± SEM for 3 experiments performed in duplicate.

supports a mechanism of lowered calcium requirement since GTPγS shifted the dose-response for stimulation of PLA$_2$ by Ca^{2+} to lower concentrations in both 3T3 cells and mast cells. However, in rod outer segments, no calcium dependence was found for PLA$_2$ stimulation by G$_t$ (Jelsema and Axelrod, 1987).

In at least some systems it is the ßγ subunit of the G protein that activates PLA$_2$. ßγ subunits purified from G$_t$ or G$_o$, when added to dark-adapted, G$_t$-depleted rod outer segments, caused marked increase in PLA$_2$ activity, while addition of an equivalent amount of G$_{t\alpha}$ subunit was with minimal effect (Jelsema and Axelrod, 1987; Jelsema et al., 1989). Complicating the interpretation, however, is the observation that G$_{o\alpha}$ also activated PLA$_2$ in rod outer segments (Jelsema et al., 1989). In membrane patches from neonatal rat atrial myocytes, addition of ßγ subunit also activated PLA$_2$ (Kim et al., 1989). In both of these systems, ßγ activation of PLA$_2$ occurred in the presence of GTPγS. Thus, an effect of the ßγ subunits to bind to and remove an inhibitory α subunit is precluded.

Instead of directly interacting with PLA$_2$, G proteins may indirectly affect the enzyme. Pertussis toxin-sensitive leukotriene receptor-mediated activation of PLA$_2$ is proposed to result from G protein initiated rapid synthesis of a PLA$_2$ stimulatory protein (Crooke et al., 1989). G protein-mediated increases in intracellular free calcium via activation of receptor-operated calcium channels or intracellular release mediated by inositol phosphates is unlikely as a general mechanism since inositol phosphate formation and arachidonate release are readily separated (Burch, 1989a, b), and certainly could not occur in the rod outer segment preparations, atrial patches, or the permeabilized, calcium-clamped cells described here. Jelsema has suggested that ßγ subunit-mediated PLA$_2$ activation may derive from interaction of the ßγ subunits with PLA$_2$ inhibitory proteins (Jelsema et al., 1989).

FUTURE DIRECTIONS

PLA$_2$ activation by G proteins has now been described in several experimental systems. As of yet, no simple reconstitution system has been developed that allows unequivocal demonstration that a G protein subunit can directly interact with a purified PLA$_2$. Until such a system is described no examination of mechanism can be attempted. The present observation suggesting that an exogenous PLA$_2$ may be activated by an endogenous G protein in permeabilized mast cells, may be a clue that provision of a proper membrane-lipid environment might allow a completely *in vitro* reconstitution system.

ACKNOWLEDGEMENT

The author wishes to thank Cheryl Sowards for preparation of the manuscript, and Julius Axelrod for many stimulating discussions concerning the regulation of PLA$_2$.

REFERENCES

Ali, H., Cunha-Melo, J. R., and Beaven, M. A., 1989, Receptor-mediated release of inositol 1,4,5-trisphosphate and inositol 1,4-bisphosphate in rat basophilic leukemic RBL-2H3 cells permeabilized with streptolysin O, *Biochim. Biophys. Acta*, 1010:88.

Axelrod, J., Burch, R. M., and Jelsema, C. L., 1988, Receptor-mediated activation of phospholipase A$_2$ via GTP-binding proteins: arachidonic acid and its metabolites as second messengers, *Trends Neurosci.*, 11:117.

Ballou, L. R., DeWitt, L. M., and Cheung, W. Y., 1986, Substrate-specific forms of human platelet phospholipase A$_2$, *J. Biol. Chem.*, 261:3107.

Banga, H. S., Simons, E. R., Brass, L. F., and Rittenhouse, S. E., 1986, Activation of phospholipase A$_2$ and C in human platelets exposed to epinephrine: role of glycoprotein IIb/IIIa and dual role of epinephrine, *Proc. Natl. Acad. Sci. USA*, 83:9197.

Bell, R. L., Kennerly, D. A., Stanford, N., and Majerus, P. W., 1979, Diglyceride lipase: a pathway for arachidonate release from human platelets, *Proc. Natl. Acad. Sci. USA*, 76:3238.

Berridge, M. J., 1983, Rapid accumulation of inositol trisphosphate reveals that agonists hydrolyze polyphosphoinositides instead of phosphatidylinositol, *Biochem. J.*, 212:849.

Bokach, G. M., and Gilman, A. G., 1984, Inhibition of receptor-mediated release of arachidonic acid by pertussis toxin, *Cell*, 39:301.

Buckingham, L., and Duncan, J., 1983, Approximate dimensions of membrane lesions produced by streptolysin S and streptolysin O, *Biochim. Biophys. Acta*, 729:115.

Burch, R. M., 1987, Protein kinase C mediates endotoxin- and zymosan-induced prostaglandin synthesis, *Eur. J. Pharmacol.*, 142:431.

Burch, R. M., 1989a, G protein regulation of phospholipase A$_2$, *Mol. Cell. Neurobiol.*, in press.

Burch, R. M., 1989b, Regulation of phospholipase A$_2$ by G proteins, in: "G Proteins and Calcium Mobilization," P. H. Naccache, ed., CRC Press, Orlando, FL.

Burch, R. M., and Axelrod, J., 1987, Dissociation of bradykinin-induced prostaglandin formation from phosphatidylinositol turnover in Swiss 3T3 Fibroblasts: evidence for G protein regulation of phospholipase A$_2$, *Proc. Natl. Acad. Sci. USA*, 84:6374.

Burch, R. M., Connor, J. R., and Axelrod, J., 1988a, Interleukin 1 amplifies receptor-mediated activation of phospholipase A$_2$ in 3T3 fibroblasts, *Proc. Natl. Acad. Sci. USA*, 85:6306.

Burch, R. M., Jelsema, C. L., and Axelrod, J., 1988b, Cholera toxin and pertussis toxin stimulate prostaglandin E$_2$ synthesis in a murine macrophage cell line, *J. Pharmacol. Exp. Ther.*, 244:765.

Burch, R. M., Luini, A., and Axelrod, J., 1986, Phospholipase A$_2$ and phospholipase C are activated by distinct GTP-binding proteins in response to α_1 adrenergic stimulation in FRTL5 thyroid cells, *Proc. Natl. Acad. Sci. USA*, 83:7201.

Burch, R. M., and Mahan, L. R., 1989, Bradykinin receptors, in: "Bradykinin Receptor Antagonists: Basic and Clinical Aspects," R. M. Burch, ed., Marcel Dekker, New York, NY.

Casey, P. J., and Gilman, A. G., 1988, G protein involvement in receptor-effector coupling, *J. Biol. Chem.*, 263:2577.

Cockcroft, S., and Gomperts, B. D., 1985, Role of guanine nucleotide binding protein in the activation of polyphosphoinositol phosphodiesterase, *Nature*, 314:534.

Crooke, S. T., Mattern, M., Sarau, H. M., Winkler, J. D., Balcarek, J., Wong, A., and Bennett, C. F., 1989, The signal transduction system of the leukotriene D$_4$ receptor, *Trends Pharmacol. Sci.*, 10:103.

Gilman, A. G., 1987, G proteins: transducers of receptor-generated signals, *Ann. Rev. Biochem.*, 56:615.

Hofmann, S. L., Prescott, S. M., and Majerus, P. W., 1982, The effects of mepacrine and p-bromophenacyl bromide on arachidonic acid release in human platelets, *Arch. Biochem. Biophys.*, 215:237.

Howell, T. W., and Gomperts, B. D., 1987, Rat mast cells permeabilized with streptolysin O secrete histamine in response to Ca^{2+} at concentrations buffered in the micromolar range, *Biochim. Biophys. Acta*, 927:177.

Irvine, R., 1982, How is the level of free arachidonic acid controlled in mammalian cells? *Biochem. J.*, 204:3.

Jelsema, C. L., 1987, Light activation of phospholipase A_2 in rod outer segments of bovine retina and its modulation by GTP-binding proteins, *J. Biol. Chem.*, 262:163.

Jelsema, C. L., and Axelrod, J., 1987, Stimulation of phospholipase A_2 activity in bovine rod outer segment by the $\beta\gamma$ subunits of transducin and its inhibition by the α subunit, *Proc. Natl. Acad. Sci. USA*, 84:3625.

Jelsema, C. L., Burch, R. M., Jaken, S., Ma, A. D., and Axelrod, J., 1989, Modulation of phospholipase A_2 activity in rod outer segments of bovine retina by G protein subunits, guanine nucleotides, protein kinases, and calpactin, in: "Extracellular and Intracellular Second Messengers in the Vertebrate Retina," v. 49, "Neurology and Neurobiology," D. A. Redburn and H. Pasanter-Morales, eds., Alan R. Liss, New York, NY.

Katada, T., Oinuma, M., and Ui, M., 1986, Two guanine nucleotide-binding proteins in rat brain serving as the specific substrate of islet-activating protein, pertussis toxin, *J. Biol. Chem.*, 261:8182.

Kim, D., Lewis, D. L., Graziadel, L., Neer, E. J., Bar-Sagi, D., and Clapham, D. E., 1989, G-protein $\beta\gamma$-subunits activate the cardiac muscarinic K+-channel via phospholipase A_2, *Nature*, 337:557.

Loeb, L. A., and Gross, R. W., 1986, Identification and purification of sheep platelet phospholipase A_2 isoforms, *J. Biol. Chem.*, 261:10467.

Majerus, P. W., Connolly, T. M., Deckmyn, H., Ross, T. S., Bross, T. E., Ishii, H., Bansal, V. S., and Wilson, D. B., 1986, The metabolism of phosphoinositide-derived messenger molecules, *Science*, 234:1519.

Nakamura, T., and Ui, M., 1983, Suppression of passive cutaneous anaphylaxis by pertussis toxin, an islet-activating protein, as a result of inhibition of histamine release from mast cells, *Biochem. Pharmacol.*, 32:3435.

Nakamura, T., and Ui, M., 1984, Islet-activating protein, pertussis toxin, inhibits Ca^{2+}-induced and guanine nucleotide-dependent releases of histamine and arachidonic acid from rat mast cells, *FEBS Lett.*, 173:414.

Nakamura, T., and Ui, M., 1985, Simultaneous inhibitions of inositol phospholipid breakdown, arachidonic acid release, and histamine secretion in mast cells by islet-activating protein, pertussis toxin, *J. Biol. Chem.*, 260:3584.

Nakashima, S., Nagata, K. -I., Ueeda, K., and Nozawa, Y., 1988, Stimulation of arachidonic acid release by guanine nucleotide in saponin-permeabilized neutrophils: evidence for involvement of GTP-binding protein in phospholipase A_2 activation, *Arch. Biochem. Biophys.*, 261:375.

Neer, E. J., and Clapham, D. E., 1988, Roles of G protein subunits in transmembrane signalling, *Nature*, 333:129.

Okajima, F., and Ui, M., 1984, ADP-ribosylation of the specific membrane protein by islet-activating protein, pertussis toxin associated with inhibition of a chemotactic peptide-induced arachidonate release in neutrophils, *J. Biol. Chem.*, 259:13863.

Prescott, S. M., and Majerus, P. W., 1981, The fatty acid composition of phosphatidylinositol from thrombin-stimulated human platelets, *J. Biol. Chem.*, 256:579.

Slivka, S. R., and Insel, P. A., 1987, α_1 Adrenergic receptor-mediated phosphoinositide hydrolysis and prostaglandin E_2 formation in Madin-Darby canine kidney cells. Possible parallel activation of phospholipase C and phospholipase A_2, *J. Biol. Chem.*, 262:4200.

Slivka, S. R., and Insel, P. A., 1988, Phorbol esters and neomycin dissociate bradykinin receptor-mediated arachidonic acid release and polyphosphoinositide hydrolysis in Madin-Darby canine kidney cells, *J. Biol. Chem.*, 263:14640.

Smith, C. D., Cox, C. C., and Snyderman, R., 1986, Receptor-coupled activation of phosphoinositide-specific phospholipase C by an N protein, *Science*, 232:97.

Stryer, L., 1987, The molecules of visual excitation, *Sci. Am.*, 257(July):42.

Sweatt, J. D., Connolly, T. M., Cragoe, E. J., and Limbird, L. E., 1986, Evidence that Na^+/H^+ exchange regulates receptor-mediated phospholipase A_2 activity in human platelets, *J. Biol. Chem.*, 261:8667.

Ui, M., 1984, Islet-activating protein, pertussis toxin: a probe for functions of the inhibitory guanine nucleotide regulatory component of adenylate cyclase, *Trends Pharmacol. Sci.*, 5:277.

Welsh, C., Dubyak, G., and Douglas, J. G., 1988, Relationship between phospholipase C activity and prostaglandin E_2 and cyclic adenosine monophosphate production in rabbit tubular epithelial cells, *J. Clin. Invest.*, 81:710.

Wightman, P. D., Humes, J. L., Davies, P., and Bonney, R. J., 1981, Identification and characterization of two phospholipase A_2 activities in resident peritoneal macrophages, *Biochem. J.*, 195:427.

LIPOCORTINS AS ANTIPHOSPHOLIPASE A2

AND ANTI-INFLAMMATORY PROTEINS

Françoise Russo-Marie

INSERM U 332
ICGM
Hôpital Cochin
27 rue du Fg St Jacques
75014 Paris, France

Introduction

The inflammatory reaction is a complex and multiparametric reaction occuring after a living organism has been injured (Ryan and Majno, 1977). It involves a cascade of events leading to cellular activation, mediator formation, enzymatic cleavage... In such a complex network, involving so many different parameters, it seemed difficult to consider a common pathway. Anyhow, the empirical discovery of the anti-inflammatory action first of aspirin and then of glucocorticosteroids opened a new way to analyse this reaction. More, the progressive understanding of the mechanism of action of these anti-inflammatory drugs (Vane, 1971; Smith and Willis, 1971, Gryglewski et al., 1975, Hong and Levine, 1976) allowed to propose that there could be a common pathway involving phospholipase A2 (PLA2) activation and the formation of its enzymatically-derived products. Indeed, these drugs interfere with phospholipase A2 activation and/or with the subsequent formation of its derivatives. Phospholipase A2 is a cellular enzyme which, upon activation, cleaves phospholipids in the sn-2 position, giving rise to arachidonic acid, the usual unsaturated fatty acid located at this position in mammalian cells and to lyso phospholipids. Arachidonic acid is the precursor of the very powerful proinflammatory mediators, prostaglandins, leukotrienes and other hydroperoxyderivatives formed by the action of enzymatic complexes present in almost all inflammatory cells, namely cyclooxygenase and lipoxygenases

Biochemistry, Molecular Biology, and Physiology of Phospholipase A₂ and Its Regulatory Factors
Edited by A. B. Mukherjee, Plenum Press, New York, 1990

(Burgoyne et al., 1987, Hamilton and Adams, 1987, Smith, 1989). These enzymes, when present, utilize all the arachidonic acid provided, so the formation of prostaglandins, leukotrienes and hydroperoxyderivatives is a function of the amount of arachidonic acid made available by phospholipase activation. This step represents the rate-limiting step in the formation of these potent proinflammatory mediators and might be a common pathway of some inflammatory processes. This hypothesis of a central role for PLA2 activation was reinforced by the fact that many proinflammatory mediators such as bradykinin (Garcia-Perez and Smith, 1984), zymosan (Tsunawaki and Nathan, 1986), immunoglobulins G (Hamilton and Adams, 1987), fMLP, the chemotactic peptide (Fradin et al, 1988; Errasfa and Russo-Marie, 1989), release arachidonic acid through the activation of phospholipase A2. Glucocorticosteroids, the potent steroidal anti-inflammatory drugs, were shown to inhibit arachidonic acid release (Gryglewsky et al., 1975; Hong and Levine, 1976) indicating that they were interfering with phospholipase A2 activation. In parallel, glucocorticosteroids were shown to inhibit prostaglandin formation in *in vitro* systems after requiring mRNA and protein synthesis (Danon and Assouline, 1978, Flower and Blackwell, 1979, Russo-Marie et al., 1979). Taken altogether, these results suggested that part of the anti-inflammatory effects of glucocorticosteroids could be mediated by the induction of protein synthesis and that this protein would be interfering with the activation of phospholipase A2.

In this paper, we shall discuss the present knowledge on the relationships existing between glucocorticosteroids and lipocortins. We shall discuss also the ability of lipocortins not only to inhibit phospholipase A2 *in vitro* but also cellular phospholipase A2 in inflammatory cells. Finally we shall present evidence that lipocortins mimick the *in vivo* anti-inflammatory effect of glucocorticosteroids.

A difficult relationship : glucocorticosteroids and lipocortins

As mentioned previously, glucocorticosteroids were reported to inhibit prostaglandin synthesis through the induction of mRNA and protein synthesis; in addition, glucocorticosteroids were reported to prevent arachidonic acid release from cell membranes (see review from Flower, 1984). It was therefore supposed that glucocorticosteroids were inhibiting phospholipase A2 through the induction of the synthesis of proteins capable of inhibiting phospholipase A2. Various attempts of characterization and purification showed the presence of proteins able to inhibit phospholipase A2 either in the supernatant of glucocorticosteroid-treated cells (Hirata, et al., 1980; Blackwell, et al., 1980; Cloix, et al., 1983, Rothhut, et al., 1983) or in the peritoneal lavage fluid of glucocorticosteroid-treated animals (Blackwell et al, 1982). The biological assay used was an *in vitro* inhibition of porcine pancreatic phospholipase A2 using either ^{3}H-oleic acid labeled E Coli membranes or ^{14}C-arachidonic acid labeled phosphatidyl-choline as a substrate. Using these assays, it was shown that the major phospholipase A2 inhibitory protein was a 40 kDa protein but proteins of 30 and 15 kDa supposed to be proteolytic fragments of the 40 kDa protein were also found. Various names were given : "macrocortin" for the proteins described and partially purified in rat macrophages (macro for macrophage and cortin for induced by corticosteroids) (Blackwell, et al., 1980), "lipomodulin" for the protein partially purified from rabbit peritoneal neutrophils (lipo for phospho"lipase' and modulin for modulating phospholipase) (Hirata, et al., 1980) and "renocortin" for the proteins found in the supernatant of rat renomedullary interstitial cells treated with dexamethasone ("reno" is reffering to renal cells) (Cloix, et al., 1983). Monoclonal antibodies raised against the different described proteins cross-reacted indicating that the proteins not only had the same biological activity but also were immunologically identical leading to the proposal of a common name of "lipocortin" (DiRosa, et al., 1984) which accounted for the phospholipase A2 inhibitory property and the control by glucocorticosteroids.

In summary, the biological results together with the ability to obtain proteins led to the hypothesis that lipocortins could then be "the anti-inflammatory messenger" of glucocorticosteroids.

Lipocortins belong to a family of constitutive intracellular proteins

The first complete purification of lipocortin was performed on proteins isolated from the rat peritoneal lavage fluid of glucocorticosteroid-treated rats (Pepinsky, et al., 1986). The biological property used to follow the purification was the inhibition of pancreatic phospholipase A2 using ^{3}H-oleic acid E. Coli labeled membranes as a substrate; it allowed the purification and partial sequencing of a 37 kDa protein. Using oligonucleotides made from the partial sequence of the protein, the cDNA of human lipocortin I was cloned and sequenced (Wallner, et al., 1986). It thus appeared that this sequenced protein was identical to the 35 kDa cellular substrate of phosphorylation of the EGF receptor tyrosine kinase moiety (Fava and Cohen, 1984; De, et al., 1986; Pepinsky and Sinclair, 1986). By searching for an additional source of proteins, Huang, et al. (1986) found in placental homogenates not only lipocortin I, the previously described protein, but another protein able to inhibit phospholipase A2 using the same E. Coli assay, exhibiting very similar biochemical properties. They named this second phospholipase A2 inhibitory protein, lipocortin II. When cloning the cDNA of this protein, it appeared that it was identical to the cellular substrate of phosphorylation of the oncogene product pp60 [v-src] and to the heavy chain of the heterodimer calpactin I (Gerke and Weber, 1984, Saris, et al., 1986) Lipocortins I and II shared 50% sequence homology between them but also with a family of proteins involved in cellular traffic, in exocytosis and which were all able to bind to acidic phospholipids in the presence of calcium. Common names were given to this family of proteins (annexins, chromobindins, calcimedins, calpactins, calelectrins, endonexins and proteins I-III) (see Pepinsky, et al., 1988 for references). These proteins were found in almost all tissues. Nevertheless their intracellular function was far for being understood. To add to the confusion, the same proteins were shown to possess anticoagulant properties (Funakoshi, et al., 1987a and 1987b; Iwasaki, et al., 1987; Grundman, et al., 1988).

In summary, in identifying a possible cytokine able to be a messenger for the anti-inflammatory and anti-phospholipase A2 properties of glucocorticosteroids, proteins were found which belonged to a family of intracellular cytosolic proteins. All these proteins were capable of inhibiting phospholipase A2 and of binding to acidic phospholipids in the presence of calcium. None of them had at its N-terminal end or within its sequence a hydrophobic stretch of aminoacids (the peptide signal sequence) allowing the passage through the ER and the secretion outside of the cell as secreted proteins normally do (Walter and Lingappa, 1986). More, the ability to be be under the genomic control of glucocorticosteroids did not appear clearly (Isacke et al., 1989). The levels of lipocortin I's mRNA increased in rat peritoneal macrophages after *in vivo* glucocorticosteroid treatment although no increase in the intracellular level of lipocortin I was detectable (Wallner, et al., 1986). Clearly, the relationship between glucocorticosteroids, phospholipase A2, cytoskeletal proteins and tyrosine kinases was far from being understood.

Lipocortins, Phospholipase A2 : a dual partnership

Lipocortins were initially reported to inhibit phospholipase A2 by a direct protein-protein interaction. Evidence for a direct inhibition came from three types of experimental results : i) In *in vitro* experiments, the inhibitor (lipocortin) was added onto a purified enzyme (porcine pancreatic phospholipase A2) for inhibiting its activity (cleavage of ^{3}H-oleic acid from labeled E.Coli membranes) suggesting a direct inhibitory effect (Rothhut, et al., 1983, Pepinsky, et al., 1986). ii) Some authors, in purifying lipocortins from biological extracts using PLA2-agarose coupled affinity columns, found lipocortins among the eluted proteins (Hirata, et al., 1982 ; Flower, et al., 1984). iii) In experiments performed on whole cells, it was suggested that lipocortin exerted a constraint on phospholipase

A2, released upon cell activation and lipocortin phosphorylation (Hirata, et al., 1984 ; Touqui, et al., 1986). These different results were interpreted as lipocortin and phospholipase A2 being coupled molecules interacting one with each other in order to control the cleavage of phospholipids in the membrane; in this putative scheme, the antiphospholipase A2 activity of lipocortin was relieved by phosphorylation through protein kinase C activation (Touqui, et al., 1986).

In summary, both biochemical data together with biological data suggested that lipocortin was a direct inhibitor of phospholipase A2 in cells.

The lipocortin-phospholipase A2 partnership involves a third partner : phospholipids

When the sequence of lipocortin I was reported and that it appeared that lipocortin I was related to proteins whose properties of binding to acidic phospholipids were well described in other systems, the issue of a possible relationship between this physical property and the reported PLA2 inhibitory effect was addressed. Using one dose of lipocortin V purified from mononuclear cells, we could show that it was inhibiting porcine pancreatic phospholipase A2 on ^{3}H-oleic acid release from labeled E. Coli membranes (Rothhut et al, 1987) as expected, but when mammalian cellular phospholipases A2 from rat platelets and rat liver mitochondriae were used on labeled phosphatidyl-ethanolamine vesicles as a substrate, the concentration of phospholipids was crucial to allow lipocortin to inhibit phospholipase A2. Inhibition occured only for low substrate concentrations, indicating that the substrate (phospholipids) was interfering with the inhibitor (Aarsman et al, 1987). Those results suggested that lipocortin could inhibit phospholipase A2 activity in two possible new ways different from the way initially described : i) at low substrate concentration, lipocortin could sequester phospholipids from phospholipase A2 whereas at high substrate concentration, enough phospholipids could be available for the enzyme which would then recover its full activity; ii) if lipocortin were bound to phospholipase A2, then a high phospholipid concentration could remove lipocortin from phospholipase A2 and allow it to recover its enzymatic activity. Our data did not allow to choose between these two possible mechanisms of inhibition. Additional experiments suggested that the first described mechanism was likely to occur : using either a zwitterionic or a negatively-charged substrate, both hydrolized by the same Naja Naja phospholipase A2, we could show that lipocortin V from human mononuclear cells was unable to inhibit phospholipase A2 activity when the substrate was zwitterionic (Radvanyi, et al., 1989; Rothhut, et al., 1987), evidencing that the substrate was interfering with lipocortin and that this interference was modifying the activity of phospholipase A2. In a very elegant study, Davidson, et al. (1987), were able to demonstrate that lipocortin I and lipocortin II inhibited pancreatic phospholipase A2 *in vitro* by substrate depletion. Using other lipocortins than lipocortin V, present in mononuclear cells, we were able to demonstrate that this property of inhibiting phospholipase A2 by interacting with the substrate was shared by all lipocortins *in vitro*. Finally, using cross-linkers in the presence or in the absence of calcium or in the presence or in the absence of phospholipids, we were always unable to find any binding of lipocortins to phospholipase A2. Taken altogether, these results suggest that lipocortins inhibit phospholipase A2 *in vitro* by sequestering the sustrate and not by a direct lipocortin-phospholipase A2 interaction.

In summary, all lipocortins bind to acidic phospholipids. This binding is probably responsible for the apparent inhibition of phospholipase A2 observed in the *in vitro* assays used. It does not exlude the possibility for another *in vivo* mechanism.

Do lipocortins have a role in controlling inflammation

All the properties reported up to now have been established in *in vitro* situations and there is no understanding of any intracellular role. Anyhow, since we and others had found lipocortins in the supernatant of dexamethasone-treated cells, (Hirata, et al., 1980; Flower, et al., 1980; Cloix, et al., 1983; Rothhut, et al., 1983) - although this extracellular presence is still unexplained -, and since these supernatants were found to possess anti-inflammatory activities (see Flower, et al., 1984), we asked whether purified lipocortins would act similarly on inflammatory cells. The inflammatory parameters that we assayed were : i) ^{3}H-arachidonic acid release from prelabeled inflammatory cells stimulated by the calcium ionophore A23187 or by fMLP, ii) prostaglandin, leukotriene et paf-acether release in cells stimulated either by the calcium ionophore A23187 or by the chemotactic agent fMLP, iii) neutrophil chemotaxis induced by fMLP, iv) superoxide production in alveolar macrophages induced by the calcium ionophore A23187. The two firts assays estimated the cellular phospholipase A2 activity since they were measuring either the release of ^{3}H-arachidonic acid itself or the formation of proinflammatory mediators derived from phospholipid cleavage in the sn-2 position. The two last assays estimated other cellular responses occuring in challenged inflammatory cells, either chemotaxis in neutrophils or superoxide production in alveolar macrophages. Both responses were shown to be secondary to phospholipase A2 activation. The different lipocortins that we assayed were lipocortin I - either from human mononuclear cells or the recombinant one, provided by Biogen

Research Corporation, and lipocortin V from mouse thymus (Errasfa et al., 1989). Both proteins gave identical results : they inhibited in a dose dependent-manner all the parameters that we measured : ^{3}H-arachidonic acid release from prelabeled cells which were stimulated either by the calcium ionophore A23187 or by the chemotactic peptide fMLP (Errasfa, et al, 1988), prostaglandin E2, leukotriene B4 and paf-acether release from inflammatory rat neutrophils stimulated with the calcium ionophore A23187 (Fradin, et al., 1989). When assayed in a microchemotaxis assembly, they inhibited dose-dependently the cell migration induced by fMLP (Errasfa, et al., 1989). On alveolar guinea-pig macrophages, they inhibited dose-dependently the production of superoxide induced by the calcium ionophore A23187 whereas they did not inhibit the PMA-induced superoxide production (Maridonneau-Parini, et al., 1989). All these effects seemed to involve at a step an inhibition of cellular phospholipase A2. It thus seems that, when added exogenously on cell membranes, lipocortins are able to provoke a modification of cellular phospholipase A2 activation. Similar data implicating an inhibition of cellular phospholipase A2 by exogenously added lipocortins had already been reported in other models : on perfused lungs (Cirino, et al., 1987) and endothelial cells in culture (Cirino and Flower, 1987). More strikingly, peptides derived from a region of high similarity between uteroglobin and lipocortin I have been reported to possess antiphospholipase A2 activity *in vitro* and anti-inflammatory properties *in vivo* (Miele, et al., 1988). On the opposite, Northrup et al., 1988, could not find any effect of lipocortin I on cellular PLA2 in inflammatory cells.

Since lipocortins were shown to be able to interfere with the activation of inflammatory cells and mainly with cellular phospholipase A2 activation, and since, the inflammatory process could be related to a permanent reactivation of cellular phospholipase A2 through the arrival of external stimuli on the cell membrane, we predicted that lipocortins could possess anti-inflammatory properties when given *in vivo* to an animal carrying an inflammatory lesion. Indeed, when injected *in vivo* in mice carrying an inflammatory lesion, they gave the same anti-inflammatory effects as an intravenous infusion of glucocorticosteroids (Errasfa and Russo-Marie, 1989), i.e., they inhibited in a dose-dependent manner both cell migration to the inflammatory site and the production of eisocanoids. While this work was in press, Cirino et al., (1989) reported that recombinant lipocortin I had acute local anti-inflammatory properties in the rat paw edema. Here again, Northrup et al., 1988, could not find any anti-inflammatory effect of lipocortin I in the rat paw edema. The reason for the opposite results of this group as compared to results from other groups does not appear clearly.

Several possible models allow to explain these effects on cellular phospholipase A2 inhibition : i) Lipocortin penetrates into the cell and inhibits phospholipase activity by binding to the phospholipidic substrate of phospholipase A2. ii) Lipocortin binds to some putative negatively charged phospholipids which are exposed on the outer face of the membrane during the inflammatory response of the cell leading to the same mechanism as in (i). iii) Lipocortin has cell surface receptors which transmit an information leading to the inhibition of phospholipase A2. iv) Lipocortin, by its ability to bind lipids, binds the products of phospholipase A2 activation (arachidonic acid, prostaglandins, leukotrienes, paf-acether) and therefore inhibits the reamplification phenomenon which is normally occuring. v) An extracellular phospholipase A2 is responsible for the activation of the cell and lipocortin inhibits it in a similar way as what was reported to occur *in vitro*. At that time, there is no clear answer on the mechanism whatsoever.

In summary, lipocortins when added exogenously on activated inflammatory cells inhibit cellular phospholipase A2 activation and when added *in vivo* to the whole animal possess potent anti-inflammatory properties.

Secretion of lipocortin I induced by glucocorticosteroids

We have been able to demonstrate that after glucocorticosteroid treatment of mononuclear cells, the only lipocortin which was found extracellularly was lipocortin I (Comera, unpublished results), raising the question of a specific secretion since among all the lipocortins present in the cell, only lipocortin I is found outside.

These results, although they do not explain how lipocortin is found outside of the cell, would reconcile the first experimental observations made by ourselves and other groups on the anti-inflammatory role of glucocorticosteroids which should be mediated by the induction of the synthesis and the release of lipocortins. A speculative scheme could be that glucorticosteroids trigger, in a way still to define, the promoter region of lipocortin I (the only lipocortin whose mRNA has been shown to be under the control of glucocorticosteroids), lipocortin I would then be found outside of the cells and would inhibit the activation of other cells by various stimuli, preventing then the amplification procedure responsible of the propagation of the inflammatory reaction. In order to be efficient, this should occur in the early stages of the inflammatory reaction which always involves activation of circulating cells such as neutrophils and monocytes. These data however are only taking into consideration the possibility that lipocortins are released and acting outside of the cell, they do not take into account the possibility that lipocortins might have a role in the control of the inflammatory reaction by an intracellular function. Indeed, it is only speculative to assign a possible role of lipocortin to control the endogenous phospholipase A2 activation. Basically, the idea would be that lipocortins would interfere with the signalling pathway in mononuclear cells, pathway which would be followed by molecules after they have been released upon cell activation by the interaction of inflammatory stimuli with the cell. From what is known about this pathway in inflammation, the most likely candidate would be the activation of cellular phospholipase A2 and the production of prostaglandins, leukotrienes and paf-acether leading to subsequent events. Lipocortins would then interfere with this activation either from outside as we have shown or from inside in a way still to define. Their intracellular role could be to interfere with phospholipids and with the cytoskeleton in order to modify the answer of the cell to inflammatory stimuli.

References

Aarsman, A.J., Mymbeck, G., Van den Bosch, H., Rothhut, B., Prieur, B., Comera, C., Jordan, L., and Russo-Marie, F. 1987, Lipocortin Inhibition of Extracellular and Intracellular Phospholipases A2 is Substrate Concentration Dependent. FEBS Lett. 219 : 176.

Blackwell, G.J., Carnuccio, R., DiRosa, M., Flower, R.J., Langham, C.S.J., Parente, L., Persico, P., 1980, Macrocortin, A Polypeptide Causing the Anti-phospholipase Effect of Glucocorticoids. Nature 287 : 147.

Blackwell, G.J., Carnuccio, R., DiRosa, M., Flower, R.J., Langham, C.S.J., Parente, L., Persico, P., Russell-Smith, N.C., and Stone, D., 1982, Glucocorticoids Induce the Formation and Release of Anti-inflammatory and Anti-phospholipase Proteins Into the Peritoneal Cavity of the Rat. Br. J. Pharmac 76 : 185.

Burgoyne, R.B, Cheek, T.R. and O'Sullivan, A.J., 1987, Receptor Activation of Phospholipase A2 in Cellular Signaling. TIBS 12 : 332.

Cirino, G., and Flower, R.J., 1987, Human Recombinant Lipocortin I Inhibits Prostacyclin Production by Human Umbilical Artery in vitro. Prostaglandins 34 : 59.

Cirino, G., Flower, R.J., Browning, J.L., Sinclair, L.K. and Pepinsky, R.B., 1987, Recombinant Lipocortin I Inhibits Thromboxane Release from Guinea Pig Perfused Lung. Nature 328: 270.

Cirino, G., Peers, S.H., Flower, R.J., Browning, J.L., and Pepinsky, R.B., 1989, Human Recombinant Lipocortin I has Acute Local Anti-inflammatory Properties in the Rat Paw Edema Test. Proc. Natn. Acad. Sci. USA, 86, 3428.

Cloix, J.F., Colard, O., Rothhut, B., and Russo-Marie, F.,1983, Characterization and Partial Purification of "Renocortins" Two Polypeptides Formed in Renal Cells Causing the Anti-phospholipase-like Action of Glucocorticoids. Br. J. Pharmacol. 79 : 313

Danon, A, and Assouline, G., 1978, Inhibition of Prostaglandin Biosynthesis Requires RNA and Protein Synthesis. Nature 273 : 552.

Davidson, F., Dennis, E.A., Powell, M., and Glenney, J.R. Jr, 1987, Inhibition of Phospholipase A2 by Lipococortins and Calpactins : An Effect of Binding to Susbstrate Phospholipids. J. Biol. Chem. 262 : 1698.

De, B.K., Misono, K.S., Lukas, T.J., Mrocskowski, B. and Cohen, S., 1986, A Calcium-dependent 35 Kilodalton Sustrate for Epidermal Growth Factor Receptor /Kinase Isolated from Normal Tissue. J. Biol. Chem. 261 : 13784.

Di Rosa, M., Flower, R.J., Hirata, F., Parente, L. and Russo-Marie, F. 1984, Nomenclature Annoucement. Anti-phospholipase Proteins. Prostaglandins 28 : 441.

Errasfa, M., Bachelet, M., and Russo-Marie, F., 1988, Inhibition of Phospholipase A2 Activity of Guinea-pig Alveolar Macrophages by Lipocortin-like Proteins Purified from Mice Lungs. Biochem. Biophys. Res.Commun. 153 : 1267.

Errasfa, M., Rothhut, B., and Russo-Marie, F., 1989, Phospholipase A2 Inhibitory Activity in Thymocytes of Dexamethasone-treated Mice. Possible Implication of Lipocortins. Biochem. Biophys. Res. Commun. 159 : 538.

Errasfa, M., and Russo-Marie, F., 1989, A Purified Lipocortin Shares the Anti-inflammatory Effect of Glucocorticosteroids *in vivo* in Mice. Br. J. Pharmac (in press).

Fava R.A., and Cohen, S., 1984, Isolation of a Calcium-dependent 35 Kilodalton Substrate for the Epidermal Growth Factor Receptor/Kinase from A-431 Cells. J. Biol. Chem. 259 : 2636.

Flower, R.J. and Blackwell, G.J., 1979, Anti-inflammatory Steroids Induce the Biosynthesis of a Phospholipase A2 Inhibitor which Prevents Prostaglandin Generation. Nature 278 : 456.

Flower, R.J. : Macrocortin and the Anti-phospholipase Proteins, 1984, Adv. Inflam. Res., 8 : 1.

Flower, R.J., Wood, J.N., and Parente, L., 1984, Macrocortin and the Mechanisms of Action of the Glucocorticoids. Adv. Inflamm. Res. 7 : 61.

Fradin, A., Rothhut, B., Poincelot-Canton, B., Errasfa, M. and Russo-Marie, F., 1988, Inhibition of Eicosanoid and Paf Formation by Dexamethasone in Rat Inflammatory Polymorphonuclear Neutrophils May Implicate Lipocortin"s". Biochim. Biopys. Acta 963 : 248.

Funakoshi, T., Heinmark, R.L., Henrickson, L.E., McMullen, B.A., and Fujikawa, K., 1987a, Human Placental Anticoagulant Protein : Isolation and Characterization. Biochemistry 26: 5572.

Funakoshi, T., Hendrickson, L.E., McMullen, B.A., and Fujikawa, K., 1987b, Primary Structure of Human Placental anticoagulant Protein. Biochemistry 26 : 8087.

Garcia-Perez, A., and Smith, W.L., 1984, Apical-Basolateral Membranes Asymmetry in Canine Cortical Collecting Tubule Cells : Bradykinin, Arginin Vasopressin and Prostaglandin E2 Interrelationships. J. Clin. Invest 70 : 63.

Grundman, U., Abel, K-J., Bohn, H., Lobermann, H., Lottspeich, F., and Kupper, H., 1988, Characterization of cDNA Encoding Human Placental Anticoagulant Protein (PP4) : Homology with the Lipocortin Family. Proc. Natl. Acad. Sci. USA 85 : 3708.

Gryglewski, R.J., Panczenko, B., Korbut, R., Grodzinska, L., and Ocetkiewicz, A., 1975, Corticosteroids Inhibit Prostaglandin Release from Perfused Mesenteric Blood Vessels of Rabbit and from Perfused Lungs of Sensitized Guinea-pigs. Prostaglandins 10 : 343.

Hamilton, T.A., and Adams, D.O., 1987, Molecular Mechanisms of Signal Transduction in Macrophages. Immunol. Today 8 : 151.

Hirata, F., Schiffman, E., Venkatasubramanian, K., Salomon, D., and Axelrod, J., 1980, A Phospholipase A2 Inhibitory Protein in Rabbit Neutrophils induced by Glucocorticoids. Proc. Natl. Acad. Sci. USA 77: 2533.

Hirata, F, Matsuda, K., Notsu, Y., Hattori, T., and Del Carmine, R., 1984, Phosphorylation of a Tyrosine Residue of Lipomodulin in Mitogen-stimulated Murine Thymocytes, Proc Natl. Acad. Sci. USA 81 : 4717.

Hirata, F, Notsu, F., Iwata, M., Parente, L., DiRosa, M., and Flower, R.J., 1982, Identification of Several Species of Phospholipase Inhibitory

Protein(s) by Radioimmunoassay for Lipomodulin. Biochem. Biophys.Res. Commun. 109 : 223.

Hong S-C and Levine, L.,1976, Inhibition of Arachidonic Acid Release from Cells as the Biochemical Action of Anti-inflammatory Steroids, Proc. Natl Acad. Sci. USA 73 : 1720.

Huang, K.S., Wallner, B., Mattaliano, R.J., Tizard, R., Burne, C., Frey, A., Hession, C., Mcgray, P., Sinclair, L.K., Chow, E.P., Browning,J.L., Ramachadran, K.L., Tang, J., Smart, J.E., and Pepinsky, R.B. 1986, Two Human 35 kd Inhibitors of Phospholipase A2 are Related to Substrates of pp60^{v-src} and to Epidermal Growth Factor Receptor/Kinase. Cell 46 : 191.

Isacke, C.M., Lindberg, R.A., and Hunter, T., 1989, Synthesis of p36 and p35 Is Increased When U-937 Cells Differentiate in Culture but Expression is not Inducible by Glucocorticoids. Mol. Cell Biol., 9 : 232.

Iwasaki, A., Suda, M., Nakao, H., Nagoya, T., Saino, Y., Arai, K., Mizoguchi, T., Sato, F., Yoshozaki, H., Hirata, M., Miyata, T., Shidara, Y. Muruta, M. and Maki, M. 1987, Structure and Expression of cDNA for an Inhibitor of Blood Coagulation Isolated from Human Placenta : a New Lipocortin -like Protein. J. Biochem 102 : 1261.

Maridonneau-Parini, I., Errasfa, M. and Russo-Marie, F., 1989, Inhibition of O$^-_2$ Generation by Dexamethasone is Mimicked by Lipcortin I in Alveolar Macrophages. J. Clin. Invest. 83, 1936.

Miele, L., Cordella-Miele, E., Facchiano, A., and Mukherjee, A.B., 1988, Novel Anti-inflammatory Peptides from the Region of Highest Similarity between Uteroglobin and Lipocortin I . Nature 325 : 726.

Northrup, J.K., Valentine-Braun, K.A., Johnson, L.K., Severson, D.L., and Hollenberg, M.D., 1988, Evaluation of the Anti-inflammatory and Phospholipase-inhibitory Activity of CalpactinII/Lipocortin I. J. Clin. Invest. 82 : 1347.

Pepinsky, R.B. and Sinclair, L.K., 1986, Epidermal Growth Factor-dependent Phosphorylation of Lipocortin. Nature 321 : 81.

Pepinsky, R.B., Sinclair, L.K., Browning, J.L., Mattaliano, R.J., Smart, J.E., Chow, E. P., Falbel, T., Ribolini, A., Garwin, J. L., and Wallner, B.P. 1986, Purification and Partial Sequence analysis of a 37 kDa Protein that Inhibits Phospholipase A2 Activity from Rat Peritoneal Exudates. J. Biol. Chem. 261 : 4329.

Pepinsky, R.B., Tizard, R., Mattaliano, R.J., Sinclair, L.E., Miller, G.T., Browning, J.L., Chow, E.P., Burne, C., Huang, K-S, Pratt, D., Wachter, L., Hession, C., Frey, A.Z., and Wallner, B.P., 1988, Five Distinct Calcium and Phospholipid Binding Proteins Share Homology with Lipocortin I. J. Biol. Chem. 263 : 10799.

Radvanyi, F., Jordan, L., Russo-Marie, F., and Bon, C., 1989, A Sensitive and Continuous Fluorimetric Assay for Phospholipase A2 Using Pyrene-labeled Phospholipids in the Presence of Serum-albumin. Anal. Biochem. 177 : 103.

Rothhut, B., Russo-Marie, F., Wood, J., DiRosa, M., and Flower, R.J., 1983, Further Characterization of the Glucocorticoid-induced Anti-phospholipase Protein "Renocortin" . Biochem Biophys. Res. Commun. 117 : 878.

Rothhut, B, Comera, C., Prieur, B., Errasfa, M., Minassian, G. N., and Russo-Marie, F., 1987, Purification and Characterization of a 32 kDa Phospholipase A2 Inhibitory Protein (Lipocortin) from Human Peripheral Blood Mononuclear Cells. FEBS Lett 219 : 169.

Russo-Marie, F., Paing, M., and Duval, D., 1979, Involvment of Glucocorticoid Receptors in Steroid-induced Inhibition of Prostaglandin Secretion. J. Biol. Chem. 254 : 8498.

Ryan G.B. and Majno, G., 1977, Acute Inflammation. A Review. Am. J. Pathol. 86 : 183.

Saris, C.J.M., Tack, B.F., Kristensen, T., Glenney, R.J., Jr and Hunter, T., 1986, The cDNA Sequence for the Protein-Tyrosine Kinase Substrate p36 (Calpactin I Heavy Chain) Reveals a Multidomain Protein with Internal Repeats. Cell 46 : 201.

Smith, W.L., 1989, The Eicosanoids and their Biomechanical Mechanisms of Action. Biochem. J 259 : 315.

Smith, J.B, and Willis, A.L., 1971, Aspirin Selectively Inhibits Prostaglandin Production in Human Platelets. Nature (New Biol) 231 : 235.

Touqui, L., Rothhut, B., Shaw, A.B., Fradin, A., Vargaftig, B.B., and Russo-Marie, F., 1986, Platelet Activation : A Role for a 40 K Anti-phospholipase A2 Protein Indistinguishable from Lipocortin. Nature 321 : 177.

Tsunakawi, S. and Nathan, C.F., 1984, Enzymatic Basis of Macrophage Activation. J. Biol Chem. 259 : 4305.

Vane J.R., 1971, Inhibition of Prostaglandin Synthesis as a Mechanism of Action of Aspirin-like Drugs. Nature (New Biol) 604 : 191.

Wallner, B., Mattaliano, R.J., Hession, C., Cate, R.L., Tizard, R., Sinclair, L. K., Foeller, C., Chow, E. P., Browning, J.L., Ramachadran, K.L., and Pepinsky, R.B., 1986, Cloning and Expression of Human Lipocortin, a Phospholipase A2 Inhibitor with Potential Anti-inflammatory Activity. <u>Nature</u> <u>320</u> :77.

Walter, P., and Lingappa, V.R., 1986, Mechanism of Protein Translocation across the Endoplasmic Reticulum Membrane. <u>Ann. Rev. Cell. Biol.</u> <u>2</u> : 499 .

BIOLOGY OF PHOSPHOLIPASE INHIBITORY PROTEINS

Fusao Hirata and Aiko Hirata

Department of Environmental Health Sciences, School of
Hygiene and Public Health, The Johns Hopkins University
615 North Wolfe Street, Baltimore, Maryland 21205

INTRODUCTION

Glucocorticoids are widely used as one of the most efficacious
medicines against various inflammatory diseases and allergic diseases
including chronic rheumatoid arthritis, lupus erythematosus and bronch-
ial asthma. The mechanism of action is attributed to the anti-inflammatory action, immunosuppressive action and anti-edematous action[1]. Of
these actions, the mode of anti-inflammation is now correlated to the
inhibition of release of arachidonic acid, a precursor of prostaglandins (PGs), hydroxyeicosatetraenoic acids (HETEs) and leukotrienes
regarded to be inflammatory mediators[2]. Since the anti-inflammatory
action of glucocorticoids with respect to the suppressive effect on the
release of eicosanoids can be blocked by inhibitors of protein synthesis
as well as mRNA synthesis[3,4], it is conceived that upon entering into a
cell, glucocorticoids are first bound to the cytosolic receptor and that
the resulting complex is transferred into the nucleus to activate the
specific genes, after all, inducing the synthesis of specific proteins[5].
Although the key enzymes and major sources of arachidonic acid still
remain to be established in a quantitative aspect, phospholipase A_2
(PLA_2) and alkylphospholipids have been considered as the important
factors[6]. It is, therefore, suggested that glucocorticoids exhibit
their anti-inflammatory action at least partly, if not exclusively,
through the induced synthesis of PLA_2 inhibitory protein(s)[7]. Attempts
have been made by several groups to isolate such a protein (a) whose
synthesis is induced by glucocorticoids, (b) that inhibits the activity
of PLA_2 in vitro and in situ and (c) that displays the anti-inflammatory
action like glucocorticoids do. Several different proteins that inhibit
PLA_2 in vitro have been identified and partially purified. Among those,
the proteins whose chemical natures are well characterized were found to
have the common feature, a capacity to bind phospholipids and Ca^{2+}. In
this communication, we would like to discuss whether those Ca^{2+}- and
phospholipid binding proteins are physiologically relevant as the second
messenger protein(s) of the anti-inflammatory action of glucocorticoids,
or whether there is any other inhibitory protein(s) for the action of
glucocorticoids.

The mechanism of phospholipase A_2 activation

Phospholipase A_2 (phosphatide 2-acylhydrolase, EC 3.1.1.4) is
widely distributed enzyme involved in various biological functions by
participating in the metabolism and turnover of phospholipids[6]. PLA_2

Biochemistry, Molecular Biology, and Physiology of Phospholipase A₂ and Its Regulatory Factors 211
Edited by A. B. Mukherjee, Plenum Press, New York, 1990

produces the precursors of biologically active lipids such as hydroxy fatty acids (HETEs), leukotrienes, prostaglandins and platelet activating factor. Since these lipid mediators profoundly influence inflammatory reactions, PLA_2 has been implicated in the pathogenesis of disorders of the cardiovascular, gastrointestinal, and pulmonary systems and skin and connective tissues, especially relating to rheumatoid arthritis, lupus erythematosus and hyperreactive airways (see other chapters in this volume). Evidence to date has shown that both membrane-associated and soluble secreted forms of PLA_2 are present and are produced by cells participating in the inflammatory reaction. The two groups, the California Biotechnology Inc. and Biogen Inc., have recently isolated the cDNA and genomic clones encoding one such non-pancreatic PLA_2 present in inflammatory rheumatoid synovial fluids[8,9]. This PLA_2 contains structural features common to all known PLA_2s. It is a secreted enzyme with a signal peptide. This characteristic is essentially similar to those of PLA_2s from snake venom and mammalian pancreas, enzymes whose structure and molecular mechanism of enzymatic action have been extensively studied. These enzymes are secreted as an inactive zymogen which, in turn, becomes fully active against glycerophospholipid dispersions following a limited proteolytic cleavage. They have a low molecular mass (12-15KDa), a strict calcium requirement for phospholipid hydrolytic activity and the presence of a histidine residue in the active site, which is irreversibly inhibited by alkylation by bromophenacyl bromide. Some recent evidences indicated the existence of calcium-independent-PLA_2s and/or PLA_2 activated by low concentrations of calcium[10]. The latter enzymes have higher molecular weights and are insensitive to bromophenacyl bromide. One such enzyme is present in brush border membranes and has been isolated after the treatment of the membrane with papain[11]. It has an apparent molecular weight of 97KDa. Although the biochemical properties are not yet characterized well, several reports described the existence of PLA_2s specific for phosphatidylinositol (PtdIns), phosphatidic acid or alkylphospholipids, respectively.

The cellular PLA_2s are a part of the deacylation-reacylation cycle, which protects cellular membranes from peroxidation damages. Recently, these enzymes are also implicated for receptor-mediated signal transduction[12,13]; arachidonate and its metabolites serve as second messengers for alteration of ion conductances via activation of protein kinase C or cyclic GMP[14,15]. At least, the activation of PLA_2 through the muscarinic receptors of cardiac pace maker can be blocked by anti-pancreas PLA_2 antibody[15]. Further, Burch and his associates reported the functional reconstitution of receptor-mediated PLA_2 activation systems, using exogenous porcine pancreas PLA_2 (see a chapter in this volume). These observations suggest that a small form of PLA_2 is coupled to various receptors. Activation of receptor-coupled PLA_2 could be achieved by adding GTP or its non-hydrolyzable analogue or by adding β/γ subunits of G protein[13,14,15]. Most of GTP and/or G protein effects are explained by enhancement of the lipase activity at low concentrations of Ca^{2+}. These effects can be seen, only when the permeabilized cells but not isolated membranes are used. Therefore, there is still a possibility that a certain type of proteases is involved in the activation of the zymosan form of PLA_2 by splitting their NH_3-terminal peptides. The detail mechanism remains to be clarified.

Phospholipase inhibitory proteins

Glucocorticoids inhibit the release of arachidonic acid, thus leading to the suppression of formation of arachidonate metabolites such as prostaglandins and leukotrienes. In certain types of tissues and cells, the production of prostaglandin and/or leukotriene can not be affected by the treatment with glucocorticoids. It has been reported that glucocorticoids do not affect phospholipase C, an enzyme which produces diacylglycerol, another source of arachidonate. In fact, the glucocorticoid treatment of astrocytoma cells brought about the inhibition of arachidonate release but not of PtdIns turnover[16]. Further, glucocorticoids do not change cellular phospholipase activity as measured by arachidonate release following protease or Ca^{2+} ionophore treatment[17]. Accordingly, it has been proposed that glucocorticoids induce the synthesis of PLA_2 inhibitory proteins(s) in the target tissues. This proposal explains well about the previous observations that inhibitors of mRNA and protein synthesis such as Actinomycin D and cycloheximide can block the suppressive effects on arachidonic acid release as well as anti-inflammatory effects by glucocorticoids[3,4]. Further, it is also consistent with the hypothesis of mechanism by which glucocorticoids act though genomic effects; the glucocorticoid-receptor complex regulates the transcription of genes, ultimately the synthesis of new proteins.

Many attempts to isolate proteins that regulate cellular PLA_2 had been made. We initially reported that one such protein purified from the glucocorticoid treated rabbit neutrophils has a molecular weight of approximately 40KDa with a pI value of 9.0 to 10.0 and is capable to bind 2 moles of Ca^{2+} [18]. A concentration of Ca^{2+} required for half maximal binding is around 1 μM. Further, this protein is glycosylated and can be phosphorylated by various kinases[19]. As the mechanism of PLA_2 inhibition, we proposed that this inhibitory protein suppresses PLA_2 by making a complex with PLA_2 rather than with phospholipid substrates. Some other groups including Flower's and Russo-Marie's groups reported that the inhibitory proteins which they isolated have the molecular weights of 15KDa and 30KDa, respectively. They did not further characterize the biochemical properties of these proteins. However, all these proteins have similar, if not identical, biological properties: (a) the synthesis of these proteins is induced by glucocorticoids, (b) preparations of these proteins, regardless of their purity, inhibit porcine pancreas and/or snake venom PLA_2 in vitro, (c) preparations of these proteins mimic some actions of glucocorticoids, especially with regard to the release of arachidonic acid and/or prostaglandin production and (d) some monoclonal antibodies raised against PLA_2 inhibitory protein prepared in one laboratory crossreact with preparations of those proteins from the other laboratories. Based upon these findings, it has been proposed that a family of these PLA_2 inhibitory proteins whose synthesis is induced by glucocorticoids be named as lipocortins[6].

The Biogen group purified one protein from human placenta whose properties are almost identical to those which our group described. They isolated cDNA clones for this protein which can inhibit porcine pancreas PLA_2 in vitro with ^{3}H-oleate labelled E. coli (phosphatidyl-ethanolamine (PtdEtn) and phosphatidylserine (PtdSer) as major phospholipids)[20,21]. The chemical natures of this protein are as follows; (a) the molecular weight of this protein is 36KDa (b) it has multiple phosphorylation sites (including one for tyrosine kinase), and a potential glycosylation site, and (c) it has 4 repeats which can bind acidic phospholipids such as PtdIns and PtdSer, and Ca^{2+}. Subsequently, it was found that this protein is the same with calpactin which was previously discovered as a major substrate for EGF (epidermal growth factor) receptor kinase. Currently, 7 related proteins are categorized in this family (see other chapters in this volume). Since oleic acid labelled E. coli (which contain PtdEtn and PtdSer) were used for the in vitro assay, the mechanism of PLA_2 inhibition was attributed to depletion of the substrate by binding phospholipids rather than to interaction of such a protein with PLA$_2$ (see other chapter). Whether the proteins of such inhibitory mechanism have some physiological relevances in intact cells is being debated (see other chapters). Further, induction of the synthesis of calpactins in the target cells by glucocorticoids has also become under question. Recent several reports described that the inhibition of prostaglandin production by glucocorticoids do not accompany the induction of the synthesis of calpactins[22,23]. We also failed to detect the increased synthesis of mRNA for calpactins by the Northern blot with anti-sense 50mer oligonucleotides, when human peripheral monocytes were incubated with dexamethasone *in vitro*. Under these conditions, dexamethasone at 1 μM inhibited [1-^{14}C]-arachidonic acid release from prelabelled monocytes elicited by fMetLeuPhe. The Biogen group described in the original report that glucocorticoids did not increase mRNA of calpactin-II in the cultured rat peritoneal macrophages, while the cells isolated from the peritoneal lavages of the animals treated with glucocorticoids contained higher levels of this mRNA[20]. Therefore, it could be speculated that glucocorticoids induce the synthesis of calpactin via unknown pathways or some factors from cells other than macrophages are responsible for this increase. This is against accumulated evidences demonstrating that glucocorticoids act directly on isolated macrophages *in vitro* with respect to suppression of the arachidonate release and/or eicosanoid production[2]. Nevertheless, Flower's and Russo-Marie's groups demonstrated that purified natural and recombinant calpactins can suppress the release of arachidonate and its metabolites, thus showing anti-inflammatory action against carrageenan induced paw edema or other immunological animal tests[24-26] (see also a chapter by Russo-Marie). The discrepancies between the observations lead us to reexamine the biological activities of classic lipomodulin and calpactins.

We purified lipomodulin and calpactin according to the previously published methods[19,27]. Briefly, human placenta were homogenized in 10 mM Tris-Cl buffer containing 5 mM Ca^{2+} and protease inhibitor cocktails. After centrifugation, the precipitates were washed with 10 mM Tris-Cl, pH 7.4, containing 0.1% Nonidet P40. The Nonidet P40 extracts were passed through a DEAE-Sepharose column. Pancreas PLA$_2$ activity was measured with ^{3}H-oleate labelled *E. coli* for calpactin and with ^{14}C-arachidonyl PtdCho for lipomodulin. Calpactins were eluted from the DEAE column with a gradient of the EDTA (0 to 20 mM). Lipomodulin was purified from the flow-through fraction using PLA$_2$-coupled agarose. The major fraction of calpactin contained 36KDa protein, while the major fraction of lipomodulin contained 40KDa protein as judged by SDS-PAGE. As previously reported, calpactins interact very weakly with PtdCho, thus leading to slight inhibition of PLA$_2$ under the present assay conditions. However, apolipoproteins such as AII inhibited strongly PLA$_2$ activity, since these proteins are able to bind PtdCho. Apolipoprotein-AII has an apparent molecular weight of 15KDa (dimmer of 7.5K under the non-reduced conditions). In addition, C reactive protein (CRP, molecular weight 24KDa) exerted strong inhibition of PLA$_2$ with PtdIns and PtdCho as substrates. Other phospholipid transfer proteins also had more or less inhibitory actions on PLA$_2$ *in vitro*. The mechanism of all these actions of apolipoprotein-AII and CRP is attributed to depletion of the phospholipid substrates by the capacity of these proteins to bind phospholipids. Their binding capacity is often dependent upon Ca^{2+}. Thus, the inhibition of phospholipase by these proteins was also Ca^{2+} dependent. In contrast, we have previously proposed that lipomodulin makes one-to-one stoichiometric complex with PLA$_2$. This hypothesis was supported by effectiveness of PLA$_2$-affinity column chromatography for purification as well as the kinetics of inhibition in which calpactin inhibits PLA$_2$ by reducing phospholipid substrates, while lipomodulin reduces PLA$_2$ activity by decreasing the amount of active PLA$_2$.

Biological action of phospholipase inhibitor proteins.

   There are several lines of evidences suggesting that arachidonate
and its metabolites act as intracellular second messengers for receptor
signal transduction by activating protein kinase C and increasing cellular cyclic GMP level. Although evidences supporting that PLA_2 is
activated directly by stimulation of receptors are not exclusive, it is
a general consensus that PLA_2 plays an important role for the release of
arachidonic acid. Therefore, PLA_2 is now proposed to be involved in
receptor-mediated cellular functions including mitogenesis, heterogenous
desensitization of certain receptors, chemotaxis and cellular differentiation[28]. If this is the case, PLA_2 inhibitory proteins can manipulate
all these events by blocking the activation of PLA_2. Availability of
pure and recombinant calpactin will facilitate investigation as to
whether a family of calpactins, Ca^{2+}-, phospholipid- and actin binding
proteins, have such actions. Along this line, Flower and his associates
recently showed several lines of evidences suggesting that calpactins
may exhibit such effects. They reported that recombinant lipocortin-I
(calpactin-II) blocks the *in vitro* release of eicosanoids from the
guinea pig perfused lung and human umbilical artery respectively[24,25].
Further, they employed this recombinant protein to an animal model of
inflammation and demonstrated that micrograms of doses of the protein,
locally administered, inhibit the eicosanoid- but not amine-dependent
component of carrageenin edema in the rat paw assay[26]. They further
speculated that calpactins are specific for PLA_2-induced inflammation
and that other glucocorticoid-induced proteins mediate the ability of
steroids with respect to the anti-inflammatory effects on the paw edema
induced by dextran, PAF, bradykinin and serotonin which can not be
blocked by calpactin. In accord with this interpretation, Jelsema has
proposed that calpactin provides tonic inhibition to cellular PLA_2 and
that the β/γ subunits of G proteins dissociated by the receptor stimulation diminish this inhibition together with actins[13].

   There had been confusion about the nature of the steroid-induced
anti-phospholipase proteins. Despite similar biological properties in
terms of various aspects, there was a disparity in their molecular
weights. Tentatively, it was interpreted by a proposal that the species
of the molecular weights with 15KDa and 30KDa are the proteolytic fragments of 40KDa species. To confirm such explanation and search the
minimal length of effective peptides, the two approaches have been
taken. First, purified calpactin was digested and individual fragments
were tested for their anti-PLA_2 activity[29]. Secondly, peptides whose
sequences match to those of cDNA clones for calpactins were synthesized
and tested for anti-phospholipase activity. The peptide (C-MT peptide)
which has 80% homology to a consensus sequence for Ca^{2+}- and phospho-
lipid-binding sites were found to be inhibitory against porcine pancreas
PLA_2[30]. Its inhibitory effects were in a Ca^{2+}- dependent manner. Although this peptide exerted its inhibitory action towards a variety of
phospholipases, PtdIns phospholipase C was most sensitive to the action
of this peptide[30]. This is probably due to its affinity for acidic
phospholipids including PtdIns. It was also the case with native
calpactin. In this sense, calpactin is PtdIns phospholipase C inhibitory protein rather than PLA_2 inhibitory protein. The C-MT peptide
could inhibit arachidonate release and chemotaxis of the rabbit neutrophils stimulated by fMetLeuPhe. Since this peptide competed with
fMetLeuPhe at their binding sites, there was no conclusion concerning
that the effects of C-MT peptide on the arachidonate release is attributed to inhibition of cellular PLA_2 or antagonism at the receptor
sites. Similarly, the fragments of CRP (calcium dependent phospholipid
binding protein in the acute phase of inflammation) are reported to have
a variety of biological activities including inhibition of O_2^- production

chemotaxis and lysozyme release[31]. Recently, Mukherjee and his associates synthesized several peptides whose sequences correspond to the region of the highest similarity between uteroglobulin and lipocortin-I[32]. The detailed description of anti-inflammatory action of these peptides is available in this volume. The mechanism of action by these peptides remains to be clarified. It is likely that these peptides are partitioned into membrane phospholipids, thus leading to alteration of physical properties of phospholipid structure in the membranes which, in turn, changes the susceptibility to hydrolytic activity. All these findings are apparently consistent with the earlier observations that calpactins are the steroid-induced phospholipase inhibitory proteins. However, calpactins are cytosolic proteins rather than membrane-bound proteins. Therefore, it is difficult to assume that these proteins enter the cells to inhibit cellular PLA_2. When we take it into consideration that calpactins have anticoagulant activity, it is conceivable that exogenously added calpactins bind to the cell surfaces to protect the cell membrane from the attack by extracellular PLA_2, if PLA_2 is secreted from inflammatory cells as in the case of pancreatic enzyme (see above). Then, questions arises as to how glucocorticoids enhance the excretion of calpactins. It may be possible to presume that cell damages or injuries by various stimulants including cytolytic action of glucocorticoids are the main forces to excrete calpactins from the cells.

Table 1. Effects of calpactin and lipomodulin on the allogeneic cytotoxic T cell generation.

		Calpactin	Lipomodulin
Antigen presentation		no effects	no effects
Proliferation	AA release	slight inhibition	inhibition
	TdR uptake	no effects	inhibition
	IL-1 mRNA	no effects	no effects
	IL-1 release	no effects	inhibition
	IL-2 mRNA	ND	ND
	IL-2 release	no effects	inhibition
Maturation		no effects	promotion
Cytotoxic reaction			
	Binding	no effects	no effects
	Lysis	slight inhibition	inhibition
	Recycling	ND	ND
Memory cells		no effects	enhanced

* ND; not determined.

In contrast to discussions favoring the roles of calpactins, recent reports described that the inhibition of eicosanoid production by dexamethasone, a potent synthetic glucocorticoids, does not accompany the increased levels (or synthesis) of calpactins[22,23]. Together with our preliminary observations that glucocorticoids do not enhance the synthesis of mRNA for calpactins in various cell lines, it has become under question that calpactins are the second messengers of glucocorticoids. To answer to such questions, we tested two preparations of phospholipase inhibitory proteins, lipomodulin and calpactin-II, on the generation of allogeneic cytotoxic T cells. We have previously reported that this system is blocked by various PLA_2 inhibitors including quinacrine, parabromophenacyl bromide and tetracaine[33]. At least the mechanism of the lysis of tumors by NK cells is akin to that of the cytotoxic reaction by cytotoxic T cells. We have previously reported that lipomodulin can block the lytic reaction of NK cells[34]. The effects of calpactin-II and lipomodulin were summarized in Table 1. The effects of lipomodulin were essentially similar to those observed with quinacrine. The inhibition of IL-1 production by glucocorticoids were previously reported to be partly attributable to the decreased stability of mRNA. When we examined the cellular and extracellular IL-1 using the radioimmunoassay, our preliminary results suggest that the main site of glucocorticoid action is the step of release of IL-1 molecule. This is in accord with our previously proposal that lipomodulin inhibits the release of cytolytic factor from NK cells, thus inhibiting tumorcidal activity of NK cells[34]. The effects of lipomodulin and dexamethasone on the synthesis of mRNA and protein molecule for lymphotoxin is currently under investigation in our laboratory. Lipomodulin inhibited the stimulatory effects of IL-1 on the TdR uptake of D10 cells (murine helper T cell clone) elicited by an anti-T cell receptor antibody. Further, it also blocked proliferation of CTLL 2 cells (murine cytotoxic T cell clone) stimulated by IL-4. These results suggest that lipomodulin is able to block not only the release but also the action of these cytokines. Our preliminary observations suggest that this protein also partially block the action of TNF-α (tumor necrotic factor). In contrast, calpactin-II at higher doses could slightly reduce the release of arachidonic acid as well as lytic process of the cytotoxic T cells, but exerted essentially no effects on the other processes. These findings strongly suggest but not necessarily prove that lipomodulin is a protein distinct from calpactin. Recently, we have isolated several monoclonal antibodies against lipomodulin. One of these antibodies was found to block the suppressive effects of glucocorticoids on the glycoconjugate production in feline trachea *in vitro*[35], suggesting that this antibody recognizes the glucocorticoid-induced protein in this tissue. However, this antibody failed to react with purified human placental calpactin-II. These findings also support our contention in which lipomodulin is a distinct protein from calpactin. We have previously reported that lipomodulin can inhibit broad spectra of phospholipases. A highly purified preparation of lipomodulin was more strictly specific for porcine phospholipase A_2, an enzyme which was used for the assay. We assumed that the difference of specificity between two preparations might be due to contaminated calpactins.

SUMMARY

Our present results showed that calpactins might be not the proteins whose synthesis is modulated by glucocorticoids, and suggest that there may be another type of PLA_2 inhibitory protein which directly interact with PLA_2 and whose synthesis is induced by glucocorticoids. Further characterization and isolation of this protein will provide the most concrete evidence in this aspect.

ACKNOWLEDGEMENT

This work is supported in part by NIH grants (ES-04802 and NS-24628).

REFERENCES

1. A.S. Fauci, D.C. Dall and J.E. Balon, *Ann. Intern. Med.*, 84:315-346 (1976).
2. F.A. Kuehl and R.W. Eagen, *Science*, 210:978-984 (1980).
3. S. Trufuji, K. Sugio and T. Takemasa, *Nature*, 280:408-410 (1979).
4. A. Danon and G. Assouline, *Nature*, 273:552-554 (1978).
5. E.B. Thompson and M.E. Lippman, *Metab. Clin. Exp.*, 23:159-202 (1974).

6. M. DiRosa, R.J. Flower, F. Hirata, L. Parente and F. Russo-Marie, *Prostaglandins*, **28**:441-442 (1984).

7. H. Van Den Bosh, *Biochem. Biophys. Acta*, **604**:191-246 (1988).

8. J.J. Seilhamer, W. Pruzanski, P. Vadas, S. Plant, J.A. Miller, J. Kloss and L.K. Johnson, *J. Biol. Chem.*, **264**:5335-5338 (1989).

9. R.M. Kramer, C. Hession, B. Johansen, G. Hayes and P. McGray, E.P. Chow, R. Tizard and R.B. Pepinsky, *J. Biol. Chem.*, **264**:5768-5745 (1989).

10. R.A. Wolfe and R.W. Gross, *J. Biol. Chem.*, **260**:7295-7303 (1985).

11. A. Gassama-Diagne, J. Fauvel and H. Chap, *J. Biol. Chem.*, **264**:9470-9475 (1989).

12. J. Axelrod, R.M. Burch and C.L. Jelsema, *Trends Neurosci.*, **11**:117-123 (1988).

13. C.L. Jelsema, *Ann. N.Y. Acad. Sci.*, **559**:158-177 (1989).

14. Y. Kurachi, H. Ito, T. Sugimoto, T. Shimizu, I. Miki and M. Ui, *Nature*, **337**:555-557 (1989).

15. D. Kim, D.L. Lewis, L. Graziadei, E.J. Neer, D. Bar-Sagi and D.E. Clapham, *Nature*, **337**:557-560 (1989).

16. J.J. DeGeorge, A.H. Ousley, K.K. McCarthy, P. Morrel and E.G. Lapetina, *J. Biol. Chem.*, **262**:9979-9984 (1987).

17. F. Hirata, E. Schiffmann, K. Venkatasubramanian, D. Salomon and J. Axelrod, *Proc. Natl. Acad. Sci. USA*, **77**:2533-2536 (1980).

18. F. Hirata, *Adv. Prostaglandins Thromboxane Leukotriene Res.*, **7**:71-78 (1980).

19. F. Hirata, *J. Biol. Chem.*, **256**:7730-7732 (1981).

20. B.P. Wallner, R.J. Mattaliano, C. Hession, R.L. Cate, R. Tizard, L.K. Sinclair, C. Foeller, E.P. Chow, J.L. Browning, K.L. Ramanchandran, R.B. Pepinsky, *Nature*, **320**:77-81 (1986).

21. K-S, Huang, B.P. Wallner, R.J. Mataliano, R. Tizard, C. Burne, A. Frey, C. Hession, P. McGray, L.K. Sinclair, E.P. Chow, J.L. Browning, K.L. Ramchandran, J. Tang, J.E. Smart and R.B. Pepinsky, *Cell*, **46**:191-196 (1986).

22. M.J. Bienkowski, M.A. Petro and L.J. Robinson, *J. Biol. Chem.*, **264**:6536-6544 (1989).

23. F. Hullin, P. Raynal, J.M.F. Ragab-Thomas, J. Fanvel and H. Chap, *J. Biol. Chem*, **264**:3506-3513 (1989).

24. G. Cirino and R.J. Flower, *Prostaglandins*, **34**:59-62 (1987).

25. G. Cirino, R.J. Flower, J.L. Browning, L.K. Sinclair and R.B. Pepinsky, *Nature*, **328**:270-272 (1987).

26. G. Cirino, S.H. Peers, R.J. Flower, J.L. Browning and R.B. Pepinsky, *Proc. Natl. Acad. Sci. USA*, **86**:3428-3432 (1989).

27. J. Soric and J.A. Gordon, *J. Biol. Chem.* **261**:14490-14495, (1986).

28. F. Hirata, in *"Anti-inflammatory Steroid Action: Basic and clinical aspects"*, eds. Schleimer, R.P., Claman, H.N. and Oronsky, A., Academic Press, NY, pp. 67-95 (1989).

29. K-S. Huang, P. McGray, R.J. Mattaliano, C. Burne, F.P. Chow, L.K. Sinclair and R.B. Pepinsky, *J. Biol. Chem.*, **262**:7639-7642 (1987).

30. Y. Notsu, S. Namiuchi, T. Hattori, K. Matsuda and F. Hirata, *Arch. Biophys. Biochem.*, **235**:195-204 (1985).

31. F.A., Robey, K. Ohura, S. Futaki, N. Fujii, H. Yajima, N. Goldman, K.D. Jones and S. Wahl, *J. Biol. Chem.*, **262**:7053-7057 (1987).

32. L. Miele, E. Cordella-Miele, A. Facchiano and A.B. Mukherjee, *Nature*, **335**:726-730 (1988).

33. S. Namiuchi, S. Kumagai, H. Imura, T. Suginoshita, T. Hattori and F. Hirata, *J. Immunol.*, **132**:1456-1461 (1984).

34. T. Hattori, F. Hirata, T. Hoffman, A. Hizuta and R.B. Herbermann, *J. Immunol.*, **131**:662-665 (1983).

35. J.D. Lungren, F. Hirata, Z. Marom, C. Logun, L. Steel, M. Kaliner and J. Shelhamer, *Am. Rev. Resp. Dis.*, **137**:353-357 (1988).

INHIBITION OF HUMAN PHOSPHOLIPASES A$_2$ BY CIS-UNSATURATED

FATTY ACIDS AND OLIGOMERS OF PROSTAGLANDIN B$_1$

R. Franson[1], R. Raghupathi[1], M. Fry[1], J. Saal[2],
B. Vishwanath[1], S.S. Ghosh[1], and M.D. Rosenthal[3]

[1]Dept. of Biochemistry and Mol. Biophysics
 Virginia Commonwealth Univ., Richmond, VA
[2]San Francisco Spine Inst., Daly City, CA
[3]Department of Biochemistry
 Eastern Virginia Medical School, Norfolk, VA

Abstract

Inhibition of human phospholipases A$_2$ by cis-unsaturated fatty acids and their oxidative metabolites and/or polymers was studied using partially purified human phospholipases A$_2$ and [1-^{14}C]oleate labelled, autoclaved E. coli as substrate. As previously reported for other phospholipases A$_2$, oleic and arachidonic acids inhibited human synovial fluid phospholipase A$_2$ with IC$_{50}$s of 15 and 30 μM respectively. Air oxidation of arachidonic acid or hydroxylation of oleic acid (12-hydroxy-oleate) substantially relieved that inhibition. Similarly, the enzymatically oxidatized metabolite of arachidonate, prostaglandin B$_1$ (PGB$_1$), did not inhibit enzymatic activity. However, prostaglandin B$_x$ (PGB$_x$), an oligomer (n=6) of PGB$_1$, was a potent inhibitor of Ca^{++}-dependent, neutral-active phospholipase A$_2$ activities. Enzymatic activity in acid extracts from human neutrophils, platelets, sperm, plasma, synovial fluid, endometrium, degenerative disc, and snake venom was inhibited by PGB$_x$ with IC$_{50}$s ranging from 0.5-7.0 μM. Inhibition was independent of substrate phospholipid concentration over a 24-fold range (5-120 μM) and PGB$_x$ quenched the tryptophan fluorescence of snake venom phospholipase A$_2$ in a dose-dependent manner. Agonist-induced (A23187) release of arachidonic acid from prelabelled human neutrophils and cultured human endothelial cells was also inhibited by PGB$_x$ with IC$_{50}$s of 3 and 20 μM, respectively. These results illustrate that oxidative reactions of cis-unsaturated fatty acids relieve their natural inhibitory activity, and polymerization of an inactive fatty acid metabolite yields a potent inhibitor of in vitro and in situ phospholipase A$_2$ activity.

Biochemistry, Molecular Biology, and Physiology of Phospholipase A₂ and Its Regulatory Factors
Edited by A. B. Mukherjee, Plenum Press, New York, 1990

219

INTRODUCTION

It is well known that cis-unsaturated, but not trans-unsaturated or saturated fatty acids inhibit phospholipase A_2 activity in vitro (1-4). In our report of 1974, we concluded that the inhibition of rabbit neutrophil phospholipase A_2 by oleic acid was an example of end product inhibition which helped to explain the nonlinear kinetic behavior observed in interfacial catalysis by these enzymes (1). More recent studies suggest that unesterified fatty acids in biological fluids and cells may act as endogenous suppressors of phospholipase A_2 activity (2-4). If cis-double bonds in a fatty acid are indeed essential for inhibition of phospholipase A_2 activity, we reasoned that free radical oxidation reactions directed at the double bond may alter inhibitory activity. In this report, we demonstrate that the inhibitory activity of cis-unsaturated fatty acids is readily relieved by oxygen free radical reactions, and that a potent inhibitor of in vitro and in situ phospholipase A_2 activity can be produced from an otherwise inactive oxygenated metabolite of arachidonic acid. Thus, fatty acids and their oxidative metabolites may play an important role not only in modulating lipolytic cascades but perhaps many other biological events affected preferentially by unesterified, polyunsaturated fatty acids (12).

MATERIALS AND METHODS

Phospholipase A_2 assay

Phospholipase A_2 activity was measured by established methods using [1-^{14}C]oleate-labelled, autoclaved E. coli as substrate (1). Reaction mixtures in a total volume of 0.5 ml contained 50 mM Hepes (pH 7.0), 5.0 mM $CaCl_2$, 4.8 x 10^8 [1-^{14}C]oleate-labelled E. coli (10.0 nmols phospholipid, 8-10,000 cpms); the time of incubation and amount of phospholipase A_2 were adjusted to optimize linear kinetics for quantitation. Fatty acids, prostaglandin B_1 (PGB_1) monomer and oligomer (PGB_x) were added as DMSO solutions, using a DMSO-enzyme control. Identical results were obtained when fatty acids were added as aqueous dispersions and when PGB_x was added as the sodium salt. Reactions were stopped and total lipids were extracted and separated as previously described (1,5). Phospholipase A_2 activity expressed as percent of enzyme control. All data are the average of at least duplicate determinations corrected for non-enzymatic hydrolysis ($\leq$ 1.5% in all experiments). Arachidonic and oleic acids were obtained from Supelco, Inc., Bellefonte, PA; ricinoleic acid and PGB_1 monomer were purchased from Sigma Chemicals, St. Louis, MO. Oligomers of PGB_1 were generously provided by Drs. G.L. Nelson (St. Joseph's University) and T.M. Devlin (Hahnemann Medical College) and the Office of Naval Research (Bethesda, MD).

Extracts of human cells and fluids

Calcium-dependent and neutral-active phospholipases A_2

were extracted from human synovial fluid, neutrophils, platelets, sperm, endometrial cells and degenerative discs by mixing equal volumes of the fluid or cell homogenates and ice-cold 0.36 N sulfuric acid. The pH was adjusted to 1.0-1.2 and the turbid mixture was stirred at 4°C for 4 hours. After centrifugation at 10,000 x g (Beckmann, JA-20 rotor), the resulting supernatant was dialyzed to pH 4.5 against 10 mM sodium acetate and the dialysate was recentrifuged to yield a supernatant enriched in phospholipase A_2 activity. Fresh human plasma was extracted in a similar manner except that the plasma-acid mixture was heated in a boiling water bath for 15 min. and cooled prior to dialysis. The phospholipase A_2-enriched supernatants obtained from acid extracts of human synovial fluid and human platelets were further purified using cation exchange chromatography (sulfopropyl Sephadex); phospholipase A_2 activity eluted from the column with 1.5 M NaCl (5,6).

Oxidation of arachidonic acid

Arachidonic acid (1 μmol) was resuspended in 10 mM sodium acetate, pH 5.5 (1.0 ml) by brief probe sonication. The tube was loosely capped and incubated at 37°C for 24 hours. Oxidation was monitored by a decrease in turbidity of the fatty acid dispersion (measured at 430 nm) and the appearance of hydroperoxides as measured by an iodometric assay (7).

Fluorescence of snake venom phospholipase A_2

To measure interaction of PGB_x with snake venom phospholipase A_2, 3.3 nmol of Naja mossambica mossambica (N. moss. mossambica) phospholipase A_2 (pI 9.6, Sigma Chemicals, St. Louis, MO) was mixed with the indicated concentrations of PGB_x-sodium salt in the presence of 50 mM Hepes (pH 7.0) and 5.0 mM $CaCl_2$ in a total volume of 1.0 ml. Samples were excited at 280 nm and emission was monitored at 340 nm. Relative fluorescence is expressed as percent of enzyme control.

Agonist-stimulated mobilization of arachidonic acid

Polymorphonuclear leukocytes were isolated by centrifugation over Ficoll-Hypaque discontinuous gradients and resuspended in phosphate buffered saline (PBS). The neutrophils were radiolabelled with [3H]arachidonic acid as described previously (8). Unincorporated [3H]arachidonate was removed by washing cells with PBS containing 25 μM fatty acid-free albumin. Aliquots of 1 x 10^6 cells in PBS-albumin were incubated for 15 min. at 37°C with or without PGB_x. Calcium ionophore A23187 was added to obtain a final concentration of 10 μM; PBS-albumin was added to the control tubes. The neutrophils were incubated for 5 min., centrifuged at 4000 x g and the radioactivity quantitated in the supernatant by liquid scintillation spectrometry.

Human endothelial cells obtained from umbilical veins were cultured in gelatin coated flasks in Hepes buffered Medium 199 supplemented with 10% fetal bovine serum, 30 μg/ml Endothelial Cell Growth Supplement and 90 μg/ml heparin. For each experiment, replicate flasks of confluent second passage cells were labelled by incubating the cells for 24 hours in complete culture medium (4.0 ml) supplemented with 0.1 μCi [^{14}C]arachidonate. Agonist-stimulated release was performed with confluent monolayers adherent to flasks. The labelled cells were washed with culture medium containing 50 μM albumin and then with albumin-free medium; the cells were incubated at 37°C for 30 min. in medium (containing 2.5 μM albumin) with or without PGB$_x$. A small volume of medium was added to provide a final concentration of 50 μM albumin with or without 10 μM A23187. After incubation for 10 min. the incubation media was removed and the radioactivity quantitated by liquid scintillation spectrometry.

RESULTS

Figure 1 illustrates the effect of cis-unsaturated fatty acids (CUFAs) and their oxidative analogues or metabolites on inhibition of neutral-active, Ca^{++}-dependent PLA$_2$ purified from human synovial fluid. As reported for other phospholipases A$_2$ (1-4), oleic and arachidonic acids inhibited synovial fluid phospholipase A$_2$ activity in a dose-dependent manner with IC$_{50}$s of 15 and 30 μM, respectively. Interestingly, hydroxylation of oleic acid at the 12-position (ricinoleic acid) attentuated the inhibitory activity. When exposed to air at 37°C, 55% of the arachidonic acid was oxidized (12) and inhibitory activity was markedly reduced. Thus, oxidative-free radical reactions, in general, relieve the phospholipase A$_2$ inhibitory activity of CUFAs.

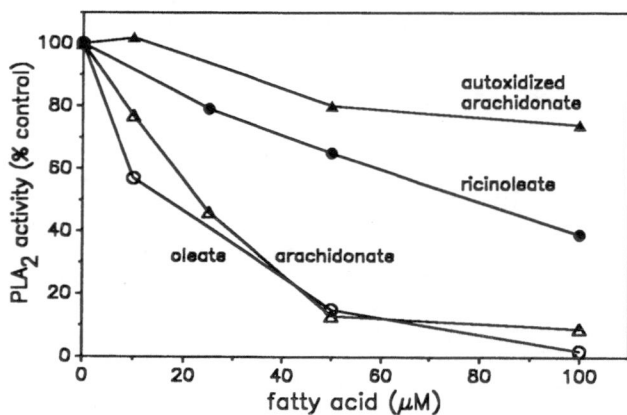

Figure 1 Inhibition of human synovial fluid phospholipase A$_2$ activity by fatty acids.

Fatty acids dissolved in DMSO (5% of reaction volume) were added at the indicated concentrations. Arachidonic acid was oxidized as described in Methods and was added directly to the reaction mixture. Control activity (in the absence of fatty acid) was in the range of 10-15% substrate hydrolysis.

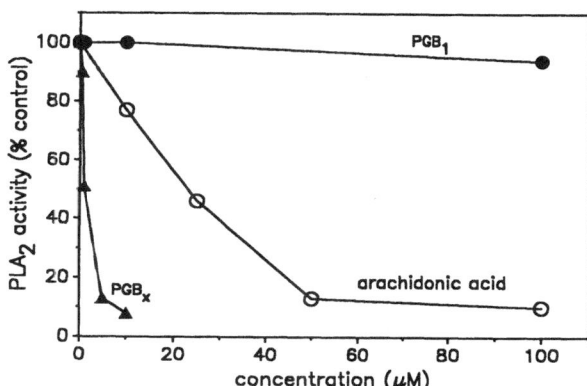

Figure 2. Inhibition of human synovial fluid phospholipase A_2
activity by PGB$_x$.

Arachidonic acid, PGB$_1$ and PGB$_x$ were dissolved in DMSO (5% of
reaction volume) and added to the reaction mixture at the
indicated concentrations. Control activities (in the absence
of inhibitor) were in the range of 10-15% substrate
hydrolysis.

As illustrated above, the oxidative conversion of
arachidonic acid to prostaglandin B_1 (PGB$_1$) resulted in an
almost complete loss of precursor inhibitory activity
(figure 2). However, when the inactive monomer, PGB$_1$, was
heated at alkaline pH, a base-catalyzed oligomer, PGBx, was
formed that was reported to have unusual biological
protective properties (9,10) and potent phospholipase A_2
inhibitory activity. PGBx inhibited human synovial fluid
phospholipase A_2 activity with an IC$_{50}$ of 1.6 μM (figure 2),
and inhibited similar activities extracted and/or partially
purified from human polymorphonuclear leukocytes, platelets,
sperm, plasma, endometrium, and degenerative disc with IC$_{50}$s
that ranged from 1.4 to 7.0 μM (Table I).

Table I   **Inhibition of Human Phospholipase A$_2$ Activities by PGBx**

Standard reaction mixtures contained 0-10 μM PGB$_x$-sodium salt. Activity is expressed as percent of enzyme control (100% ranged from 0.8-1.6 nmol fatty acid released/reaction). Enzyme samples were prepared as described in Methods.

Human Source	Inhibition of PLA$_2$ IC$_{50}$ (μM)
PMN	5.1
Platelet	2.4
Sperm	2.2
Plasma	7.0
Synovial Fluid	1.6
Endometrium	1.9
Degenerative Disc	1.4

Inhibition was independent of substrate phospholipid concentration over a 24-fold range, from 5 to 120 μM E. coli phospholipid phosphorus (Table II).

Table II   **Inhibition of Human Platelet PLA$_2$ Activity as a Function of Substrate Concentration**

Standard reaction mixtures contained enzyme, 10 μM PGB$_x$-sodium salt and the indicated concentrations of autoclaved E. coli phospholipid phosphorus.

[Substrate Phospholipid] (μM)	% Inhibition PLA$_2$
5	59
10	53
15	57
30	61
60	58
120	54

Because the phospholipase A$_2$ inhibitory activity of PGBx was independent of substrate phospholipid concentration, we examined the ability of PGBx to interact with snake venom phospholipase A$_2$ by fluorescence spectroscopy (figure 3). PGBx quenched the fluorescence of N. moss. mossambica phospholipase A$_2$ (pI 9.6) in a dose-dependent manner. Relative fluorescence was decreased 50 % at a molar ratio of 1.5 (PGBx/PLA$_2$), demonstrating that PGBx associates with the purified snake venom phospholipase A$_2$. PGBx inhibited in vitro snake venom phospholipase A$_2$ activity with an IC$_{50}$ of 0.5 μM (not shown).

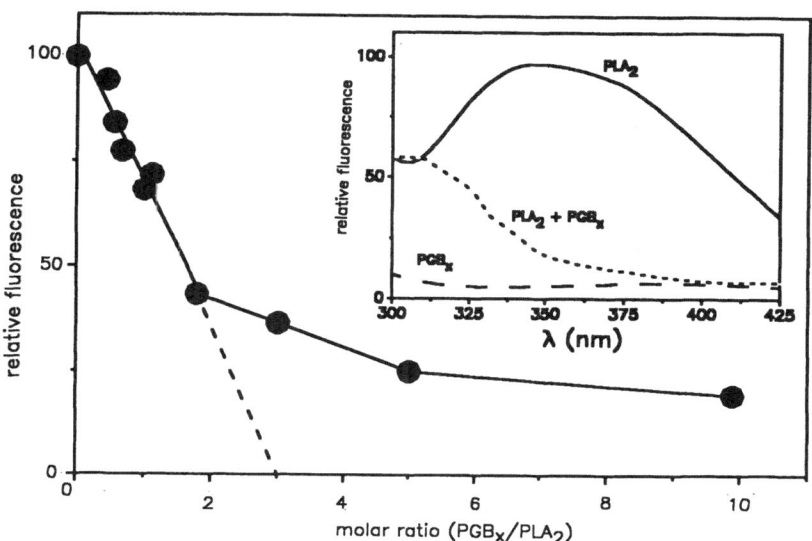

Figure 3. Effect of PGB$_x$ on the fluorescence of **N. moss. mossambica** phospholipase A$_2$.

Mixtures in a total volume of 1.0 ml contained 3.0 nmol of the enzyme, 50 mM Hepes, pH 7.0 and 5.0 mM CaCl$_2$ with and without PGB$_x$-sodium salt. Samples were excited at 280 nm and emission intensity measured at 340 nm. <u>Inset</u> : Relative intensity of the emission spectra of 3.0 nmol enzyme in the absence and presence of 10 μM PGB$_x$-sodium salt.

In addition to inhibition of <u>in</u> <u>vitro</u> phospholipase A$_2$ activities, PGBx inhibited agonist-induced release of arachidonic acid from prelabelled human PMNs and cultured human endothelial cells (figure 4). Release of arachidonic acid from both cell types was inhibited a dose-dependent manner; the IC$_{50}$s were 3 and 20 μM, respectively. PGBx had no effect on basal level release of arachidonic acid at the concentrations tested. Thus, PGBx not only inhibits phospholipase A$_2$ activity <u>in</u> <u>vitro</u> but also affects an <u>in</u> <u>situ</u> lipolytic cascade in which phospholipase A$_2$ is a prominent participant (17).

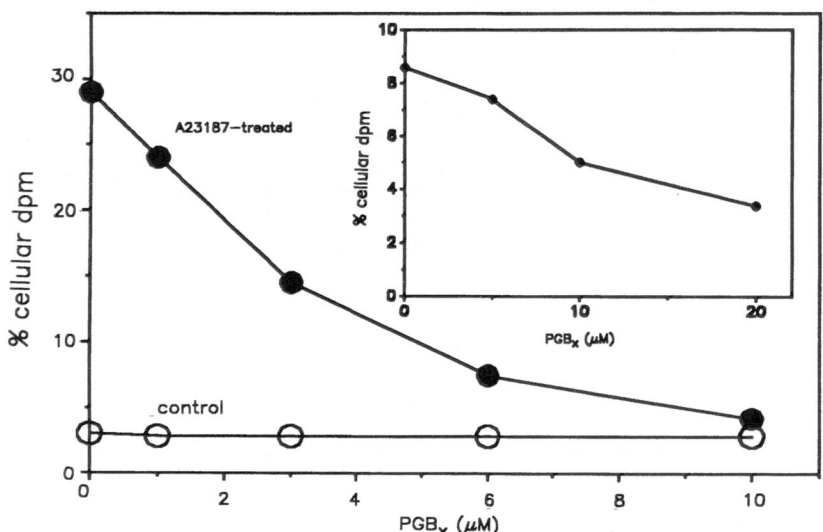

Figure 4. Effect of PGB_x on A23187-stimulated release of arachidonate from human neutrophils and vascular endothelial cells (inset).

Aliquots of prelabelled cells were incubated with PGB_x-sodium salt as described in Methods. Radioactivity released into the media was quantitated and expressed as percent of total cellular [^{3}H]arachidonate. <u>Inset</u> : Vascular endothelial cells prelabelled with [^{14}C]arachidonate were treated with PGB_x as described in Methods. Fatty acid release was calculated as radioactivity released in the presence of A23187 minus that in control flasks and expressed as a percent of total cellular [^{14}C]arachidonate.

DISCUSSION

These results demonstrate that *cis*-unsaturated fatty acids (CUFAs) inhibit human phospholipase A_2 activities and oxidative modification of CUFAs relieves their inhibitory effect. Prostaglandin B_x (PGB_x), an oligomer ($n=6$) of the oxygenated metabolite of arachidonic acid, prostaglandin B_1 (PGB_1), inhibits neutral-active, Ca^{++}-dependent phospholipases A_2 in vitro and arachidonate mobilization in situ.

Cis- but not *trans*-unsaturated fatty acids inhibit a broad spectrum of neutral-active, Ca^{++}-dependent phospholipases A_2 (2-4). In addition to the *cis*-double bond, a free carboxyl group is necessary for inhibition, since methylated fatty acids have little or no effect on phospholipase A_2 activity in vitro (2, 3). Both the physical structure imparted by the *cis* and *trans* nature of fatty acid double bonds and the chemical environment of the *cis*-double bond play an important role in determining the inhibitory activity of these fatty acids. Thus, hydroxylation of oleic acid at the C-12 position markedly reduces inhibition. Similar results were observed when arachidonic acid was oxidized either enzymatically to PGB_1 or via free radical attack during autoxidation (figure 1). Like PGB_1, other cyclooxygenase metabolites of arachidonic acid are reported to lack phospholipase A_2 inhibitory activity (4). Chang et al., (11) concluded that the phospholipase A_2 inhibitory potencies of the lipoxygenase metabolites, 5-, 12-, and 15-hydroxyeicosatetraenoic acids were dependent on the position of the hydroxyl group. These data emphasize the complex role of the *cis*-double bond in the inhibition of phospholipase A_2 activity by CUFAs.

Inhibition by CUFAs was observed at levels below their critical micellar concentration, indicating that interactions with monomers and not the detergent properties of micelles affect enzymatic activity. In this regard, there is indirect evidence to suggest that CUFAs interact directly with the phospholipase A_2 (3, 4). Ballou and Cheung (3) noted that phospholipase A_2 activity was unaffected by increasing substrate concentrations. Moreover, Lister et al., (4) reported that arachidonic acid behaved as a competitive inhibitor of the enzyme. In recent studies, we have demonstrated direct association of a snake venom phospholipase A_2 with CUFAs by sedimentation analysis as well as fluorescence (12). While these studies suggest that inhibition of phospholipase A_2 activity by CUFAs could be mediated by direct association with the enzyme, Jain et al., (13) have recently demonstrated that some lipophilic inhibitors may reside in the lipid bilayer and compete with the substrate phospholipid for binding to the enzyme. These observations illustrate the difficulty in interpretation of mechanistic studies with phospholipase A_2 inhibitors in interfacial catalysis.

Enzymatic oxidation of arachidonic acid to PGB_1 drastically reduces the inhibitory activity of the CUFA (figure 2). Oligomerization of PGB_1 produced a polymer, PGB_x, capable of inhibiting a wide variety of Ca^{++}-

dependent phospholipases A_2 _in vitro_ (table 1). Inhibition of enzymatic activity by PGB_x was unaffected over a 24-fold range of substrate concentration (table 2). PGB_x may inhibit phospholipases A_2 by interacting directly with the enzyme. In this regard, PGB_x quenched the fluorescence of N. moss. mossambica venom in a dose-dependent manner (figure 3). Thus, PGB_x is unlike non-specific inhibitors such as quercetin (14) and calpactin (15) which inhibit phospholipase A_2 activity by binding to the substrate. PGB_x inhibits _in situ_ phospholipase A_2 activity, reducing arachidonate mobilization in both human polymorphonuclear leukocytes and endothelial cells at concentrations comparable to _in vitro_ inhibition. Modulation of cellular and extracellular phospholipases A_2 (16) and the bioactive transmitters generated by this catalytic event (17) may be the mechanism by which PGB_x exerts its membrane protective effects (9, 10).

Initial results on the inhibitory effects of fatty acids on Ca^{++}-dependent phospholipase A_2 activity were used to explain the non-linear kinetics characteristic of interfacial catalysis by suggesting that the fatty acids functioned as end product inhibitors (1). Subsequently, high concentrations (0.1-12 mM) of endogenous unesterified fatty acids in human platelets and neutrophils were reported, suggestive of their potential role as endogenous inhibitors (2, 3). Oxidative modification of CUFAs may regulate their biological potencies (18, 19). Unsaturated fatty acids are highly susceptible to free radical attack. Extensive oxidation occurs via a self-propagating chain reaction forming hydroperoxides and hydroxides (20), thus altering the chemical environment of the cis-double bond. In pathological conditions such as ischemia, free radical reactions are promoted by acidosis and peroxidation of unesterified CUFAs may lead to an "apparent activation" of phospholipase A_2 activity _in vivo_. In this regard, we recently reported that when arachidonic acid was autoxidized, it could no longer inhibit the neutral-active, Ca^{++}-dependent phospholipase A_2 isolated from rat heart (21). Our results are consistent with the hypothesis that lipid peroxidation and activation of phospholipases contribute to irreversible cell injury.

FUTURE STUDIES

The results of these experiments indicate that more careful mechanistic studies are required to evaluate suppressors as well as activators of cellular and extracellular phospholipases A_2. It is clear that CUFAs could function as one class of endogenous "switches" to up or down regulate lipolytic activity. In the future more attention will be focussed on enzyme-targeted effectors by monitoring inhibitor-protein interaction. Ultimately, correlating modulation of physical and enzymatic properties upon enzyme-inhibitor interaction with altered biological responses, will provide badly needed clues as to the role of these enzymes in normal and pathologic processes.

Acknowledgement This work was initiated by stimulating conversations with Dr. William Regelson and was supported by Phoenix Advanced Technology Inc., the Office of Naval Research and Virginia's Center for Innovative Technology.

REFERENCES

1. R.C. Franson, P. Patriarca and P. Elsbach, Phospholipid metabolism in phagocytic cells. Acid and alkaline phospholipases A associated with rabbit polymorphonuclear leukocyte granules, J. Lipid Res. 15:380, 1974.

2. F. Märki and R.C. Franson, Endogenous suppression of neutral active and calcium-dependent phospholipase A_2 in human polymorphonuclear leukocytes, Biochim. Biophys. Acta 879:149, 1986.

3. L.R. Ballou and W-Y. Cheung, Inhibition of human platelet phospholipase A_2 activity by unsaturated fatty acids, Proc. Natl. Acad. Sci. 82:371, 1985.

4. M. Lister, R. Deems, Y. Watanabe, R. Ulevitch and E. Dennis, Kinetic analysis of the calcium-dependent, membrane-bound, macrophage phospholipase A_2 and the effects of arachidonic acid, J. Biol. Chem. 263:7506, 1988.

5. B.S. Vishwanath, A.A. Fawzy and R. Franson, Edema-inducing activity of phospholipase A_2 purified from human synovial fluid and inhibition by aristolochic acid, Inflammation, 12:549, 1988.

6. R. Raghupathi and R. Franson, Isolation of a calcium-dependent phospholipase A_2 from human platelets, Fed. Proc. 45:1560, 1987.

7. F.J.G.M. van Kuijk, G.J. Handelman and E.A. Dratz, Consecutive action of phospholipase A_2 and Glutathione peroxidase is required for reduction of phospholipid hydroperoxides and provides a convenient method to determine peroxide values in membranes, J. Free Rad. Biol. Med. 1:421, 1985.

8. M.D. Rosenthal, B.S. Vishwanath and R. Franson, Effects of aristolochic acid on phospholipase A_2 activity and arachidonate metabolism of human neutrophils, Biochim. Biophys. Acta 1001:1, 1989.

9. E.T. Angelakos, R.L. Riley, and B.D. Polis, Recovery of monkeys after myocardial infarction with ventricular fibrillation. Effects of PGB_x, Physiol. Chem. Phys. 12:81, 1980.

10. B.D. Polis, E. Polis and S. Kwong, Protection and reactivation of oxidative phosphorylation in mitochondria by a stable free-radical prostaglandin polymer, Proc. Natl. Acad. Sci. 76:1598, 1979.

11. J. Chang, E. Blazek, A.F. Kreft and A.J. Lewis, Inhibition of platelet and neutrophil phospholipases A_2 by hydroxyeicosatetraenoic acids, Biochem. Pharmacol. 34:1571, 1985.

12. R. Raghupathi and R. Franson, Effect of autoxidation of fatty acids on their inhibition of phospholipase A_2 activity, Biochim. Biophys. Acta , in press, 1989.

13. M.K. Jain, W. Yuan and M. Gelb, Competitive inhibition of phospholipase A_2 in vesicles, Biochem. 28:4135, 1989.

14. A.A. Fawzy, B.S. Vishwanath and R. Franson, Inhibition of human non-pancreatic phospholipases A_2 by retinoids and flavonoids. Mechanism of action, Agents and Actions 25:395, 1988.

15. F. Davidson, E. Dennis, M. Powell and R. Glenney, Jr., Inhibition of phospholipase A_2 by 'lipocortins' and calpactins, J. Biol. Chem. 262:1695, 1987.

16. R. Franson and M.D. Rosenthal, Oligomers of prostaglandin B_1 inhibit _in vitro_ phospholipase A_2 activity, _Biochim. Biophys. Acta_, (in press), 1989

17. M.D. Rosenthal and R. Franson, Oligomers of prostaglandin B_1 inhibit arachidonic acid mobilization in human neutrophils and endothelial cells, _Biochim. Biophys. Acta_ (in press), 1989.

18. P. Needleman, J. Turk, B.A. Jakschik, A.R. Morrison and J.B. Lefkowith, Arachidonic acid metabolism, _Ann. Rev. Biochem._ 55:69, 1986.

19. Y. Kurachi, H. Ito, S. Tsuneaki, T. Shimizu, I. Miki and M. Ui, Arachidonic acid metabolites as intracellular modulators of G-protein gated cardiac K^+ channels, _Nature_ 337:555, 1989.

20. E.N. Frankel, Lipid oxidation, _Prog. Lipid Res._ 19:1, 1980.

21. R. Franson, L.K. Harris and R. Raghupathi, Fatty acid oxidation and myocardial phospholipase A_2 activity, _Mol. Cell. Biochem._ 88:155, 1989.

ACTIVATION OF PHOSPHOLIPASE A[2] IN RHEUMATOID ARTHRITIS

John S. Bomalaski[1] and Mike A. Clark[2]

[1]V.A. Medical Center, Medical College of Pennsylvania
University of Pennsylvania, Philadelphia, PA
[2]Washington University School of Medicine, St. Louis, MO
(Current address: Schering-Plough Research, Bloomfield, NJ)

Rheumatoid arthritis (RA), a common and often disabling systemic disease with a predilection for joints, is characterized by an inflammatory and proliferative reaction of synovial cells associated with infiltration of immunocompetent cells and fluid into the synovial tissue, as well as the destruction of articular cartilage. Prostaglandins and related eicosanoids are thought to be important mediators and regulators of these immune and inflammatory responses (1,2,3). For example, prostaglandin E_2 induces bone resorption, and leukotriene B_4 stimulates vasodilitation and chemotaxis (1,4). Increased quantities of eicosanoids are produced by rheumatoid synovium in both organ and cell culture (2,5,6) and by freshly isolated or cultured peripheral blood monocytes isolated from RA patients as compared to cells obtained from normal donors (7,8,9,10). In addition, high concentrations of eicosanoids have been shown to be present in rheumatoid synovial fluid (2,11,12). Because eicosanoids are important mediators of this disease, numerous investigators have sought to understand the mechanisms of enhanced eicosanoid biosynthesis in this illness. The rate-limiting step in eicosanoid biosynthesis is the release of the precursor fatty acid from membrane phospholipids (4,13,14,15,16,17). Once liberated, the unsaturated fatty acid, usually arachidonic acid, is then oxygenated to form prostaglandins, leukotrienes and related lipid metabolites. As a first step toward the biochemical characterization of this human disease, we examined the uptake and incorporation of unsaturated fatty acids into the phospholipids of cultured peripheral blood monocytes from human volunteers. We chose monocytes as a cell system because of the evidence implicating the monocyte as a major producer of eicosanoids in the inflamed joint tissue (3). The fatty acids utilized in these studies were linolaic acid, a common dietary polyunsaturated fatty acid, dihomagamolinolaic acid, the precursor of the one series prostaglandins, and arachidonic acid, the precursor of the major prostaglandins and leukotrienes (7). Cultured monocytes (macrophages) from patients with RA exhibit enhanced uptake of the unsaturated fatty acids as compared to normal cells (Table 1).

Next we identified the major phospholipids into which these fatty acids were incorporated. Irrespective of the fatty acid used, the majority of the labeled material appeared to be incorporated into phosphatidylcholine, both in cells from patients on treatment with nonsteroidal antiinflammatory drugs as well as those patients who had received no treatment (Table 2), suggesting that rheumatoid cells have enhanced fatty acid turnover. We next stimulated these cells with a variety of agents and observed marked release of arachidonic acid from the phosphatidylcholine (Table 3).

TABLE 1

MONOCYTE-MACROPHAGE FATTY ACID UPTAKE*

Fatty Acid	Day of Culture			
	3	5	7	12
Arachidonic	145**	146**	290**	190**
Linoleic	94	188**	160**	218**
Dihomogammalinoleic	90	140**	197**	273**

* Expressed as percent of control uptake. Cultured cells were exposed to radiolabeled fatty acid overnight. The remaining fatty acid washed away with phosphati buffered saline. The cells were then lysed with Triton X-100 and aliquots were taken for scintillation spectrophotometry. Uptake was corrected for total cellular protein.

** $p < 0.05$.

TABLE 2

INCORPORATION OF FATTY ACIDS INTO PHOSPHATIDYLCHOLINE*

Rheumatoid Patient	Arachidonic	Linoleic	Dihomogammalinolenic
Untreated	145**	43**	82
Treated	80	80	109

* Expressed as percent of control

** $p < 0.05$.

TABLE 3

RELEASE OF FATTY ACID FROM PHOSPHATIDYLCHOLINE*

Fatty Acid	Rheumatoid (On Therapy)	Rheumatoid (No Therapy)
Arachidonic	150**	142**
Linoleic	148**	171**

* Expressed as percent of control. Cells were stimulated with the calcium ionophore A23187, (1μM).

** $p < 0.05$.

mechanism that explains the increased eicosanoid production observed in this disease. Several other mechanisms have been proposed that would explain the enhanced phospholipase activity found in patients having RA. Auto-antibodies to lipocortin, a protein which inhibits phospholipase activity, are found in serum from RA patients (21). This has been proposed to decrease the amount of functional lipocortin which may result in enhanced phospholipase enzyme activity. Another possible mechanism would be the presence of increased levels of a phospholipase activating protein.

The venom from stinging insects, sea anemones and snakes (22,23,24,25) contains a number of phospholipase enzyme activating proteins. These venom components act locally to produce an inflammatory action characterized by warmth, redness, pain, swelling and loss of function. Similar symptoms of inflammation are noted in the rheumatoid joint. Mellitin, a major component of bee venom, is a well characterized phospholipase A_2 activating protein (15,25,26,27,28,29). Mellitin not only stimulates phospholipase A_2, but also causes cells to release eicosanoids (25). To determine if a similar protein may be found in mammalian, we made antibodies to mellitin and using these antibodies affinity purified a mammilian protein that activated phospholipase A_2 (30,31). We have termed this phospholipase activating protein PLAP. Biochemical characterization of PLAP indicated that this mammalian protein stimulated phospholipase A_2 only when phosphatidylcholine was used as a substrate, it had no effect on phospholipase A2 when either phosphatidyl-ethanolamine or phosphatidylinositol was used as a substrate in the enzyme reactions. This protein had a molecular mass of approximately 28,000 Kd as determined by SDS-PAGE.

TABLE 4

IMMUNO DOT BLOTS OF PLAP IN SYNOVIAL FLUID*

Arthritis	N	Dot Blot Value	P Value
Rheumatoid	56	1.88 ± 1.19	< 0.001
Osteoarthritis	27	0.44 ± 0.66	-
Gout	18	1.17 ± 0.99	< 0.025
Pseudogout	12	0.75 ± 0.87	> 0.050
Seronegative Spondyloarthropathy	11	0.46 ± 0.69	> 0.050

* Dot blot analysis was performed in a double blind manner on synovial fluid samples. The blots were read and scored 0-3 (positive). To determine the level of significance, student's t-test was used.

We next examined the synovial fluid isolated from a large number of patients having a diverse spectrum of joint diseases. The presence of PLAP in these fluid samples was assayed using immuno dot blotting techniques (Table 4). The dot blots were scored 0-3 positive. Synovial fluid from RA patients was found to contain significantly more PLAP than synovial fluid from patients with osteoarthritis (a noninflammatory arthropothy) or other forms of inflammatory joint disease. Of the synovial fluids tested, 81% contained detectable levels of PLAP. Synovial fluid samples obtained from patients having gout also contained significantly more PLAP than those samples isolated from osteoarthritic patients (P < 0.05). Thus RA and gout synovial fluid was found to contain a PLAP-like molecule as determined by antibody techniques. To further characterize the PLAP found in these fluid samples, we initiated a biochemical purification strategy designed to isolate this protein. PLAP was isolated from the synovial fluid using immuno affinity chromatography followed by gel filtration chromatrography. To assess the purity of the final product we analyzed the samples using SDS gel chromatrography. The human PLAP was found to be 28,000 molecular weight and also had an apparent specificity for phospholipase A2 when phosphatidyl-choline was used as a substrate. Thus the PLAP isolated from human synovial fluid appeared to be biochemically similar, to the protein isolated from bovine and murine sources.

To assess the role of PLAP in joint arthropathese, experimental animals were injected in the joint capsules with purified PLAP protein. After 14 hours the animals were necropsied and the joint tissues were examined by standard histological techniques. A dose dependent destructive arthropathy was observed in these experimental animals.

In order to characterize the cellular response to PLAP, we examined the effect of the addition of exogenous PLAP on macrophage and neutrophil cells. PLAP readily induced human peripheral blood neutrophil aggregation (Table 5) as well as chemokinesis (32). Neutrophils treated with purified PLAP were found to release lisosomal enzymes as determined by assays for betaglucuronodase as well as neutrophil metaloprotease activity (Table 6). PLAP was also examined for its ability to stimulate neutrophils to produce superanion, a toxic metabolite of oxygen that is produced by neutrophils in the inflammatory lesion. The results from these experiments are illustrated in Table 7. All of these effects were achieved by concentrations of PLAP that were not toxic to the cells as determined by LDH or tripan blue dye exclusion. These data support the concept that PLAP may not only be generated in the inflammatory lesion but may also play an important role in the perpetuation of the inflammatory response.

TABLE 5

EFFECT OF PLAP ON NEUTROPHIL AGGREGATION[a]

Treatment	% of Maximal Aggregation
Buffer	11 ± 6
PLAP (5 U/ml)	35 ± 5
Melittin (10 µg/ml)	39 ± 2
LTB_4 (10^{-8} M)[b]	59 ± 3

a   Data expressed as mean percent $\pm$ SD of maximal aggregation from three experiments.

b LTB_4 (leukotriene B_4). p < 0.01 (buffer vs PLAP and melittin); p < 0.01 (buffer vs LTB_4); p < 0.05 (LTB_4 vs PLAP and melittin).

TABLE 6

	β-Glucuronidase[a]	Metalloproteinase[b]
Control	1.0	20
PLAP 1 Unit	1.5	25
PLAP 10 Units	4.0	30
PLAP 50 Units	8.0	41
Melittin (5 μg/ml)	4.0	65

a $nmol/10^6$ cells in 20 min.

b   o/o of total activity released in 20 min.

PLAP induction of human neutrophil β-glucuronidase release. Normal human peripheral blood neutrophils in PBS with 0.5 mM calcium. 1 nM magnesium, and 1 g glucose/l at pH = 7.4 (PICM solution) at 2×10^6 cells m 0.2 ml were treated with PLAP or melittin as noted above. Similar inductino of enzyme release was observed when the cells were suspended in only PBS. Assays were performed in triplicate and are presented as the mean value; SD were less than the size of the symbol presented. Data obtained by using PLAP that was boiled for 10 min was not different from data obtained by using untreated cells. Data was from a representative experiment of which three were performed, all in triplicate: $p < 0.05$ (PLAP 1 U/ml vs control); $p < 0.01$ (PLAP 5 U/ml vs control). PLAP induction of human neutrophil metalloproteinase release. Normal human peripheral blood neutrophils were prepared as described in Figure 1. The total metalloproteinase activity released digested 75 μg azocoll/m 0.05 ml supernatant from 2×10^6 cells in 0.2 ml buffer. Data are from a representative experiment of which three were performed. $p < 0.05$ (PLAP 1 U/ml and 10 U/ml vs control); $p < 0.001$ (Melittin 5 μg/ml vs control).

TABLE 7

PLAP INDUCES NEUTROPHIL SUPEROXIDE RELEASE[a]

Treatment	Superoxide Production
Buffer	2.1 ± 0.2
PLAP (1 U/ml)	4.3 ± 0.1
PLAP (10 U/ml)	5.0 ± 0.2
Melittin (5 μg/ml)	5.8 ± 0.5
fMLP (10^{-8} M)	10.4 ± 1.0

a   Expressed as mean nanomoles of cytochrome c reduced per 2×10^6 neutrophils $\pm$ SD of samples prepared in triplicate. Superoxide dismutase prevented the majority of cytochrome c reduction in all experimental situations. $p < 0.05$ (Buffer vs PLAP and melittin); $p < 0.01$ (Buffer vs fMLP).

The mechanisms responsible for enhanced phospholipase enzyme activity in RA are only beginning to be understood. We have shown that cells from rheumatoid patients incorporate more eicosanoid precursor fatty acids compared to normal cells, thus indicating that the phospholipids are turning over more rapidly in these rheumatoid cells (7). The phospholipase enzymatic activities also appeared to be activated or enhanced in cells isolated from patients with rheumatic disease. The enhanced phospholipase A2 activity found not only in the cells but also in the patients having rheumatoid arthritis could be a result of enhanced expression of the phospholipase A2 enzyme itself or alternatively may result from an enhanced production of the phospholipase activating protein we have termed PLAP.

REFERENCES

1.   Bomalaski JS, Wiliamson PK, Zurier RB.   Prostaglandins and the inflammatory response. Clin Lab Med 1983; 3:695-717.

2.   Klickstein LB, Shapleigh C, Goetzl EJ. Lipoxygenation of arachidonic acid as a source of polymorphonuclear leukocyte chemotactic factors in synovial   fluid and tissue in   rheumatoid arthritis and spondyloarthritis. J Clin Invest 1980; 66:1166-1170.

3.   Krane SM, Goldring SR, Dayer J-M.   Interactions among lymphocytes, monocytes, and other synovial cells   in the rheumatoid synovium. Lymphokines 1982; 7:75-136.

4.   Samuelsson B, Dahlen S-E, Lindgren JA, Rouzer CA, Serhan CN. Leukotrienes and lipoxins: Structures, biosynthesis and biological effects. Science 1987; 237:1171-1175.

5.   Robinson DR, Tashjian AH, Levine L.   Prostaglandin stimulated bone resorbtion by rheumatoid synovium: a possible mechanism for bone destruction in rheumatoid arthritis.   J Clin Inves 1975; 56:1181-1188.

6.   Salmon JA, Higgs GA, Vane JR, Bitensky L, Chayen J, Henderson B, Cashman B.   Synthesis of arachidonate cyclooxygenase products by rheumatoid and   non-rheumatoid synovial lining in nonproliferative organ culture. Ann Rheum Dis 1983; 42:36-39.

7.   Bomalaski JS, Goldstein CS, Dailey AT, Douglas SD, Zurier RB. Uptake of fatty acids and their mobilization from phospholipids in cultured monocyte-macrophages from rheumatoid arthritis patients. Clin Immunol Immunopathol 1986; 39:198-212

8.   Dayer J-M, Trentham DE, David JR, Krane SM. Collagens stimulate the production of mononuclear cell factor and   prostaglandins (PGE_2) by human monocytes.   Trans Assoc Am Phycisians 1981; 93:326-335.

9.   Dayer J-M, Trentham DE, Krane, SM. Collagens act as ligands to stimulate human moncytes to produce mononuclear cell factor and prostaglandins (PGE_2). Coll Relat Res 1982; 2:523-540.

10.   Seitz M, Deimann W, Gram N, Hunstein W, Gemsa D. Characterization of blood mononuclear cells of rhematoid arthritis patients. I. Depressed lymphocyte proliferation and enhanced prostanoid release from monocytes. Clin Immunol Immunopathol 1982; 25:405-416.

11. Chang J, Gilman S, Lewis AJ. Interleukin -1 activates phospholipase A_2 in rabbit chondrocytes: a possible signal for IL-1 action. J Immunol 1986; 136:1283-1287.

12. Robinson DR, McGuire MB, Levin L. Prostaglandins in the rheumatic diseases. Ann NY Acad Sci 1974; 256:318-329.

13. Dennis EA. Phospholipases. Enzymes 1983; 16:307-353.

14. Dennis EA. Phospholipase A_2 mechanism--inhibition and role in arachidonic acid cascade. Drug Dev R 1987; 10:205-220.54.

15. Flower RJ, Blackwell GJ. The importance of phospholipase A_2 in prostaglandin biosynthesis. Biochem Pharmacol 1976; 25:285-291.

16. Higgs GA, Flower RJ, Vane JR. A new approach to anti-inflammatory drugs. Biochem Pharmacol 1979; 28:1959-1961.

17. Waite M. Approaches to th estudy of mammaliam cellular phospholipases. J Lipid Res 1985; 26:1379-1388.

18. Bomalaski JS, Clark MA, Zurier RB. Enhanced phospholipase activity in mononuclear phagocytes from patients with rheumatoid arthritis. Arthritis Rheum 1986; 29:312-318.

19. Bomalaski JS, Clark MA, Douglas SD, Zurier RB. Enhanced phospholipase A_2 and C activities of peripheral blood poly-morphonuclear leukocytes from patients with rheumatoid arthritis. J Leuk Biol 1985; 38:649-654.

20. Pruzanski W, Vadas P, Stefanski E, Urowitz MB. Phospholipase A_2 activity in sera and synovial fluids in rhematoid arthritis and osteoarthritis: its possible role as a proinflammatory enzyme. J Rheumatol 1985; 12:211-216.

21. Hirata F, del Carmine R, Nelson CA, Axelrod J, SSchiffman E, Warabi A, deBlas A, Nirenberg M, Magnaniello V, Vaughn M, Kumagi S, Green I, Decker JL, Steinberg AD. Presence of autoantibody for phospholipase inhibitory protein, lipomodulin, in patients with rheumatic diseases. Proc Natl Acad Sci USA 1981; 78:3190-3194.

22. Haberman E. Bee and wasp venoms. Science 1972; 117:314-322.

23. Hessinger DA, Lenhoff HM. Membrane structure and function: mechanism of hemolysis induced by nematocyst venom: roles of phospholipase A_2 and direct lytic factor. Arch Biochem Biophys 1976; 173:603-613.

24. Rozengurt E, Gelehrter TD, Legg A, Pettican P. Melittin stimulates Na entry, Na-K pump activity and DNA synthesis in quiescent cultures of mouse cells. Cell 1981; 23:781-788.

25. Shier WT. Activation of high levels of endogenous phospholipase A_2 in cultured cells. Proc Natl Acad Sci USA 1979; 76:195-199.

26. Bernheimer AW, Rudy B. Interactions between membranes and cytolytic peptides. Biochem Biophys Acta 1986; 864:123-141.

27. Dufource RJ, Smith ICP, Dufource J. Molecular details of melittin-induced lysis of phospholipid membranes as revealed by deuterium and phosphorus NMR. Biochemistry 1986; 25:6448-6455.

28. Kurihara H, Kitajima K, Senda T, Jujita H, Nakajima T. Multiglandular exocytosis induced by phospholipase A_2 activators, melittin and mastoparan, ion rat anterior pituitary cells. Cell Tissue Res 1986; 243:311-316.

29. Weissmann G, Hischhorn R, Krakauer K. Effect of melittin upon cellular and lysosomal membranes. Biochem Pharmacol 1969; 18:1771-1775.

30. Clark MA, Chen M-J, Crooke ST, Bomalaski JS. Tumor necrosis factor (cachectin) induces phospholipase A_2 activity and the synthesis of a phospholipase A_2 activating protein (PLAP) in endothelial cells. Biochem J 1988; 250:125-132.

31. Bomalaski JS, Baker DG, Brophy L, Resurreccion N, Spilberg I, Clark MA. A phospholipase A_2 - activating protein (PALP) stimulates human neutrophil aggregation and release of synovial enzymes superoxide and eicosanoides. J Immunol 1989; 142: 3957-3962.

SOLUBLE PHOSPHOLIPASE A$_2$ IN HUMAN PATHOLOGY: CLINICAL-LABORATORY INTERFACE

W. Pruzanski and P. Vadas

Inflammation Research Group, University of Toronto
The Wellesley Hospital, Toronto, Ontario
Canada

INTRODUCTION

The arachidonic acid cascade and its end products eicosanoids, play an important role in the metabolism of phospholipids and in a variety of mechanisms related to hemodynamic homeostasis, inflammatory reactions, phagocytic activity and others (1). Arachidonic acid cascade is initiated enzymatically by a group of phospholipases, phospholipase A$_2$ (PLA$_2$) being one of the pivotal triggering enzymes. PLA$_2$ is located in at least 2 subcellular sites, as membrane bound and as cytosolic enzymes (2). Whereas the former acts intracellularly, the latter has the potential to be released from the cell upon a variety of stimuli (3). In humans, such soluble extracellular enzyme of pancreatic origin has indeed been found in the blood (4).

Until a few years ago nothing was known about the physiological and pathological roles of soluble extracellular non- pancreatic PLA$_2$, whereas the roles of intracellular PLA$_2$ including membrane remodelling, destruction of microorganisms, metabolism of surfactant-associated phospholipids, digestion of dietary phospholipids, and release of fatty acids, have been well documented (2,5,6) (Table 1).

TABLE 1

ROLES OF INTRACELLULAR PHOSPHOLIPASE A$_2$

* MEMBRANE REMODELLING

* DIGESTION OF BACTERIA

* METABOLISM OF PULMONARY SURFACTANT PHOSPHOLIPIDS

* INTESTINAL DIGESTION OF DIETARY PHOSPHOLIPIDS

* RELEASE OF FATTY ACIDS

At the present time, the physiological role(s) of extracellular PLA$_2$ have not yet been elucidated, however there is substantial

Biochemistry, Molecular Biology, and Physiology of Phospholipase A₂ and Its Regulatory Factors
Edited by A. B. Mukherjee, Plenum Press, New York, 1990

239

evidence that excessive concentrations of extracellular PLA$_2$ may initiate and/or propagate inflammation (3) and cause cellular damage (3). Extracellular PLA$_2$ was also found to modulate various aspects of phagocytic activity (7), and vascular tone and permeability (8,9) (Table 2).

TABLE 2

ROLES OF EXTRACELLULAR PHOSPHOLIPASE A$_2$

* INITIATION AND PROPAGATION OF INFLAMMATION

* CELLULAR DAMAGE

* MODULATION OF CHEMOTAXIS, PHAGOCYTOSIS AND SUPEROXIDE GENERATION

* MODULATION OF VASCULAR TONE

* ENHANCEMENT OF VASCULAR PERMEABILITY

* IMPACT ON T-CELL FUNCTIONS (?)

Substantial evidence has been found that in humans both type I and type II extracellular PLA$_2$ are synthesized, pancreatic and synovial fluid PLA$_2$'s being the respective examples. Extracellular non-pancreatic PLA$_2$ was found in the culture supernatant of human synoviocytes (10,11), and chondrocytes (12). Type II PLA$_2$ was also found in platelets (13), placenta (14), cartilage (15), peritoneal cells and peritoneal fluid (16), spleen, (17), and other organs . It was also found in the culture supernatants of the rat calvaria osteoblasts (18).

Research into biological role of PLA$_2$ has started when it was found that the induction of local inflammatory reaction in experimental animals caused release of hyperemia- inducing activity (HIA) which was detected in the lymph draining inflammatory sites (19). This vasoactive factor was found to be different from plasminogen activator and from prostaglandins and was subsequently identified as phospholipase A$_2$ (8). Furthermore, in rabbits intradermal injection of PLA$_2$ purified from crotalus atrox venom, induced hyperemia with kinetics similar to those of HIA (8). Two potential sources of the above PLA$_2$ have been identified (20). Enzyme activity was detected in the supernatant of con A- stimulated macrophages from afferent lymph of sheep and in supernatants of aggregated platelets (20). In several animal species PLA was also found to be released from PMN's (21).

PLA$_2$ IN SYSTEMIC INFLAMMATION (Table 3)

Since PLA$_2$ was found to be vasoactive and to induce a rapid and marked hypotension with the suppression of the mean arterial blood pressure (MABP) after intravenous administration (8,9,22), it was of interest to investigate whether PLA$_2$ plays a role in the pathogenesis of endotoxin induced shock. It was found indeed that intravenous administration of endotoxin in rabbits reproduces hemodynamic picture of gram negative septic shock and that endotoxin administration causes the rise in circulating PLA$_2$. The magnitude of the increase of PLA$_2$ was directly related to the fall in MABP (9). Pretreatment of animals with glucocorticoids abrogated hypotension as well as the rise of PLA$_2$.

When PLA $_2$ purified from the blood of rabbits with endotoxic shock, was infused into healthy rabbits, it caused hypotension similar to that induced by endotoxin itself. PLA $_2$ inactivated with pBPB did not induce hypotension (9).

TABLE 3

HIGH ACTIVITY OF PHOSPHOLIPASE A $_2$ IN HUMAN DISEASE

DISEASE	SITE
ARTHRITIS	SERUM, SYNOVIAL FLUID, WBC, (?)
COLLAGEN VASCULAR DISEASES	SERUM
PANCREATITIS	SERUM
PERITONITIS	PERITONEAL FLUID AND CELLS
SEPSIS AND SHOCK	SERUM
ARDS	SERUM AND ALVEOLAR FLUID
RENAL FAILURE	SERUM

Marked increase in circulating PLA$_2$ found in endotoxin- treated animals prompted to investigate the role of PLA$_2$ in humans who suffer from gram negative septic shock (GNSS). In retrospective (23) and prospective (24) studies of GNSS, the circulating PLA$_2$ was found to be markedly elevated during the hypotensive phase of sepsis. PLA$_2$ activity correlated significantly with the duration and severity of circulatory collapse (p < 0.001). In contrast, in patients with cardiogenic shock due to myocardial infarction, PLA$_2$ activity remained normal. PLA$_2$ in GNSS was found to be of extra- pancreatic origin (23,25). Recently a group of volunteers was injected intravenously with endotoxin and the level of tumor necrosis factor (TNF) and hemodynamic effects were measured (26,27). Marked elevation of TNF in the peripheral blood was noted after 90 to 120 minutes ,followed by fever, tachycardia and increase in epinephrine level. Testing of the sera from both the arterial and venous compartments for PLA$_2$ activity has shown that a very marked increase in the activity of this enzyme took place, the peak of the activity occurring at least 4 hours after the peak of TNF (Pruzanski, Wilmore and Vadas, unpublished). There were no significant arterial - venous PLA$_2$ gradients in any volunteer.

PLA$_2$ IN LOCALIZED INFLAMMATION

Following the findings of high circulating PLA $_2$ activity in systemic inflammatory processes such as sepsis (23,24), the relevance of PLA$_2$ to localized inflammation was explored. Two clinical models were especially suitable for this purpose, namely peritonitis and arthritis. In 9 patients with acute bacterial peritonitis, marked increase in PLA$_2$ activity was detected in both peritoneal fluids (1034 to 18,674 U/ml) and in the serum (952 to 33,924 U/ml)(PLA$_2$ in normal serum ranged from 130 to 691 U/ml). It seemed that serum PLA $_2$activity derived not excl- usively from the diffusion from peritoneal cavity, since in some instances serum PLA $_2$activity markedly exceeded that in the peritoneal fluid (28). Peritoneal fluid and peritoneal cell associated PLA $_2$ have been isolated and compared. Whereas peritoneal fluid PLA $_2$was found to be a calcium dependent enzyme, the enzyme isolated from PMN's was only partially inhibited by calcium chelators (28). The source of peritoneal fluid PLA$_2$ in man is still unknown, however it has been shown that it does not derive from bacterial or pancreatic sources (28). Northern blots of RNA extract of cells from patients with acute bacterial peritonitis showed high levels of mRNA transcript of the Type II (non-pancreatic) PLA$_2$ (29).

Very high levels of PLA$_2$ were detected in synovial fluids in inflamed joints (30). (Table 4). Twenty-five percent of patients with rheumatoid arthritis were also found to have high PLA$_2$ activity in the blood (31). (Fig 1) It has been shown that the activity of rheumatoid arthritis significantly correlated to the level of serum PLA$_2$ (31). (Fig 2) When remission was induced in active rheumatoid patients, their serum PLA$_2$ activity markedly diminished (31).

TABLE 4

SYNOVIAL FLUID PHOSPHOLIPASE A$_2$

DIAGNOSIS	NO	RANGE (U/ml)	MEAN $\pm$ SD (U/ml)
RHEUMATOID ARTHRITIS	147	2261 – 43266	16430 $\pm$ 8493
OSTEOARTHRITIS	41	1886 – 19690	11108 $\pm$ 6164
PSORIASIS	15	2016 – 26804	11251 $\pm$ 8000
MONOARTHRITIS	11	4289 – 13892	8923 $\pm$ 3045
GOUT	5	4161 – 27189	17939 $\pm$ 9012

Similar observations have been made in children with chronic juvenile arthritis (32). The highest serum PLA$_2$ activity was detected in the systemic form of juvenile chronic arthritis, in which the inflammatory features are most prominent, whereas in the pauciarticular form in which inflammation is much less pronounced, the activity of PLA$_2$ was much lower, although still increased above the normal level. In those children who achieved remission, PLA$_2$ activity markedly decreased (32).

Synovial fluid PLA$_2$ was purified and characterized as a calcium dependent, type IIa enzyme with molecular weight of 13,940 and 124 amino acids in the mature peptide (29,33,34). This enzyme was cloned and its active transcription was found in 2 different inflammatory cell sources, namely in human synovial tissue and in human cell pellet from a peritoneal exudate in peritonitis (29). The identical enzyme was also detected in human platelets (35). More detailed biochemical and enzymatic characterization of synovial fluid PLA$_2$ has shown that 2 and occasionally 3 peaks with PLA$_2$ activity can be resolved by preparative HPLC (36). The most abundant peak A was found to be present in rheumatoid, osteoarthritic and psoriatic synovial fluids. In all 3 another peak-B with a longer retention time was detected. Peak-B was relatively low in rheumatoid arthritis and high in osteoarthritis. Psoriatic synovial fluids contained an intermediate quantity of peak-B.

Both peaks showed optimal activity with DOC/phosphatidyl choline mixed micelles and both were more active with PC substrate than with PI, however peak-A exhibited higher activity with PE than with PC. In contradistinction to peak-A, peak-B activity was stimulated by 0.1% DOC and by 0.5 M Tris buffer (29,36). Recently the human type IIa PLA$_2$ gene was mapped to chromosome 1. (37).

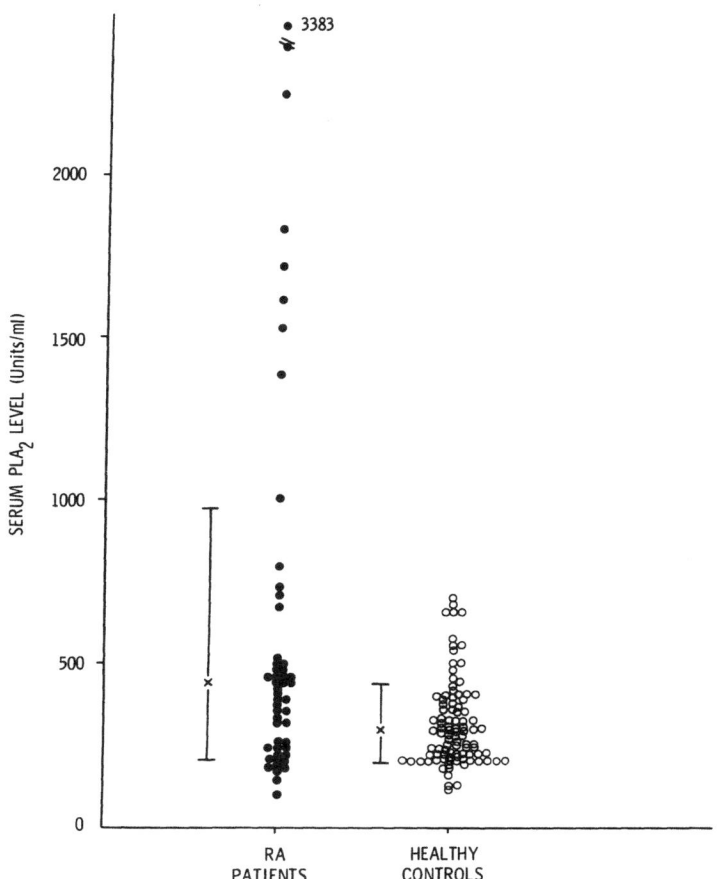

Fig 1.   Serum activity of phospholipase A2 in patients with rheumatoid
          arthritis. Reproduced by permission from the J. Rheumatol.

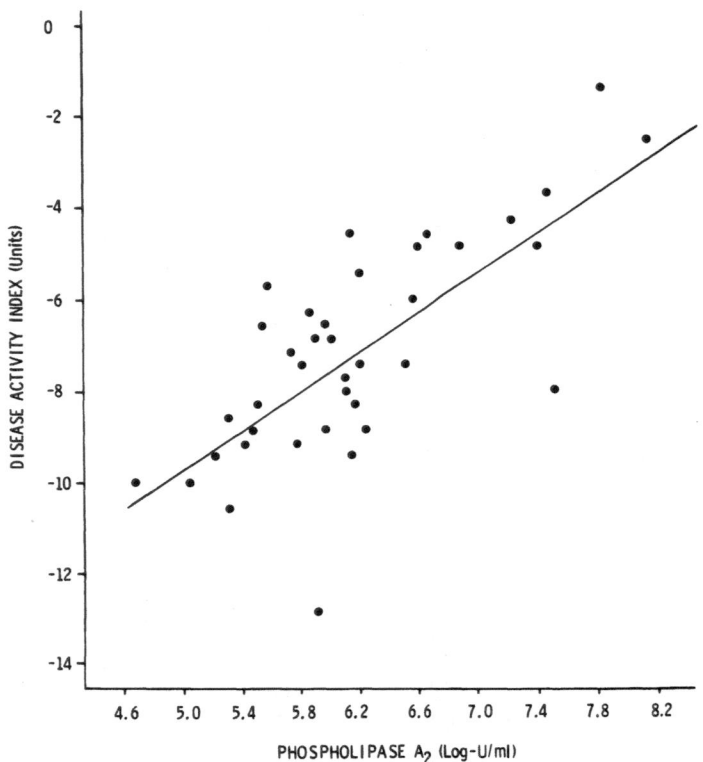

Fig 2.   Correlation of serum phospholipase A$_2$ with clinical activity index (Three variables combined – Landsbury index, number of swollen joints and duration of morning stiffness) in rheumatoid arthritis.   (p < 0.01)

The source(s) of synovial fluid PLA_2 has not yet been identified. Synthesis and release of PLA_2 from human synoviocytes from both rheumatoid and osteoarthritis have been well documented by Gilman (10) and confirmed by us (11). However cultures of synovial cells seem to have quite a variable rate of PLA_2 release and some cultures were inactive. Much more consistent was synthesis and release of PLA_2 from rat fetal calvarial osteoblasts (18). Those cells consistently synthesize and release PLA_2 in culture. Marked enhancement of PLA_2 release was achieved by exposing the cells to Il-1 and TNF and more so to a combination of both (38). (Fig.3) The release was completely inhibited by cycloheximide (38). Cultures of human articular chondrocytes have also been shown to release PLA_2 extracellulary (12). (Fig.4) A very high concentration of PLA_2 was detected in human articular cartilage (15). Deep layers of the cartilage invariably contained more PLA_2 per mg of protein than the superficial ones and rheumatoid cartilage contained less PLA_2 than osteoarthritic cartilage. Nasal septum cartilage contained very little PLA_2.. There was no relationship between the content of PLA_2 and that of muramidase. Human articular cartilage was found to be calcium dependent with optimum activity at pH 7.0 - 7.5.

High activity of PLA_2 was also found to be associated with white cells floating in the synovial fluid (39). Synovial fluid cells contained 10 fold or more PLA_2 than peripheral blood PMN's (368 $\pm$ 243 versus 31 $\pm$ 15 U/ml per 5×10^6 cells). When peripheral blood PMN's were incubated with purified synovial fluid PLA_2, the enzyme adsorbed to the surface of the cells in a concentration dependent manner (39). Thus, the studies showing increased activity of PLA_2 PMN's must take into consideration adsorption from the milieau.

INDUCTION OF INFLAMMATION BY PLA_2

Both type I and type II PLA_2 have been found to induce inflammatory reaction in experimental animals (40,41). Intracutaneous injections of PLA_2 induced dose and time dependent inflammatory reaction that involved the skin, subcutaneous tissue and cutaneous blood vessels, and ranged from simple hyperemia to vasculitis and muscle necrosis. Inactivation of the enzyme by pBPB attenuated or abolished induction of inflammatory process (40). Similar inflammatory reaction was observed when PLA_2 was injected into subcutaneous air pouch in the rat (Pruzanski, et al. Unpublished). In this model, in addition to marked tissue reaction, there was also accumulation of the inflammatory exudate in the pouch. When purified synovial fluid PLA_2 was injected intraarticularly, a time and dose dependent inflammatory process was observed (41). Whereas a single injection caused an acute inflammatory reaction, multiple injections led to the proliferative process with synovial lining cell replication and marked inflammatory infiltrate. The picture resembled that of rheumatoid arthritis (41). Inactivation of PLA_2 markedly attenuated the inflammatory process.

The above observations left little doubt that extracellular type II PLA_2 is highly pro- inflammatory both systemically and locally. The impact of this enzyme on the induction and propagation of inflammatory process may be two-fold. PLA_2 may exert both the direct irritative (non-specific?) impact on the tissues and cells and the indirect one by triggering the arachidonic acid cascade and excessive synthesis of pro-inflammatory eicosanoids, PAF and lysophosphatides. In this context the influence of 2 groups of agents, namely phospholipase activating proteins (PLAP 's) (42) and phospholipase inhibitory proteins (PLIP's) (43) seem to assume special importance. These agents may either inhibit PLA_2 (PLIP) by blocking its interaction with the substrate or activate it (PLAP).

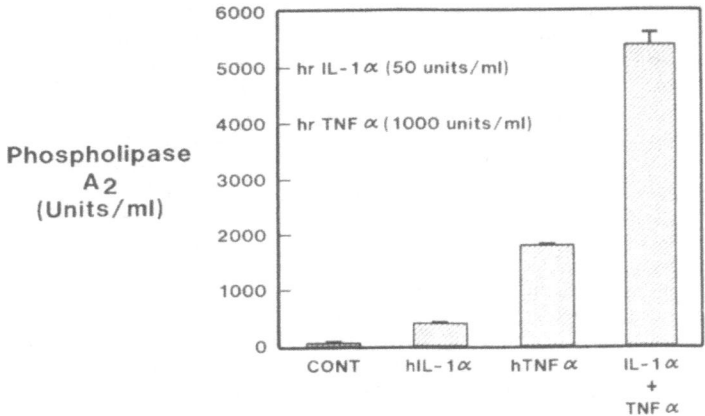

Fig 3. Extracellular phospholipase A$_2$ activity in the supernatants
from cultured rat calvaria osteoblasts.
CONT – Control; Unstimulated cultures
hrIl-1α – human recombinant Interleukin 1α
hrTNFα – human recombinant tumor necrosis factorα

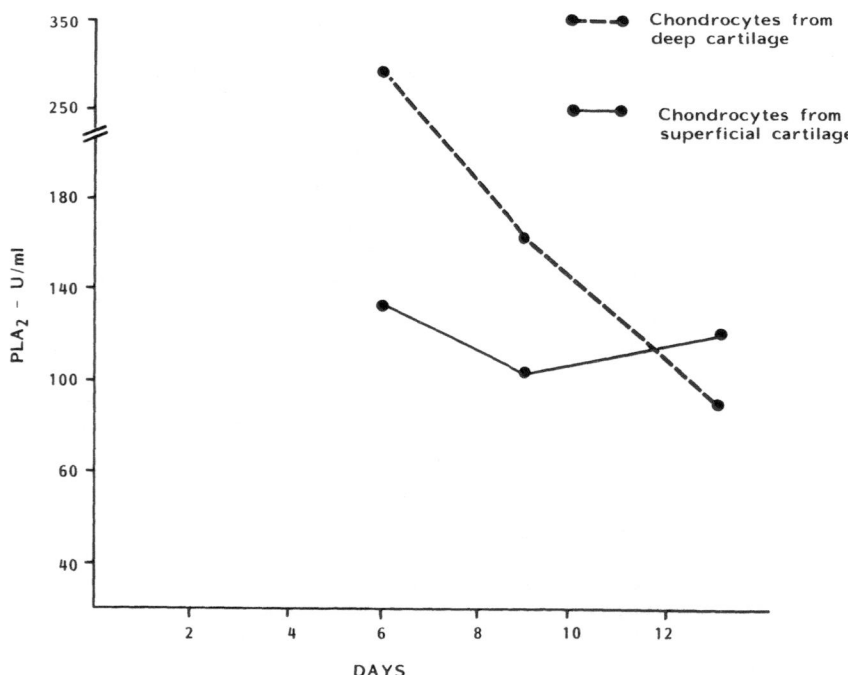

Fig 4.   Extracellular phospholipase A$_2$ activity in the supernatants from
cultured human articular chondrocytes.
DAYS – DAYS OF CULTURE.   UNSTIMULATED CELLS

246

In order to assess the impact of PLA_2 on mammalian cells, normal peripheral blood PMN's and monocytes have been chosen as a target (7). PMN's were preincubated with purified type II synovial fluid-derived PLA_2 in concentrations similar to those encountered in the synovial fluid or in the blood of patients with gram negative septic shock, or with juvenile rheumatoid arthritis. Type II PLA_2 markedly inhibited adhesiveness, chemotaxis and intracellular bactericidal activity (ICBA) of human PMN's. The latter was most probably caused by the impact of PLA_2 on the superoxide generation mechanism. PLA_2 caused spontaneous burst of energy in the cells and, as a consequence, subsequent inability of PMN's to generate sufficient amounts of energy to kill intracellular microorganisms. PLA_2 also reduced ICBA of human monocytes. Purified Naja naja type I PLA_2 did not have similar impact on phagocytes.

PROPOSED ROLE OF PHOSPHOLIPASE A_2 IN INFLAMMATION

The discovery of proinflammatory activity of PLA_2, the dependence of its synthesis and release on cytokines and its subsequent role in the initiation of arachidonic acid cascade and in the metabolism of PAF and lysophosphatides, make it possible to propose a new scheme of inflammatory process. The sequence of events that leads to the initiation and propagation of inflammatory reaction might be divided into several stages (Fig 5).

STAGE 1   Attack by noxious agent.

STAGE 2   Release of cytokines and autocoids by responding cells.

STAGE 3   Induction of synthesis and secretion of PLA_2.

STAGE 4   Direct cytotoxic impact of PLA_2, and acceleration of arachidonic acid cascade, excessive synthesis of proinflammatory eicosanoids, PAF and lysophosphatides.

In this staging several additional factors may play an important pathogenetic role. In which stage PLAP's and PLIP's are synthesized and exert biological action is still unclear. What is the role of glucocorticoids in various stages is still undecided. It seems however that in order to counteract the inflammatory impact of PLA_2 a very early exposure to or down-regulation by steroids is imperative. Most probably steroids are biologically active immediately after stage 1 and before the initiation of stage 2. Administration of steroids after that point cannot stop the continuation of proinflammatory cascade.

The future directions of research into proinflammatory PLA_2 are quite evident at the present time. More knowledge is necessary about the synthetic mechanisms of PLA_2 at both transcriptional and translational levels. The intricate relationship between the insulting factors, cytokines and PLA_2 has to be further explored and the PLIP-PLAP system has to be better defined. Needless to say the investigation of PLA_2 per se, including its specific structure and crystallographic features as well as production of monoclonal antibodies and inhibitors will be required to control this enzyme. Finally, laboratory- clinical correlations have to be expanded and the exact roles of extracellular type II PLA_2 in human physiology and pathology have to be defined. Only then will the appropriate place of extracellular PLA_2 in biology be found.

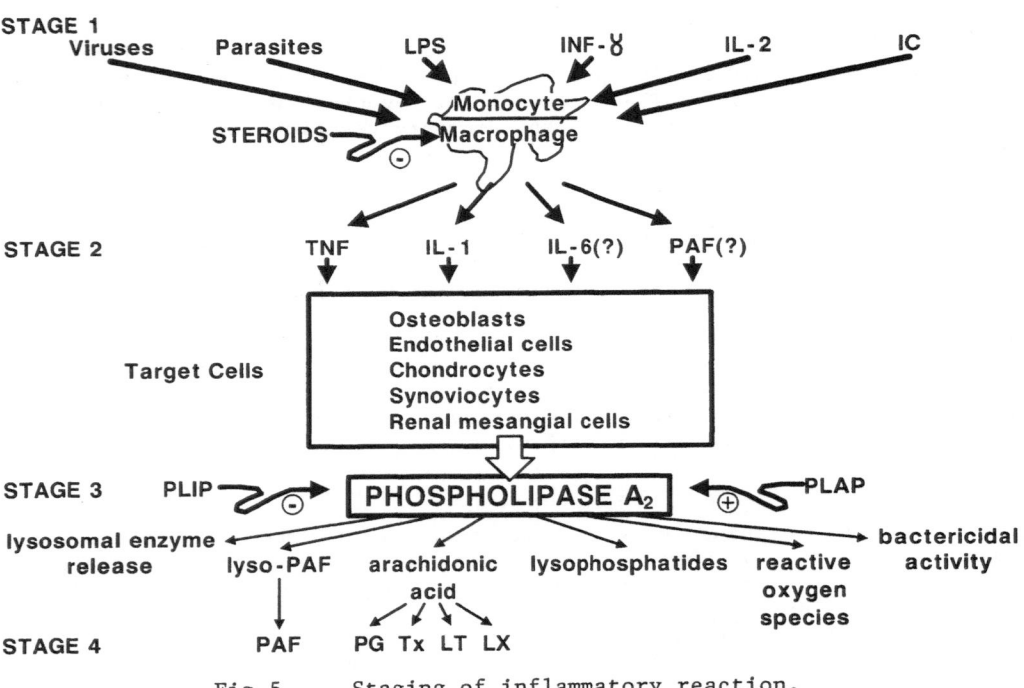

Fig 5.   Staging of inflammatory reaction.

ACKNOWLEDGEMENT

We would like to acknowledge the invaluable contribution of Mrs. E. Stefanski, Mr. M. Wloch, Drs. J. Seilhamer, L. Johnson, J. Edelson, D. Wilmore   E. Bogoch, V. Fornasier, J. Aubin, and others in preparation of this paper.

REFERENCES

1. Marcus, A.J., Eicosanoids: Transcellular Metabolism, in: "Inflammation: Basic principles and Clinical Correlates". Ed. J. Gallin, I.M. Goldstein, R. Snyderman, pp129-137 Raven Press, Ltd. New York, 1988.
2. Van den Bosch, H., Intracellular phospholipases A. Biochim Biophys Acta 604: 191, 1980.
3. Vadas, P. and Pruzanski, W. Role of Secretory Phospholipases A$_2$ in the Pathobiology of Disease.Lab Invest 55: 391, 1986.
4. Sternby, B., Akerstrom, B. Immunoreactive pancreatic colipase, lipase and phospholipase A$_2$ in human plasma and urine from healthy individuals. Biochim Biophys Acta 789: 164, 1984.
5. Van den Bosch, H. Phosphoglyceride metabolism. Ann Rev Biochem 43: 243, 1974.
6. Verheij, H.M., Slotboom, A.J., de Haas, G.H. Structure and function of phospholipase A$_2$, Rev Physiol Biochem Pharmacol 91: 92, 1981.
7. Pruzanski, W., Saito, S., Stefanski, E., Vadas, P. The modulation of human phagocytic function by rheumatoid synovial fluid phospholipase A 2 . 2nd International Conference on LEUKOTRIENES and PROSTANOIDS in HEALTH and DISEASE. Jerusalem, Israel, Oct. 1988.
8. Vadas, P., Wasi, S., Movat, H., Hay, J. Extracellular phospholipase A$_2$ mediates inflammatory hyperemia. Nature 273: 583, 1981.
9. Vadas, P., Hay, J. Involvement of circulating phospholipase A$_2$ in the pathogenesis of the hemodynamic changes in endotoxin shock in rabbits. Can J Physiol Pharmacol 61: 561, 1983.
10. Gilman, S.C., Chang, J., Zeigler, P.R. Uhl, J., Mochan, E. Interleukin-1 activated phospholipase A$_2$ in human synovial cells. Arthr. Rheum. 31: 126, 1988.
11. Vadas, P., Pruzanski, W., Kim, J., Jacobs, H. and Stefanski, E. Studies of cellular source of proinflammatory enzyme phospholipase A$_2$ in rheumatoid synovial fluid. Arthritis and Rheumatism. 28, No 4 (Supplement) April, 1985.
12. Pruzanski W., Vadas, P., Stefanski, E., Seilhamer, J., Johnson, L.K. Phospholipase A$_2$ mediates inflammation in arthritis. 7th Internat. Congress of Immunology, Berlin, July/August, 1989.
13. Kramer, R.M. Pepinsky, R.B. Purification and characterization of phospholipase A$_2$ from human platelets. J. Cell. Biochem. Suppl. 12E: 49, 1988.
14. Jordan-Tankou L., Russo-Marie, F., Purification and characterization of phospholipase A$_2$ from human placenta. J. Cell. Biochem. Suppl. 12E: 48, 1988.
15. Pruzanski, W., Bogoch, E., Stefanski, E. and Vadas, P. Phospholipase A2 in human cartilage. International Conference "Lipoproteins and Phospholipases". Paris, France, Sept. 1988.
16. Vadas, P., Pruzanski, W., Stefanski, E., Johnson, L., Seilhamer, J., Mustard, R.J., Bohnen, J. Phospholipase A$_2$ in acute bacterial peritonitis in man. in: Cell Activation and Signal initiation: Receptor and phospholipase control of inositol phosphate, PAF and eicosanoid production. A.R. Liss Inc. p311-316, 1989.
17. Kanda, A., Ono, T., Yoshida, N., Tojo, H., Okamoto, M. The primary structure of a membrane-associated phospholipase A$_2$ from human spleen. Biochem. Biophys. Res. Comm. 163: 42, 1989.

18. Vadas, P., Pruzanski, W., Sos, A., Melcher, A., Jacobs, H. and Cheong, T. Synthesis and secretion of soluble phospholipase A_2 from cultured osteoblasts. Arthr. Rheum. 30 (Suppl) S-65, 1987.

19. Vadas, P., Wasi, S., Movat, H.Z., Hay, J.B. A novel vasoactive product and plasminogen activator from afferent lymph cells draining chronic inflammatory lesions. Proc. Soc. Exper. Biol. Med. 161:82 1979.

20. Vadas, P., Hay, J. The release of phospholipase A2 from aggregated platelets and stimulated macrophages of sheep. Life Sci. 26: 1721, 1980.

21. Traynor, J.R., Authi, K.S. Phospholipase A2 of lysosomal origin secreted by polymorphonuclear leukocytes during phagocytosis or on treatment with calcium. Biochim Biophys Acta. 665:571, 1981.

22. Slotta, K.H. Vick, J.A. and Ginsberg, N.J. Enzymatic and toxic activity of phospolipase A. In: Toxins of plants and animal origin. Ed.: A.DeVries and E. Kochva. Gordon and Breach, New York p401-418, 1981.

23. Vadas, P., Pruzanski, W., Stefanski, E., Mustard, R., Bohnen, J., Fraser, I. and Andrews, D. The pathogenesis of hypotension in septic shock. The contributory role of circulating phospholipase A_2. Crit. Care. Med. 16:1, 1988.

24. Vadas, P., Pruzanski, W., Stefanski, E., Ruse, J., Farewell, V., Mclaughlin, J., Bombardier, C. Concordance of endogenous cortisol and phospholipase A_2 levels in gram negative septic shock: A prospective study. J. Lab. Clin. Med. 111: 584, 1988.

25. Vadas, P., Pruzanski, W. and Stefanski, E. Extracellular phospholipase A_2 :causative agent in circulatory collapse of septic shock? Agents and Actions. 24: 320, 1988.

26. Michie, H.R., Manogue, K.R., Spriggs, D.R., Revhaug, A., O'Dwyer, S., Dinarello, C.A., Cerami, A., Wolff, S.M., and Wilmore, D.W. Detection of circulating tumor necrosis factor after endotoxin administration. New Engl. J. Med. 318: 1481, 1988.

27. Suffredini, A.F., Fromm, R.E., Parker, M.M., Brenner, M., Kovacs, J.A., Wesley, R.A., and Parrillo, J.E. The cardiovascular response of normal humans to the administration of endotoxin. New Engl. J. Med. 321: 280, 1989.

28. Vadas, P., Pruzanski, W., Stefanski, E., Johnson, L., Seilhamer, J., Mustard, R., Bonen, J. Phospholipase A2 in acute bacterial peritonitis in man. J. Cell Biochem. In Press. 1989.

29. Seilhamer, J.J., Plant, S., Pruzanski, W., Schilling, J., Stefanski, E., Vadas, P. and Johnson, L.K. Multiple forms of phospolipase A2 present in arthritic synovial fluid. Purification and preliminary characterization. J. Biochem. 106:730, 1989.

30. Pruzanski, W., Vadas, P. and Stefanski, E. Activity of phospholipase A_2 in sera and synovial fluids in arthritis. J. Rheumatol. 12:211, 1985.

31. Vadas, P., Pruzanski, W. and Stefanski, E. Characterization of extracellular phospholipase A_2 in human synovial fluids. Life Sci. 36: 579, 1985.

32. Silverman, E., Pruzanski, W., Laxer, R., Albin-Cook, K., Stepanovic, E. and Vadas, P. Correlation of phospholipase A_2 and disease activity in juvenile rheumatoid arthritis (JRA). Arthr. Rheum. 30 (Suppl) S-127, 1987.

33. Vadas, P., Pruzanski, W. and Stefanski, E. Characterization of extracellular phospholipase A2 in human synovial fluids. Life Sci. 36: 579, 1985.

34. Stefanski, E., Pruzanski, W., Sternby, B. and Vadas, P. Purification of a soluble phospholipase A2 from synovial fluid. J. Biochem. 100: 1297, 1986.

35. Kramer, R.M., Hession, C., Johansen, B., Hayes, G., McGray, P., Chow, E.P., Tizard, R. and Pepinski, R.B. Structure and properties of a

human non-pancreatic phospholipase A_2 . J. Biol. Chem. 264: 5768, 1989.

36. Seilhamer, J., Vadas, P., Pruzanski, W., Plant, S., Stefanski, E. and Johnson, L. Synovial fluid phospholipase A_2 in arthritis. In: Therapeutic Approaches to Inflammatory diseases. p129-136. Ed. A.J. Lewis, N.S. Doherty, N.R. Ackerman, Elsevier, New York, 1989.

37. Johnson, L.K., Frank, S., Vadas. P., Pruzanski, W., Marian-Scardina, J., Lusis, A.J. and Seilhamer, J.J. Localization and evolution of two human phospolipase A_2 genes and two related genetic elements. in: "Phospholipase A_2 : role and function in inflammation". Eds. P.Y. Wong and E. Dennis. Plenum Press, New York, 1989.

38. Vadas, P., Pruzanski, W. Phospholipase A_2 activation is the pivotal step in the effector pathway of inflammation. Sent for publication.

39. Pruzanski, W., Vadas, P., Kim, J., Jacobs, H. and Stefanski, E. Phospholipase A_2 activity associated with synovial fluid cells. J. Rheumat. 15: 791 1988.

40. Vadas, P., Pruzanski, W. and Fornasier, V. Inflammatory reaction induced by phospholipase A_2 in rabbits. J. Invest. Dermatol. 86: 380, 1986.

41. Vadas, P., Pruzanski, W., Kim, J. and Fornasier, V. The proinflammatory effect of intra-articular injections of soluble human and venom phospholipase A_2. Am. J. Pathol. 134: 807, 1989.

42. Clark, M.A., Conway, T.A., Shorr, R.G.L. and Crooke, S.T. Identification and isolation of a mammalian protein which is antigenically and functionally related to the phospholipase A stimulatory peptide mellitin. J. Biol. Chem. 262: 4402 , 1987.

43. Flower, R.J. Background and discovery of lipocortins. Agents Actions 17: 255, 1985.

CONTRIBUTORS

Dr. Tore Abrahamsen
Pediatric Oncology Branch
Division of Cancer Therapy
National Cancer Institute
Bethesda, Maryland 20892

Dr. Corrado Baglioni
State University of New York at Albany
1400 Washington Avenue
Albany, New York 12222

Dr. John D. Bell
Department of Biochemistry and
Pharmacology
University of Virginia
Charlottesville, Virginia 22908

Dr. Rodney L. Biltonen
Department of Biochemistry
and Pharmacology
University of Virginia
School of Medicine
Charlottesville, Virginia 22908

Dr. John S. Bomalaski
V.A. Medical Center
Medical College of Pennsylvania
University of Pennsylvania
Philadelphia, Pennsylvania

Dr. Ezio Bonvini
Laboratory of Cell Biology
FDA, Bldg. 29, Room 225
HFB 420
Bethesda, Maryland 20892

Dr. Clara Brando
Laboratory of Cell Biology
FDA, Bldg. 29, Room 225
HFB 420
Bethesda, Maryland 20892

Dr. Ronald Burch
NOVA Pharmaceutical
Corporation
6200 Freeport Center
Baltimore, Maryland 21224-2788

Dr. Giovanni Camussi
Cattedra di Nefrologia Sperimentale
Dipartimento di Biochimica e Biofisica
Prima Facoltá di Medicina
Universitá di Napoli
Via Costantinopoli, 16
80138 Napoli, Italy

Dr. Mike Clark
Schering Plough
60 Orange Street
Bloomfield, New Jersey 07003

Dr. Robert Conroy
Department of Molecular Genetics
Hoffmann-LaRoche,Inc
Building 102
340 Kingsland
Nutley, New Jersey 07110

Dr. Eleonora Cordella-Miele
Section on Developmental Genetics
Human Genetics Branch, NIH
Bethesda, Maryland 20892

Dr. Robert Crowl
Department of Molecular Genetics
Hoffmann LaRoche
Building 102,
340 Kingsland, Nutley
New Jersey 07110

Dr. Raymond A. Deems
Department of Chemistry
University of California
San Diego,
La Jolla, California 92093

Dr. G.H. de Haas
Department of Biochemistry
State University of Utrecht
CBLE, University Center De Uithof
Padualaan 8
3584 CH Utrecht
The Netherlands

Dr. Edward Dennis
Department of Chemistry
University of California
San Diego, La Jolla,
California 92093

Dr. Richard C. Franson
Department of Biophysics,Box 614
Medical College of Virginia
Richmond, Virginia 23298

Dr. M. Fry
Department of Biochemistry and
Molecular Biophysics
Medical College of Virginia
Richmond, Virginia 23298

Dr. S. S. Ghosh
Department of Biochemistry and
Molecular Biophysics
Virginia Commonwealth University
Richmond, Virginia

Dr. Michael Hanson
FDA, Bldg. 29, Room 225
Bethesda, Maryland 20892

Dr. Theodore L. Hazlett
Department of Chemistry
University of California,
San Diego, La Jolla,
California 92093

Dr. Thomas R. Heimburg
Department of Biochemistry
and Pharmacology
University of Virginia
Charlottesville, Virginia

Dr. Robert Heinrikson
Biopolymer Chemistry Unit
The Upjohn Company
Kalamazoo, Michigan 49001

Dr. Fusao Hirata
Department of Environmental Health
Sciences
School of Hygine and Public Health
The Johns Hopkins University
Baltimore, Maryland 21205-2179

Dr. Aiko Hirata
Department of Environmental Health
Sciences, School of Hygine and Public
Health, The Johns Hopkins University
Baltimore, Maryland 21205-2179

Dr. Thomas Hoffman
Laboratory of Cell Biology
FDA, Building 29
Room 225 HFB 420
8800 Rockville Pike
Bethesda, Maryland 20892

Dr. Ferenc J. Kezdy
Biopolymer Chemistry Unit
The Upjohn Company
Kalamazoo, Michigan 49001

Dr. Yoo Jin Kim
FDA, Building 29, Room 225
Bethesda, Maryland 20892

Dr. O. P. Kuipers
Department of Biochemistry
State University of Utrecht
CBLE, University Center De Uithof
Padualaan 8
3584 CH Utrecht
The Netherlands

Dr. Brian Lathrop
Department of Biochemistry and
Pharmacology
University of Virginia
Charlottesville, Virginia 22908

Dr. Crystal Lee
Laboratory of Cell Biology
FDA, Bldg. 29, Room 225
Bethesda, Maryland 20892

Dr. Elaine F. Lizzio
Laboratory of Cell Bioilogy
FDA, Building 29, Room 225
8800 Rockville Pike
Bethesda, Maryland 20892

Dr. Lucio Miele
Section on Developmental Genetics
Human Genetics Branch, NIH
Bethesda, Maryland 20892

Dr. Anil B. Mukherjee
Section on Developmental Genetics
Human Genetics Branch,
Bldg. 10, Room 9S242
Bethesda, Maryland 20892

Dr. Yu-Ching Pan
Department of Protein Biochemistry
Hoffman-LaRoche
Building 102
340 Kingsland,
Nutley, New Jersey 07110

Dr. N. Pattabiraman
Code 6030
Laboratory for the Structure of Matter
Naval Research Laboratory
Washington, D.C. 20375

Dr. W. Pruzanski
University of Toronto
The Wellesley Hospital
160 Wellesley Street East
Toronto, Ontario
Canada M4Y1J3

Dr. Joseph Puri
FDA,
Bldg.29,Room 225
Bethesda, Maryland 20892

Dr. R. Raghupathi
Department of Biochemistry and
Molecular Biophysics
Virginia Commonwealth University
Richmond, Virginia 23298

Dr. M. D. Rosenthal
The Department of Biochemistry
Eastern Virginia Medical School
Norfolk, Virginia 23501

Dr. Francoise Russo-Marie
Department of Biochemistry
Stanford University
Medical Center
Stanford, California 94305-5307

Dr. J. Saal
San Fransisco Spine Institute
Daly City,
California

Dr. Cheryl Stoner
Department of Molecular Genetics
Hoffmann-LaRoche,Inc.
Nutley, New Jersey 07110

Dr. Tim Stoller
Department of Molecular Genetics
Hoffmann-LaRoche, Inc.
Nutley, New Jersey 07110

Dr. Ciro Tetta
Laboratorio di Immunopatologia
Cattedra di Nefrologia
Università di Torino
Torino, Italy

Dr. Karen Ting
FDA, Laboratory of Cell Biology
Bldg. 29,Room 225
Bethesda, Maryland 20892

Dr. Peter Vadas
Inflammation Research Group
University of Toronto,
The Wellesley Hospital
160 Wellesley Street East
Toronto, Ontario Canada M4y1J3

Dr. C.J. van den Bergh
Department of Biochemistry
State University of Utrecht
CBLE, University Center De Uithof
Padualaan 8
3584 CH Utrecht
The Netherlands

Dr. H.M. Verheij
Department of Biochemistry
en Lipide Enzymologie
Padualaan 8
Postbus 80054
3508 TB Utrecht
The Netherlands

Dr. B. Vishwanath
Department of Biochemistry and
Molecular Biophysics
Commonwealth University of Virginia
Richmond, Virginia 23298

Dr. Mosley Waite
Department of Biochemistry
The Bowman Gray School of
Medicine, Wake Forest University
300 South Hawthorne Road
Winston-Salem,
North Carolina 27103

Dr. Keith Ward
Code 6030
Laboratory for the Structure
of Matter, Naval Research
Laboratory,
Washington, D.C. 20375-5000

INDEX

(PL = Phospholipase)

Inflammation (continued)
  name explained, 161
  and PLA$_2$, 203
  as process, biological, 156-157
  stages, four, of, 247-248
Inositol phosphate, 49, 126
Inositol trisphosphate, 132
Interaction, molecular
  nearest-neighbor association, 90
  percolation point, 90-91
Interface
  activation, 57-60
    of phosphatidylcholine, 59-60
    theories, foru, 58-59
  and PLA$_2$, 37-47
  quality of, 38
  recognition site hypothesis, 67
Interleukin-1, 131, 246
  recombinant, 246
Ionomycin, see Calcium ionophore
    A23187
Isopeptide bond, 114-117

Keratinocyte, 106
Killer cell, natural, 217
  and tumor lysis, 216

Lahesis muta venom and PLA$_2$, 43
Laki-Lorand factor, see Trans-
    glutaminase
Lecithin, 56, 59, 78
  -cholesterol acyl transferase, 43
  structure, 79
Leukocyte, polymorphonuclear, 221,
    223
Leukotriene, 125, 131, 197, 198,
    203, 211-213
  B$_4$, 231, 234
Linoleic acid, 232
Lipase
  hepatic, 6-7
  pancreatic, 43
Lipocortin, 11, 142-147, 151, 154,
    161, 213, 216, 233
  as anti-inflammatory protein,
    197-210
  as anti-PLA$_2$, 197-210
  and autoantibody, 233
  and eicosanoid inhibition of, 161
  identification of, 199
  as inhibitor of PLA$_2$, 146, 147
  phosphorylation, 201
  and PLA$_2$, inhibition of, 161
  purification, 199
Lipomodulin, 199
  is lipocortin, 199
  purification, 214
  and T-cell, cytotoxic, 216
5-Lipoxygenase, 125, 131, 197
Lipopolysaccharide, 125
Lipoprotein lipase, 43

Liver PLA$_2$
  in lysosome, 4
  in mitochondrion, 4
Lupus erythematosus, 211
Lymphokine, 131
Lymphotoxin, 217
Lysine, 56-57
Lysolecithin, 45, 78
Lysophosphatide, 245, 247, 248
Lyso (bis) phosphatidic acid, 6
Lysophosphatidylcholine, 110
Lysophospholipase
  and surface dilution kinetics, 52
Lysosome and PLA$_2$, 4, 5, 248
Lysozyme, 115

Macrocortin, 199
  is lipocortin, 199
Macrophage, 120, 141, 162, 164,
    167, 168, 170, 248
  inhibition by lipocortin, 155
  -monocyte fatty acid uptake, 232
  permeabilization, 190
  and PLA$_2$, 190
Manoalide and lysine, 56
Mast cell, 191-192
Mellitin, 233-235
  and eicosanoid, 233
  and PLA$_2$, 233
Metalloproteinase, 234, 235
Membrane
  probe, 88
  structure and PLA$_2$, 88-90, 100
Methanol, 16
Michaelis-Menten kinetics, 113
Mitochrondrion of liver
  and PLA$_2$, 4
MK-886, 131-134
fMLP, 198, 202, 203
Mojave toxin, 34, 35
Monoarthritis and PLA$_2$, 242
Monocyte, human, 125-136, 231, 247,
    248
  and arachidonic acid release, 125
  and eicosanoid production, 129
  isolation, 127
  -macrophage fatty acid uptake,
    232
  and PLA$_2$ activation, 125-131
  release, 128
  and superoxide release, 131
Monte Carlo calculation, 91
Myocyte, 193
Myoglobin, 115

Naja melanoleuca, venom, 55, 75, 77
  N. mossambica mossambica, 221, 224
  N. naja atra venom, 56
  N. naja naja venom, 117

Phospholipase (continued)
PLA$_2$ (continued)
and calcium, 8, 49, 53, 54,
   66, 67, 71-74, 85, 95-99,
   220
ion-binding site, 28-31
and cell death, 121
and catalysis, interfacial, 37
clustering effect, 9
in cobra venom, 33, 54, 117,
   201
models, 60-61
in collagen vascular disease,
   241
and deamination, 114
and deletion of surface loop
   (62066), 74-77
and detergent, 53
and 1, 2-diacylglycerol-3-
   sulfate, 53
as dimer is active, 38-40, 43-
   44
dimerization of, 120
and dimiristoylphosphatidylcholine, 112
and dioctanoyllecithin
   hydrolysis by, 72, 73
and dipalmitoylphosphatidylcholine, 85-100
in diseases, human, listed, 241
and disulfide bridges, 65
and DNA
   blot analysis, 181-183
   clones, 173-184
   homology, 176
   recombinant, pancreatic, 65-81
and *Escherichia coli*, 68-71,
   112, 115
expression as a fusion protein,
   69-71
in yeast, 70-71
and fatty acid, unsaturated,
   219-230
features, common, 106
in fluid
   inflammatory, 212
   peritoneal, 9
   rheumatoid, 212
   synovial, 179-180, 212
and fluorescence, 54, 86-90, 97,
   221, 224, 225
functions, 50, 142, 173
and glucocorticosteroid, 198
and gout, 242
and G-protein regulation of,
   185-195
group
   I, 8, 41
   II, 8, 41
and histidine, 56
homology between DNAs, 176

Phospholipase (continued)
PLA$_2$ (continued)
hydrolysis and computer simulation,
   86-90
- immunoprecipitation analysis, 180
- inactivation by radiation, 55
and inflammation, 203
induced, 245-247
systemic, 240, 247
inhibition of, 11, 50-64, 137-
   161, 164-166, 198, 201, 211-218
interface activity, 37-47
isoforms, 183
kinetics, 113, 115
of *Lahesis muta*, 43
at lipid-water interface, 51-52
and lipocortin, 161, 199
of liver mitochondrion, 4
and lysine, 56-57
and lysosome, 4, 5
of macrophage (RAW264.7), 190
mechanism of catalysis proposed
   65-81
mechanism of DNA, recombinant
   techniques, 65-81
- in mellitin, 233
in membrane, 88-90, 100, 173
in mitochondrion, hepatic, 24
and monoarthritis, 242
in monocyte, 125-136
as monomer is inactive, 39
and neurotoxicity, 33-34
and NH$_2$ group, 57
and nucleotide dequence, 176-178
and oleic acid, 220, 222
in organs, human, 10
in osteoarthritis, 242
pancreatic, 5, 7, 24, 25, 30, 53,
   65-81, 106-113
in pancreatitis, 241
in pathology, human, 212, 239-
   251 *see* Inflammation
in peritonitis, bacterial, 241
and phosphatidylethanolamine
   vesicle as substrate, 201
and placenta, human, 173-184
and prostaglandin B, 219-230
and protein, inhibitory, 211-218
and proton-relay system for
   catalysis, 8
in psoriasis, 242
purification of, 242
and pyridoxalphosphate, 57
and radiation inactivation, 55
in rabbit, 9, 10
in rat, 9, 10
in rattlesnake venom, 24, 39, 40,
   97, 117
reaction, 49
recognition, interfacial, 37